the Good Earth

Introduction to Earth Science

David McConnell
North Carolina State University

David Steer
University of Akron

Contributions by
Catharine Knight
University of Akron

Katharine Owens
University of Akron

THE GOOD EARTH: INTRODUCTION TO EARTH SCIENCE, FOURTH EDITION

Published by McGraw-Hill Education, 2 Penn Plaza, New York, NY 10121. Copyright © 2018 by McGraw-Hill Education.
All rights reserved. Printed in the United States of America. Previous editions © 2015, 2010, and 2008. No part of this publication
may be reproduced or distributed in any form or by any means, or stored in a database or retrieval system, without the prior
written consent of McGraw-Hill Education, including, but not limited to, in any network or other electronic storage or
transmission, or broadcast for distance learning.

Some ancillaries, including electronic and print components, may not be available to customers outside the United States.

This book is printed on acid-free paper.

1 2 3 4 5 6 7 8 9 LMN 21 20 19 18 17

ISBN   978-0-07-802288-3
MHID   0-07-802288-6

Chief Product Officer, SVP Products & Markets: *G. Scott Virkler*
Vice President, General Manager, Products & Markets: *Marty Lange*
Vice President, Content Design & Delivery: *Betsy Whalen*
Managing Director: *Thomas Timp*
Director of Digital Content: *Philip Janowicz, Ph.D.*
Brand Manager: *Michael Ivanov, Ph.D.*
Director, Product Development: *Rose Koos*
Product Developer: *Jodi Rhomberg*
Marketing Manager: *Noah Evans*
Market Development Manager: *Tamara L. Hodge*
Digital Project Analyst: *Patrick Diller*
Digital Product Developer: *Joan Weber*
Director, Content Design & Delivery: *Linda Avenarius*
Program Manager: *Lora Neyens*
Content Project Managers: *Laura Bies, Rachael Hillebrand, & Sandra Schnee*
Buyer: *Laura M. Fuller*
Design: *Tara McDermott*
Content Licensing Specialists: *Carrie Burger & Lori Slattery*
Cover Image: ©Ron Watts/Getty Images
Compositor: *SPi Global*
Printer: *LSC Communications*

All credits appearing on page or at the end of the book are considered to be an extension of the copyright page.

Library of Congress Cataloging-in-Publication Data

Names: McConnell, David, 1959 February 5- | Steer, David Nathan, 1960- |
  Knight, Catharine. | Owens, Katharine, 1942-
Title: The good Earth : introduction to earth science / David McConnell,
  North Carolina State University, David Steer, University of Akron ;
  contributions by Catharine Knight, University of Akron, Katharine Owens,
  University of Akron.
Description: Fourth edition. | New York, NY : McGraw-Hill Education, [2018] |
  Includes index.
Identifiers: LCCN 2016042222 | ISBN 9780078022883 (alk. paper)
Subjects: LCSH: Earth sciences—Textbooks.
Classification: LCC QE26.3 .G66 2018 | DDC 550—dc23 LC record available at https://lccn.loc.gov/2016042222

The Internet addresses listed in the text were accurate at the time of publication. The inclusion of a website does not indicate
an endorsement by the authors or McGraw-Hill Education, and McGraw-Hill Education does not guarantee the accuracy
of the information presented at these sites.

mheducation.com/highered

Brief Contents

Contents

Preface

Teaching earth science can be viewed as content instruction, covering the principles of science and earth systems. But can it also be considered as an opportunity to *engage* students in the nature of scientific inquiry?

> *A traditional science instructor concentrates on teaching factual knowledge, with the implicit assumption that expert-like ways of thinking about the subject come along for free or are already present. But that is not what cognitive science tells us. It tells us instead that students need to develop these different ways of thinking by means of extended, focused, mental effort.*
>
> Carl Wieman
> Nobel Prize winner

© StockTrek/Getty Images

For many, the wonder of Earth and its features is enough to drive learning. For these happy few, a readable book with lots of attractive photographs is almost all that is required. *But for many—in fact most—learning takes more than pretty words and pictures.* Providing high-quality teaching is the most cost-effective, tangible, and timely effort that geoscience instructors can make to improve student engagement, increase attendance, and add majors.

But how do we do that? There is extensive literature describing what effective teaching looks like, but most science instructors have not had access to these articles and books. Further, few of us were ever explicitly taught the components of good teaching. Instead, we were left to figure it out for ourselves on the basis of our classroom experiences as students.

The Good Earth was published to support both the traditional earth science class **and** to serve as an accessible resource for instructors seeking to apply effective teaching strategies to enhance learning.

The Good Earth *Difference*

We wrote *The Good Earth* to support an active learning approach to teaching and to provide the necessary resources for instructors moving through the transition from passive to active learning. Like you, we want our students to walk away from this course with an appreciation for science and the ability to make life decisions based on scientific reasoning.

Our goal was to write a book that was engaging for students but that also included resources that illustrated for instructors how to use teaching practices that have been shown to support student learning. The materials and methods discussed in the text and the accompanying *Instructor's Manual* have been tried and tested in our own classes. Our research shows that the integration of the materials and pedagogy provided in this book not only improved students' understanding of earth science as measured by standardized national tests, but it can also improve students' logical thinking skills by twice as much as a typical "traditional" lecture class. Such methods are overwhelmingly preferred by students and increase student attendance and satisfaction with the course. Finally, a significant point for us is that these methods make teaching class more fun for the instructor.

> *I love the voice the authors use. Reading the text is like listening to a very intelligent but down-to-earth friend explain a difficult topic. The authors are excellent at organizing and presenting the material. . . . The illustrations are superior to other texts in all ways.*
>
> *Patricia Hartshorn*
> *University of Michigan–Dearborn*

Student-Centered Research

The Good Earth can be used as a text for a traditional, teacher-centered lecture-based course. In fact, we have taken great care to write a book that students would find more engaging than a typical text. But the greatest benefit will come when the book is used as part of an active-learning, student-centered course. For some instructors, it may simply be a matter of adding some of our exercises to an existing active-learning class environment. For others, the book and accompanying materials will give them an opportunity to add components as they gradually change their pedagogy. If you want a more interactive class, try one or all of the following three recommendations based on research findings:

1. Students learn key concepts better when they have opportunities to actively monitor their understanding during class. Rather than just standing up and talking, the instructor can break lectures into segments separated by brief exercises to make sure that students understand concepts before moving on. Students' understanding must be frequently challenged to provide an opportunity to identify misconceptions and replace them with improved, more realistic models.

   The Good Earth includes hundreds of Checkpoint exercises that can also be used as handout-ready PDF files (located on the text website along with answer keys). Practice makes perfect: the more opportunities students have to assess their learning and to practice the application of new skills, the better their performance. If you are concerned about reduced time for lecture, we have found that an emphasis on fostering deeper understanding and less content coverage in lecture, combined with greater student responsibility for reading,

produced no decrease in content knowledge attainment and improved student comprehension of key concepts. Some exercises can be assigned as homework, and the answer key in the back of the book can help students to assess their self-directed-learning.

2. Students become better learners when we challenge them to answer questions that require the use of higher-order thinking skills (for example, analysis, synthesis, evaluation). Brain research shows that people become smarter when they experience cognitive challenges. However, it is important not to throw students into the deep end without any help. Instead, instructors need to step through a series of problems of increasing difficulty (scaffolding) so that they can train students to correctly apply their newly acquired thinking skills.

   Therefore, we have carefully created a series of color-coded **Checkpoint** exercises for each section of every chapter. The exercises are pitched at four skill levels: basic, intermediate, advanced, and superior, to give students and instructors an opportunity to scaffold student understanding of key concepts. The questions represent four levels of Bloom's taxonomy. Blue and green questions typically are comprehension and application-level questions. Yellow and red checkpoints typically require analysis, synthesis, or evaluation skills. It is not necessary to complete all the exercises; instructors can select the exercises that are most appropriate for their learning goals.

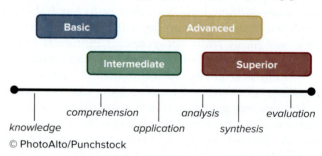

© PhotoAlto/Punchstock

> *This was kind of a neat idea, and the questions [Checkpoints] do get quite challenging at higher orders. I feel these are good things for students to do while studying, with the idea that if they understand the higher order questions they will understand concepts better for exams. I thought these checkpoints have some very well-formulated questions in the chapters I reviewed.*
>
> *Swarndeep Gill*
> *California University of Pennsylvania*

> *I like the fact that the authors are mindful and well versed in science education research and pedagogy. This aspect of the author's background is evident in the design of the Checkpoint questions.*

> *The use of Concept Maps and Venn Diagrams is fairly cutting edge for introductory Earth Science textbooks that I am familiar with. This is probably the most innovative aspect of this book and distinguishes it from similar texts, even though the content is presented very similarly to other texts.*
>
> *Jeffrey Templeton*
> *Western Oregon University*

Sort ...

✓ Checkpoint 11.1

✓ basic	✓ advanced
✓ intermediate	✓ superior

Sort the following 12 terms into six pairs of terms that most closely relate to one another. Explain your choices.

groundwater	plants	transpiration
stream	ice	infiltration
rainfall	precipitation	water vapor
gas	meltwater	runoff

Match the lettered responses ...

✓ Checkpoint 7.22

✓ basic	✓ advanced
✓ intermediate	✓ superior

Rock Cycle Diagram

The following diagram illustrates some of the interactions of the rock cycle. Match the lettered responses to the blank ovals on the diagram. (*Note:* Some letters are used more than once.) Example: If you believe that metamorphic rock is converted to magma by cementation and compaction, enter "a" in the top left oval.

 a. Cementation and compaction (lithification)
 b. Heat and pressure
 c. Weathering, transportation, deposition
 d. Cooling and solidification
 e. Melting

✓ Checkpoint 6.19

✓ basic	✓ advanced
✓ intermediate	✓ superior

Venn Diagram: Shield Volcanoes, Stratovolcanoes, and Cinder Cones

Use the Venn diagram provided here to compare and contrast the three principal types of volcanoes. Place the number corresponding to features unique to each type in the larger areas of the circles; note features they share in the overlap area in the center of the image. Five items are provided; identify at least 12 more.

 1. Associated with subduction zones
 2. Have a triangular shape in profile
 3. Example: Mount Hood, Oregon
 4. Mild eruptions
 5. Intermediate-silica magma

Compare and contrast ...

Stratovolcanoes   Shield Volcanoes

Cinder Cones

Evaluate the five most important factors ...

✓ Checkpoint 12.12

✓ basic	✓ advanced
✓ intermediate	✓ superior

Groundwater Evaluation Rubric

You are asked to help locate a new aquifer that will supply your town with water. In examining the potential sites, you recognize that several different factors will influence groundwater availability and at no single site are all of the factors optimal. You decide to create a scoring scheme to evaluate the most important factors that will influence the availability of groundwater. The location that scores the highest according to the rubric will be selected for the well field. One factor is included as an example in the table below; identify five more.

Factors	Poor (1 point)	Moderate (2 points)	Good (3 points)
Depth to water table	Deep	Intermediate	Shallow

I have to compliment you on putting together Checkpoint 3.3. This was probably the best evaluation tool I have seen for determining whether a student really understands the meaning of the words we use to describe the scientific methods (hypothesis, prediction, etc.).

Neil Lundberg
Florida State University

3. Knowledge is socially constructed and people learn best in supportive social settings. Students do not enter our classrooms as empty vessels to be filled with knowledge. Instead, they actively construct mental models that assimilate new information with previous experiences. This construction of knowledge happens most readily when students work in small collaborative groups (three to four students), where they can talk and listen to peers as they build their understanding of new concepts. Students must be provided with opportunities to be self-reflective about their learning and to help them learn how to learn. Our research confirmed that students in classes where small groups worked to solve challenging problems outperformed students in classes where they worked on the same problems independently.

© PhotoAlto/Punchstock

It is set up very user friendly and will make it easy for instructors to create an interactive learning environment. Also, the way the chapters and questions are laid out, students will know exactly what they should be getting from the chapter and how to test their knowledge and skills.

Jessica Kapp
University of Arizona

Whether you choose to use informal groups ("turn and talk to your neighbors") or formal groups determined by experiences (for example, number of science classes, scores on pretests, academic rank), collaborative learning is a powerful mechanism for maintaining attendance, increasing student-instructor dialogue, and enhancing learning. The Checkpoint exercises (especially advanced and superior level) and conceptests (conceptual multiple choice questions) provided with the book will give you many assignments that you can use as the basis for group work.

For detailed information regarding concept maps, Venn diagrams, Bloom's taxonomy, assessment, and so forth, please consult the *Instructor's Manual*.

Tools for Teaching and Learning Science Literacy

Science can be thought of in three ways: as a body of knowledge, as the processes that people employ to explain the universe, and as a set of attitudes and values possessed by those who "do science." This latter aspect is often overlooked in college science textbooks. For each chapter of *The Good Earth,* the *Instructor's Manual* gives suggestions for incorporating into class discussion science attitudes and values such as open-mindedness, skepticism, persistence, and curiosity.

Additionally, the discussion of the **scientific method** is woven throughout the text. We emphasize three scientific themes throughout the text: 1) scientific literacy, 2) earth science and human experience, and 3) the science of global change. Numerous examples of human interaction with Earth serve as introductions to each chapter. Each chapter includes examples of the connection between science and technology, and builds on a context or event familiar to the student. We believe that links to students' past knowledge and experience are essential foundations upon which to build deeper understanding.

In addition to the theme of global change permeating the text, we devote a full chapter to the topic and do not duck the tough issues related to it. We use data and evidence to help students build their own understanding and assist them to realize that *"Much of what lies ahead for the good Earth is up to us. Know, care, act."*

I am pleased to see the final chapter on global change; most students assume that climate change is a political debate, so it is nice to see a textbook that discusses the science behind the news.

Bryan C. Wilbur
Pasadena City College

Ways to Direct Learning

Rather than put key vocabulary terms in bold, we put **key concepts** in bold font. Our rationale is that conceptual understanding is the goal; vocabulary terms alone may not lead to the understanding that we desire. Research suggests that listing key terms encourages the memorization of those terms, rather than the understanding of the associated concepts—rather like learning words in a foreign language but being unable to put together a sentence. To make students fluent in science, we chose to focus on a vocabulary that builds students' conceptual understanding of major ideas in earth science. These ideas were recommended by standards-setting groups, such as the American Association for the Advancement of Science (AAAS).

Students can use the Checkpoint surveys to self-evaluate their comprehension of the major concepts in the section. Self-evaluation is a life skill that persists far longer than the evaluation imposed by an outside party (that is, the instructor). We believe in ongoing assessment tied to each key concept while ideas are still fresh. In contrast, other texts may provide tools for assessment only at the end of the chapter, after all of the content has been covered.

National Committee on Science Education Standards and Assessment
National Research Council

LEARNING SCIENCE IS AN ACTIVE PROCESS. Learning science is something students do, not something that is done to them. In learning science, students describe objects and events, ask questions, acquire knowledge, construct explanations of natural phenomena, test those explanations in many different ways, and communicate their ideas to others. Science teaching must involve students in inquiry-oriented investigations in which they interact with their teachers and peers.

FOCUS AND SUPPORT INQUIRIES. Student inquiry in the science classroom encompasses a range of activities. Some activities provide a basis for observation, data collection, reflection, and analysis of firsthand events and phenomena. Other activities encourage the critical analysis of secondary sources—including media, books, and journals in a library.

ENCOURAGE AND MODEL THE SKILLS OF SCIENTIFIC INQUIRY, AS WELL AS THE CURIOSITY, OPENNESS TO NEW IDEAS, AND SKEPTICISM THAT CHARACTERIZE SCIENCE.

USE MULTIPLE METHODS AND SYSTEMATICALLY GATHER DATA ON STUDENT UNDERSTANDING AND ABILITY. Because assessment information is a powerful tool for monitoring the development of student understanding, modifying activities, and promoting student self-reflection, the effective teacher of science carefully selects and uses assessment tasks that are also good learning experiences.

Often students have some fundamental knowledge of earth science and, when reminded, are able to apply this information to the introduction of new concepts. Each chapter includes a **Self-Reflection Survey** to promote awareness of personal experiences.

Self-Reflection Survey: Section 1.1

Respond to the following questions as a means of uncovering what you already know about Earth and earth science.

1. Which of the following earth science phenomena ha[s] experienced? Which would you most like to exper[ience]? Can you think of three more things to add to the li[st]?

    *   A volcanic eruption
    *   A glacier
    *   A river in flood
    *   A cave system
    *   An underground mine
    *   A canyon
    *   An earthquake
    *   An erosional coastline (rocky cliffs)
    *   A depositional coastline (beaches)
    *   A hot desert
    *   A continental divide
    *   Rock layers with fossils
    *   A big, assembled dinosaur skeleton
    *   A meteor shower or comet
    *   The aurora borealis (the northern lights)
    *   A meteorite crater
    *   A mountain range over 3,000 meters (over 10,000 feet) in elevation
    *   The top of a cloud

2. What three questions about Earth would you like be able to answer by the end of this course?

Self-Reflection Survey: Section 17.1

Answer the following questions as a means of uncovering what you already know about global change.

1. Respond to the following questions taken from recent CNN and Gallup polls, and compare your answers to those of other respondents. (See footnote to compare responses.*)

    i. Which of the following statements comes closest to your view of global warming?

        a. Global warming is a proven fact and is mostly caused by emissions from cars and industrial facilities such as power plants and factories.

        b. Global warming is a proven fact and is mostly caused by natural changes that have nothing to do with emissions from cars and industrial facilities.

        c. Global warming is a theory that has not yet been proved.

        d. Unsure.

    ii. In thinking about the issue of global warming, sometimes called the *greenhouse effect,* how well do you feel you understand this issue?

        a. Very well.

        b. Fairly well.

        c. Not very well.

        d. Not at all.

Visuals are of great importance for understanding earth science concepts. *The Good Earth* features two-page **Snapshots** to emphasize an important concept in every chapter.

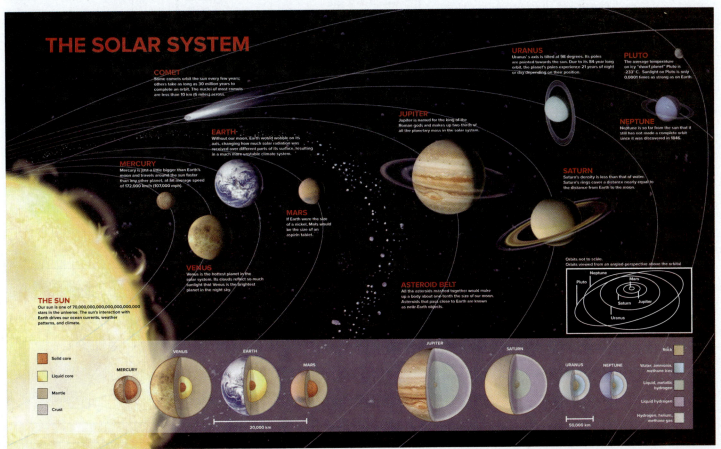

We frequently hear complaints that students don't get the **Big Picture** and become lost in the vocabulary or in trying to memorize facts. We responded to this concern by connecting a chapter- opening "Big Picture" question and photo to the end-of-chapter summary, titled **The Big Picture,** to help students link the key concepts before moving to a new chapter.

the big picture

When Mount St. Helens began rumbling in 1980, teams of scientists rushed to the mountain with truckloads of instruments to monitor the activity. Still, the May 18 eruption came as a surprise. Despite the experience of the scientists and the sophistication of the devices they deployed, little detailed information on the eruptive history of the volcano had been gathered beforehand and few monitoring instruments had been collecting data. That is no longer the case. In the past quarter-century, scientists have made a concerted effort to place a variety of instruments around the volcano, and even in space, to monitor every rumble and movement. Even with what they know today, it is unlikely that volcanologists would have predicted the precise time of the May 18 eruption. But they would have known enough to have more vigorously encouraged the authorities to move people farther from the volcano itself, dramatically reducing the loss of life.

Educating the public is an important factor in reducing the effects of hazards such as volcanoes. Education should provide a scientifically literate population with the necessary skills to critically respond to scientists' assertions. Deciding what evidence to dismiss and what to pay attention to might mean the difference between life and death for those who live in the shadow of an active volcano. The people living near Mount St. Helens in 1980 weighed the evidence and the accompanying call to action. Some heeded the call to evacuate, while others ignored the evidence provided by the volcanologists, chose to hold their ground, and paid for their decision with their lives.

Mount St. Helens is one of only a few US volcanoes with such a high degree of monitoring. However, the US Geological Survey plans to create a National Volcano Early Warning System that would identify the most threatening volcanic hazards, including the number of people and the extent of property endangered. A preliminary assessment of volcanic threat identified more than 50 volcanoes as high-threat or very-high-threat sites and recommended that each volcano have an extensive network of monitoring equipment to identify the first signs of unrest. Few such networks are currently deployed, and some of these volcanoes have no monitoring systems at all.

One of the volcanoes in the very-high-threat group is Mount Rainier, pictured looming over Tacoma, Washington, at the beginning of this chapter. At 4,392 meters (14,410 feet), Mount Rainier is the tallest and most imposing volcano in Washington. It is located about 70 kilometers (43 miles) southeast of Tacoma. What questions would you ask if you lived in Tacoma?

Historical records indicate that Mount Rainier does not erupt with the frequency of Mount St. Helens. The distance of the peak and the prevailing westerly winds make it unlikely that tephra would ever reach Tacoma. In addition, lava flows and pyroclastic debris would not extend beyond the foot of the mountain, staying tens of kilometers short of Tacoma. Still, large lahars have the potential to reach the northern suburbs of the city and enter neighboring Puget Sound. Even if Tacoma is safe, many smaller towns lie in stream valleys just a 10-minute trip from the volcano by lahar. It is the residents of towns such as Ashford, Packwood, and Orting (Figure 6.33) who need an early warning system for volcanoes.

Figure 6.33 Lahar hazards associated with Mount Rainier, Washington.
USGS

d other young mountain belts. south-central China, at the east- ished whole towns, killed more y 5 million homeless. The unrest continues; Earth at this very moment is shifting, rumbling, building, and decaying. We must carefully observe and prepare.

Volcanoes and Mountains: Concept Map

Complete the following concept map to evaluate your understanding of the interactions between the earth system and volcanoes and mountains. Match the following interactions with the lettered labels on the figure, using the information from this chapter.

they are formed by movements on faults, movements that generate damaging earthquakes. Building a mountain range like the Himalayas involves thousands of faults that generate millions of earthquakes. Unfortunately, major earthquakes are still common

Interaction	Letter
Eruption melted ice on Nevado del Ruiz to cause fatal lahars.	
Sulfur dioxide blocks incoming sunlight.	
Added water causes partial melting of mantle.	
Volcanoes add CO_2 and sulfur dioxide to atmosphere.	
Commercial airlines are at risk from tephra clouds.	
Solar radiation heats Tibetan plateau.	
Rain strips CO_2 from atmosphere.	
Krakatau eruption generated massive tsunami.	
Tephra is carried downwind over cities.	
Some 500 million people are in risk zones for volcanoes; trees are knocked down.	
Industrial materials are swept into rivers and lakes from mudflows.	
Monsoon rains result from air rising over Himalayas.	
Weathering processes break down rocks in mountains.	
Instrumentation of volcanoes.	

Numerous diagrams, photos, and tables support visual processes and concepts.

Figure 6.15 Hawaiian lava. **a.** A lava tube transports hot, fluid, low-viscosity basalt lava toward the front of a lava flow on Kilauea volcano, Hawaii. **b.** Walter's Kalapana Store and Drive-in was burned and buried within a few weeks in 1990 as lava from the Kilauea volcano invaded communities along the southern coast of Hawaii. Note the height of the original sign. How deep is the lava at this location?
6.15a: USGS, 6.15b (top): USGS; 6.15b (middle): USGS; 6.15b (bottom): USGS;

Figure 8.15 Recently discovered Tiktaalik fossil. **a.** This is a transitional fossil between fish and amphibians. The fossil was discovered on Ellesmere Island, Canada, in 375 million-year-old rocks. Several individuals were found, some up to nearly 3 meters (9 feet) long. **b.** A re-creation of what Tiktaalik may have looked like in life.
8.15a: © Ted Daeschler/VIREO; 8.15b: Zina Deretsky, National Science Foundation

To further aid in the understanding of earth processes, many figures include a simple drawing to portray a **Geologist's View.**

Geologist's View
Figure 5.5 Signs of movement on a fault. Movement on a 44-kilometer-long (27-mile-long) fault caused the Hebgen earthquake in Montana in 1959. **a.** The fault broke the surface near a ranch (background). **b.** The fault can be followed for several kilometers along the south flank of Kirkwood Ridge in the center of the image.
5.5a: USGS, 5.5b: USGS

How Is This Text Organized?

The Good Earth covers the primary topics included in other earth science texts. However, there are a few notable differences in its content compared to other textbooks.

The Good Earth begins with an introduction (Chapter 1), then takes up the topic of astronomy (Chapters 2, 3), and moves on to solid earth (Chapters 4, 5, 6, 7, 8) and the surficial processes (Chapters 9, 10, 11, 12), which overlap with the hydrosphere (Chapters 11, 12, 13), before dealing with the atmosphere (Chapters 14, 15, 16) and finishing with a wrap-up chapter on global change (Chapter 17) that incorporates elements of all the previous chapters.

Astronomy is dealt with early in the text (Chapters 2 and 3) from the context of Earth's position in space. By beginning with Earth's place in the universe, we give students a "big picture," set the context for looking at the uniqueness of this planet in contrast to our neighbors in space, and hopefully, inspire a bit of wonder in the reader. In both chapters, we grab the reader's attention by emphasizing space from a human perspective. We believe this provides a more appealing beginning to an earth science class than the traditional several weeks spent discussing minerals, rocks, and weathering. Chapter 2, in particular, guides students to see methods that scientists employ as they build our knowledge of the planet and its place in the universe.

Plate tectonics appears early (Chapter 4). We introduce this important unifying concept at the beginning of the text and then use it as a foundation to introduce other solid earth topics (for example, earthquakes, volcanoes). Because an understanding of plate tectonics is pivotal to all the content that follows in subsequent chapters, we revisit this concept several times in subsequent chapters, thereby showing students the interrelationships among the other solid earth topics, such as rock formation, earthquakes, and volcanoes.

Driven by recent research findings, we have chosen to emphasize some topics that are discussed briefly or not at all in other earth science texts. We have included chapters on the threat of a collision with near-Earth objects (Chapter 3), Earth's climate system (Chapter 16), and global change (Chapter 17). In addition, the continuing debate about the teaching of creationism in the public schools has lead us to address this topic head-on in our treatment of geologic time (Chapter 8).

New in This Edition

Updated material in this edition includes references to recent natural hazards such as the Oso landslide in Washington and frequent earthquakes in Oklahoma; new data on topics such as US energy use, near earth objects (NEOs), changing global temperatures and sea ice distribution; and references to significant events such as landing a probe on a comet.

There are many structural changes evident in this edition. The e-text is now easily used on mobile devices as well as tablets and PCs.

Additional updates to this edition include:

- Many photos have been updated and changed to be more current and better illustrate concepts
- References and discussions have been updated to include:
  - More complete descriptions of geo-scientific ways of thinking and habits of mind
  - Additional concepts related to the characteristics of complex systems and the Earth as a system
  - More data and discussion related to extra-solar planets
  - A narrative related to the 2014 landing of the Philae on a comet
  - More examples that illustrate the impacts of earthquakes on society including a 2015 study related to Los Angeles
  - Recent examples of the impact of rivers on society including the 2015 floods in Texas, Oklahoma and Death Valley and 2016 floods in the Mississippi River watershed
  - New information related to the Flint Michigan water crisis related to lead in the water supply
  - Updated statistics related to renewable energy and electrical generation from wind power
  - Added information related to global cloud cover and the impact of clouds on reflecting solar radiation
  - Updated information on weather hazards and more recent examples of weather-related deaths from recent heat waves, floods and other extreme weather events
  - Updates on changes to sea ice extent in the Arctic and the impacts on local and global ecosystems
  - Updated data related to "healing" of the ozone hole and global trends in CFC concentrations in the atmosphere
  - Expanded discussion of global sea level changes and temperature changes
  - Additional information relate to global emissions and how they compare to the past
- Figures have been updated and/or replaced throughout the text to better illustrate key concepts and to provide updated data.

©Getty Images/iStockphoto

McGraw-Hill Connect®
Learn Without Limits

Connect is a teaching and learning platform that is proven to deliver better results for students and instructors.

Connect empowers students by continually adapting to deliver precisely what they need, when they need it, and how they need it, so your class time is more engaging and effective.

> 73% of instructors who use **Connect** require it; instructor satisfaction **increases** by 28% when **Connect** is required.

Analytics

Connect Insight®

Connect Insight is Connect's new one-of-a-kind visual analytics dashboard—now available for both instructors and students—that provides at-a-glance information regarding student performance, which is immediately actionable. By presenting assignment, assessment, and topical performance results together with a time metric that is easily visible for aggregate or individual results, Connect Insight gives the user the ability to take a just-in-time approach to teaching and learning, which was never before available. Connect Insight presents data that empowers students and helps instructors improve class performance in a way that is efficient and effective.

Mobile

Connect's new, intuitive mobile interface gives students and instructors flexible and convenient, anytime–anywhere access to all components of the Connect platform.

Connect's Impact on Retention Rates, Pass Rates, and Average Exam Scores

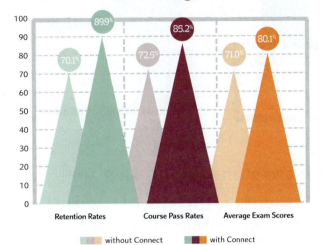

	without Connect	with Connect

> Using **Connect** improves retention rates by **19.8%**, passing rates by **12.7%**, and exam scores by **9.1%**.

Impact on Final Course Grade Distribution

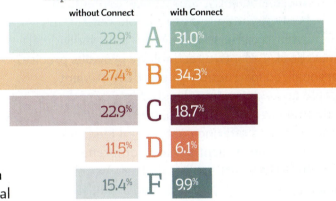

without Connect		with Connect
22.9%	A	31.0%
27.4%	B	34.3%
22.9%	C	18.7%
11.5%	D	6.1%
15.4%	F	9.9%

> Students can view their results for any **Connect** course.

Adaptive

More students earn **A's** and **B's** when they use McGraw-Hill Education **Adaptive** products.

SmartBook®

Proven to help students improve grades and study more efficiently, SmartBook contains the same content within the print book, but actively tailors that content to the needs of the individual. SmartBook's adaptive technology provides precise, personalized instruction on what the student should do next, guiding the student to master and remember key concepts, targeting gaps in knowledge and offering customized feedback, and driving the student toward comprehension and retention of the subject matter. Available on tablets, SmartBook puts learning at the student's fingertips—anywhere, anytime.

Over **8 billion questions** have been answered, making McGraw-Hill Education products more intelligent, reliable, and precise.

www.mheducation.com

ACKNOWLEDGMENTS

The authors would like to express their appreciation for family, friends, colleagues, and students who provided encouragement throughout the writing process for the book. In particular, we would like to thank Tom Angelo who guided us through a detailed course on teaching and learning, and gave us many of the tools that helped us link together our teaching goals with appropriate learning exercises. We will always be grateful for what we learned under his thoughtful instruction. In addition, we are grateful for the enthusiasm and support of the McGraw-Hill development and production teams whose names appear on the copyright page and the support of all the reviewers who helped improve the quality of the text and illustrations.

We would like to thank the following individuals who wrote and/or reviewed learning goal-oriented content for **LearnSmart.**

Northern Arizona University, Sylvester Allred
Roane State Community College, Arthur C. Lee
State University of New York at Cortland, Noelle J. Relles
University of North Carolina at Chapel Hill, Trent McDowell
University of Wisconsin—Milwaukee, Tristan J. Kloss
University of Wisconsin—Milwaukee, Gina Seegers Szablewski
Elise Uphoff

Reviewers

Special thanks and appreciation go out to all reviewers for their constructive suggestions, new ideas, and invaluable advice.

Broward College, Neil M. Mulchan
Bucks County Community College, Cristina Ramacciotti
Cal State University–Northridge, Doug Fischer
California University of Pennsylvania, Swarndeep Gill
Central Connecticut State University, Kristine Larsen
Central Michigan University, Karen S. Tefend
Cerritos College, Tor BjÖrn Lacy
Charleston Southern University, Peter B. Jenkins
C. W. Post-Long Island University, Vic DiVenere
East Los Angeles College, Randall J. Adsit
Eastern Michigan University, Maria-Serena Poli
Eastern Michigan University, Steven T. LoDuca
Emporia State University, Susan Aber
Florida State College, Jacksonville, Betty Gibson M.Ed.
Florida State College, Jacksonville, Rob Martin
Florida State University, Neil Lundberg
Georgia Institute of Technology, L. Gregory Huey
Heartland Community College, Robert L. Dennison
Hillsborough Community College, Marianne O'Neal Caldwell

Indian River State College, Paul A. Horton
Ivy Tech Community College of Indiana, Donald L. Eggert
Johnson State College, Dr. Leslie Kanat
Kingsborough Community College, Cyrena Anne Goodrich
Laredo Community College, Glenn Blaylock
Laredo Community College, Sarah M. Fearnley
Lock Haven University, Thomas C. Wynn
Methodist University, Dr. John A. Dembosky
Middle Tennessee State University, Dr. Clay Harris
Middle Tennessee State University, Mark Abolins
Middle Tennessee State University, Melissa Lobegeier
Missouri State University, Jill (Alice A.) Black
Montgomery County Community College, George Buchanan, P.G.
Murray State University, George W. Kipphut
Murray State University, Haluk Cetin
North Carolina A & T State University, Godfrey A. Uzochukwu
Northern Oklahoma College, Eugene A. Young
Northwest Missouri State University, C. R. Rohs
Northwest Missouri State University, Jeffrey D. Bradley
The Ohio State University, Lindsay Schoenbohm
Pasadena City College, Bryan C. Wilbur

Pensacola Junior College, Kathleen Shelton
Purdue University, Lawrence W. Braile
San Jose State University, Paula Messina
Santa Ana College, Claire M. Coyne
Santiago Canyon College, Debra Ann Brooks
Sierra College, Alejandro Amigo
Southwestern Illinois College, Stanley C. Hatfield
St. Petersburg College, Paul G. Cutlip
St. Petersburg College, William C. Culver
State College of Florida, Jay C. Odaffer
State University of New York, College at Potsdam, Michael C. Rygel
University of Arkansas, Fort Smith, Christopher Knubley
University of Dayton, Heidi S. McGrew
University of Dayton, Michael R. Sandy
University of Indianapolis, Thomas L. Chamberlin
University of Michigan–Dearborn, Patricia Hartshorn
University of North Carolina—Greensboro, Jeffrey C. Patton
University of North Carolina—Greensboro, Michael Lewis
University of North Florida, Jane MacGibbon
Western Oregon University, Don Ellingson
Western Oregon University, Jeffrey Templeton

about the authors

The original version of The Good Earth was a product of a team of educators from the geosciences, science education, and cognitive psychology whose combined expertise created this text to teach essential earth science content in an engaging and cognitively supportive way. We wish to thank our colleagues Kathie Owens, Cathy Knight, and Lisa Park for their contributions to the textbook through the first two editions. The writing team was reduced to the two principal authors starting with the third edition of the book.

David McConnell grew up in Londonderry, Northern Ireland, and was hooked on geology when he took his first course in high school with an inspirational teacher. His earliest geological exercises involved examining rocks along the rugged coastlines of Ireland. He graduated with a degree in geology from Queen's University, Belfast, before moving to the US to obtain graduate degrees from Oklahoma State and Texas A&M Universities. David spent much of his career at the

© Katherine Ryker

University of Akron, Ohio, where he met David Steer, beginning a research partnership that eventually resulted in the book you are now holding. David relocated to North Carolina State University to build a geoscience education research group that continues to examine how to improve the student learning experience in large general education science classes.

David has taught a dozen different courses from introductory geoscience classes to advanced graduate courses. He has received several teaching awards, and he and his collaborators and graduate students have made many presentations and published articles on their educational research. When pressed for some personal information, David will tell you that he loves collecting vinyl records, is way too attached to Tottenham Hotspur football club, and enjoys spending weeks each summer hiking trails through a mountain range somewhere.

David Steer was fascinated with rocks as a child in Ohio. That interest was nurtured by his participation in a National Science Foundation–sponsored geology field camp for high school students that took him to the Black Hills of South Dakota. David's plan to become a geologist had to wait when he accepted an appointment to West Point and then served for a decade as an Army Corps of Engineers officer. While in the military, David attended Cornell University, earning a Master's of Engineering degree. He was then assigned

Courtesy of David Steer

to West Point Military Academy, where he taught physics. After leaving the service, David returned to Cornell University to pursue his early geological interests at the Ph.D. level, albeit in the field of geophysics. He began his appointment at the University of Akron in 1999.

Several years ago, David began employing student-centered learning techniques in his large introductory earth science classes. He has extensive experience in using conceptual questions, physical models, and other active learning techniques. His education research, allowing him to identify at-risk students very early in the course so that effective intervention can occur, has produced scholarly publications in the *Journal of Geoscience Education* and numerous national and regional conference presentations. David has been recognized for his extensive research and teaching scholarship at the institutional and national levels. He and David McConnell were recognized together as National Association of Geoscience Teachers Distinguished Speakers and travel the country making presentations about their educational research.

On a more personal note, David frequently experiments with using golf clubs as seismic energy sources and travels the country with his family with a goal of visiting every national park in the continental United States. David brings military discipline to the team and is one of the principal geo-science content writers. David made this comment about his participation: "Writing this text has been both rewarding and humbling. That endeavor constantly reminded me how much I still have to learn about our planet."

Contributing Authors

© Dr. Walter J. Kuleck

Catharine Knight originally hails from Minneapolis, Minnesota. Cathy has become an expert in effective teaching and learning, and in cognitive support of learning for humans. Her research has focused on making the science of cognitive development accessible, practical, and applicable to teachers and instructors in the real world.

She has devoted more than 25 years to the study of how students learn and develop, and how instructors can effectively teach, given the characteristics of both students and the concepts and content to be learned.

Courtesy Lisa Park

Katharine Owens or Kathie, as she's called informally, is the other education member of the team. Kathie says that being a member of *The Good Earth* writing team is one of the highlights of her long career in education. She reports that her interest in science began when she watched the *Apollo 8* astronauts circle the moon and greet everyone on "the good Earth" from their vantage point millions of miles away.

Kathie is convinced that how a subject is taught is equally as important as what is taught and that, if the instructor's methods make the content dull and boring or the students are not challenged to think through the content, much is lost.

The Good Earth

Although we have long understood Earth's position in space, the unique nature of our planet was not fully appreciated until we were able to look at our home from some distance. The astronauts aboard the *Apollo 8* spacecraft were the first people to travel to the moon and were the first to glimpse our home planet from distant space. This view of Earth, commonly known as "Earthrise," was one of the most well-known images of the twentieth century. The photograph was taken by astronaut William Anders during *Apollo 8's* fourth orbit of the moon on Christmas Eve 1968. (The original image was actually rotated so that the moon's surface was near-vertical and to the right of Earth.) A few hours after snapping the photograph, the Apollo crew read the first 10 verses of the book of Genesis during a broadcast to Earth. At the end of the reading, Commander Frank Borman closed communications with ". . . Merry Christmas, and God bless all of you, all of you on the good Earth." For many at home, those early views of the planet from the inky darkness of space illustrated the unique wonders of the fragile environment we share on spaceship Earth.

—*Frank Borman*

"The materials of science are the materials of life itself. Science is part of the reality of living; it is the what, the how, and the why of everything in our experience. It is impossible to understand man without understanding his environment and the forces that have molded him physically and mentally. The aim of science is to discover and illuminate truth."

—Rachel Carson, marine biologist

1

Introduction to Earth Science

Chapter Outline

the big picture

Tsunami coming onshore following the Tohoku earthquake, Japan, 2011.

See The Big Picture box at the end of this chapter for the full story on this image.

1.1 Earth Science and the Earth System

Chapter Learning Outcomes

- Students will evaluate claims in a science-based argument.
- Students will describe the relationships among science, society, and government.
- Students will recognize that Earth is a complex system of interacting rock, water, air, and life.

The fourth-largest earthquake ever recorded buckled the seafloor east of Japan, at 2.46 P.M. on March 11, 2011. Within seconds the tremor was registered on the dense array of instruments, and alerts were issued to millions of citizens throughout the nation. Residents experienced severe shaking within a couple of minutes. Soon a tsunami warning was issued for coastal locations, giving people about 15 minutes to make it to safety. As earthquake-damaged structures blocked streets, people fled on foot, seeking higher ground. Katherine Heasley, an American teaching high school English in the harbor city of Kamaishi, didn't have classes that day. The long-lasting shaking alerted her to the need to move to a safer location. This is how she described what happened next:

An official-looking man was at the head of the column, directing us up the steep road to a recreation center. . . . From the parking lot, I could see the estuary, the port and the long spit of land ending in a pier that separated the two. The first thing that concerned me was that the river was running very rapidly out to sea. Usually, even during low tide, the river was sluggish where it met the bay. The second thing that concerned me was that it wasn't the only thing running fast. Many of the larger fishing boats were absolutely speeding out of the harbor.

Perhaps fifteen minutes after we got to the rec center, the tide turned. Kamaishi has a very sheltered bay, with the deepest breakwater in the world. The tsunami couldn't come in as the classic white-topped wave. But nothing could stop that much water. The water began flowing back, reversing the river's flow. It rose faster than I'd have thought possible. Soon, the pier was swamped. Those of us at the rec center watched helplessly as the water picked up an entire parking lot of cars on the pier and threw them into a building. And the water kept coming. It rose over the marina where the small fishing boats were kept. Then it swamped the area under the highway. Then it flowed into the parking lot of the keisatsu, or the main police station, just down the street from my apartment. And it kept. On. Coming. It threw the police cars into the keisatsu building. It began to hit houses, and the noise was indescribable as it lifted them off their foundations. It flowed across the highway and started rising up the hill we'd come up. There was a deafening crash as, across the bay, the derrick they used to load ships with Nippon Steel products was knocked down. Think about that: A derrick that regularly lifted tons of steel was no match for the tsunami.

The water had turned into some kind of malevolent beast. Some days later, I thought about the Japanese Godzilla mythos, how a creature rises from the depths of the sea and destroys whole cities on land. I can't help but think I now know where that comes from in the collective unconscious of the Japanese people. They depend on the sea; they came across it to find their land, and many make their living off it. But once in a while, the sea comes to take back from the land. When that happens, there is nothing anyone can do except run. The tsunami broke down seawalls and ignored breakwaters. Nothing could stop it. Nothing.

Heasley, Katherine "Personal Account of Tsunami and Aftermath in Japan," www.Oregonlive.com, April 13, 2011

Much of what happened in Japan was eerily similar to what had occurred just seven years earlier in southesast Asia (Figure 1.1). The third-largest earthquake ever recorded buckled the seafloor west

a.

b.

Figure 1.1   Travel times for recent tsunami. **a.** Time taken for 2011 tsunami to travel to locations along the margins of the Pacific Ocean. **b.** Time taken for 2004 tsunami to travel to locations along the margins of the Indian Ocean.
Heasley, Katherine "Personal Account of Tsunami and Aftermath in Japan," www.Oregonlive.com, April 13, 2011; 1.1a: Adapted from http://serc.carleton.edu/details/images/27439.html; 1.1b: Adapted from http://www.ngdc.noaa.gov/hazard/icons/2004_1226.jpg

of Sumatra, Indonesia, at 7:58 A.M. on the day after Christmas 2004. Hundreds of miles away, just before 10 A.M., Penny Smith saw her 10-year-old daughter, Tilly, staring at the sea off Mai Kao beach on the north end of Phuket Island, Thailand. The Smiths were enjoying a warm-weather vacation away from the cold, damp English winter. Tilly noticed that instead of the steady rhythm of breaking waves, the water was flowing down the beach—moving out to sea and not coming back in again. As the beach grew in size, everyone stood to watch, wondering just what strange natural phenomenon they were observing. Tilly knew. She remembered a recent school lesson about tsunami, a series of fast-moving, potentially dangerous ocean waves. Soon Tilly was yelling at her parents to get off the beach.

Meanwhile, about 30 kilometers (16 miles) south, Rick Von Feldt was passing the popular Patong beach in a taxi on his way to his hotel. He noticed people staring out to sea and wondered what they were looking at. A whale? Maybe some dolphins? He too noticed that the water had receded. Soon he was standing on an outdoor balcony of his hotel with a commanding view of the beach below and the bay beyond. He could see boats grounded on the beach and people milling about as if they were not quite sure what to do. This is how he described what happened next on his blog:

Suddenly—in front of our eyes—the bay begin(s) to fill. Rapidly. As if someone had turned on giant faucets—and it just seemed to rush in. In about 10–15 minutes this entire HUGE bay filled in. And then we saw the swell. It was perplexing—because the day was nice and sunny—and we could not figure out—or actually believe what was happening in front of our lives. But it did. The swell came—and we saw it rushing over the wall. And it kept rising—and went higher than the palm trees along the edge. Local people started to cry—for they knew what it meant. We all stood there, stunned. People came running up the road— shrieking. "Water—the water" they were crying. The water receded slightly—and then, again with a vengeance. Rushed

forward—rose again—and the 18 feet wall rolled over the front of the beach—the shops and everything in its path. We stood there in disbelief—not understanding WHY—but realizing that one of the most awful things that could happen—just had. But it wasn't over. It just keep coming and coming. It would recede— and then come again—rushing over the seawall.

Von Feldt, Rick "Surviving the Tsunami-Part 2-The First Hour," www.phukettsunami.blogspot.com, December 27, 2004

In both examples, images of the tsunami's destruction were soon being broadcast around the globe. The world looked on in horror as images of whole towns were wiped off the land (Figure 1.2). Piles of debris could be traced around the rim of the Indian Ocean or up and down the east coast of Japan. Before 2004, few would have anticipated the potential for tsunami linked to massive earthquakes as only three of these mega-earthquakes had ever been recorded worldwide. These types of destructive tsunami are so unusual in the modern era that earth science textbooks written 20 years ago would have briefly mentioned tsunami as part of the chapter on earthquakes, but here we are leading off the book comparing and contrasting two recent dramatic examples. As you can see from the first-hand descriptions given here in Section 1.1, there were similarities to the way the tsunami came on land, sweeping up all before them. However, there were also contrasts in the consequences of these events, mainly as a result of the awareness of the local populations and the warning systems that were in place.

Even though the earthquakes were of almost the same magnitude, the Japanese Tohoku earthquake resulted in about one-tenth of the casualties of the Indian Ocean earthquake. More than 230,000 people were missing or dead following the Indian Ocean tsunami; in contrast the death toll from the Japanese event was closer to 21,000 (Table 1.1). Much of this can be ascribed to the extensive earthquake and tsunami warning systems that are

Figure 1.2 Destruction in Banda Aceh, Sumatra, Indonesia, from the Indian Ocean tsunami. The shoreline has been heavily eroded, and buildings have been almost completely destroyed.
1.2 (before): © DigitalGlobe/ZUMA Press/Newscom; 1.2 (after): © DigitalGlobe/ZUMA Press/Newscom

Table 1.1	Human Toll and Economic Costs of Selected Recent Earthquakes				
Date	Earthquake Location	Earthquake Magnitude	Deaths	Displaced	Economic Losses ($ Billions)
April 25, 2015	Gorkha, Nepal	7.8	8,964	2,800,000	5
August 3, 2014	Wenping, China	6.1	617	12,000	6
September 24, 2013	Awaran, Pakistan	7.7	825	300,000	2
August 11, 2012	Northwestern Iran	6.4	306	36,000	0.6
February 22, 2011	Christchurch, New Zealand	6.3	185	70,000	12
March 11, 2011	Tohoku, Japan	9.0	20,896	131,000	300
February 27, 2010	Maule, Chile	8.8	525	1,500,000	15–30
January 12, 2010	Haiti	7.0	316,000	1,300,000	7–14
September 30, 2009	Sumatra, Indonesia	7.5	1,117	451,000	2.3
May 12, 2008	Sichuan, China	7.9	87,587	>5,000,000	86
October 8, 2005	Kashmir, Pakistan	7.6	86,000	2,500,000	5
December 26, 2004	Sumatra, Indonesia	9.1	230,000	1,700,000	4.5
January 16, 1995	Kobe, Japan	6.9	5,502	31,000	200
January 17, 1994	Northridge, California	6.7	72	22,000	20

in place in Japan. In contrast, in 2004 there was no tsunami early warning system in the Indian Ocean. The lack of effective communication systems in poor regions meant that thousands were killed along the coasts of India and Sri Lanka several hours after waves had already obliterated coastal towns in Indonesia. Economic losses were estimated to be more than 20 times higher in Japan than in Indonesia, largely as a consequence of the much more developed and industrialized Japanese coastline. The 2011 earthquake and tsunami produced the costliest natural disaster in history (~$300 billion). The resulting damage shut down industries for weeks, leading to a global slowdown in automobile manufacturing. The tsunami flooded the Fukushima nuclear power plant, resulting in both a meltdown in three of its six nuclear reactors and the release of dangerous radioactive gases. The disaster put a serious dent in the prospect of a nuclear-powered future as governments in many nations started backing away from plans for building new power plants. This destruction occurred in a country that is better prepared to deal with earthquakes than any nation on the planet. Much of the Japanese coastline is protected by 10-meter-high (33-foot-high) seawalls, specifically to protect against tsunami. Unfortunately, the seawalls were designed to protect against waves produced by earthquakes of smaller magnitude, the type and size that happen quite frequently along the coast of Japan.

Are you smarter than a fifth-grader such as Tilly Smith? What type of natural disaster are you most likely to experience in your lifetime? Are you ready? Will you, like Tilly, know what to do? How much responsibility do individuals have to protect themselves? How many resources should be devoted to trying to predict these events? What type of information should government agencies collect to protect their citizens from such hazards? How can this information be quickly communicated to those at risk? If we are to successfully answer these questions, we must understand

the science of how Earth works and how to effectively integrate scientific findings with the needs of society. Scientists collect data on natural phenomena such as tsunami, but it is often politicians (and, indirectly, the people who elect them) who determine what actions should be taken to protect the public.

Your Introduction to Earth Science

In the remainder of this chapter, we will define *science* and describe how it is done by trained people doing basic research using an array of skills. These scientists ask questions and analyze data; some create maps or collect fossils, while others use sophisticated technologies to search for oil and gas, track ocean currents, or measure changes in the chemistry of Earth's atmosphere. We will explain the principles that scientists use to conduct investigations that weave together data collected from experiments and observations of the natural world. We will discuss the principal roles of the earth sciences in our lives, from finding ways to protect us from natural hazards to investigating the implications of climate changes for the future of humanity.

We finish the chapter by introducing the concept of global change and humans' impact on Earth. Global change is an idea

✔ **Checkpoint 1.1**

Good questions often produce answers that lead to yet more questions. Review the following statement and suggest some related questions that could clarify or expand the topic.

Students who work together in groups often learn more than students in the same class who work alone.

Self-Reflection Survey: Section 1.1

Respond to the following questions as a means of uncovering what you already know about Earth and earth science.

1. Which of the following earth science phenomena have you experienced? Which would you most like to experience? Can you think of three more things to add to the list?
    -   A volcanic eruption
    -   A glacier
    -   A river in flood
    -   A cave system
    -   An underground mine
    -   A canyon
    -   An earthquake
    -   An erosional coastline (rocky cliffs)
    -   A depositional coastline (beaches)
    -   A hot desert
    -   A continental divide
    -   Rock layers with fossils
    -   A big, assembled dinosaur skeleton
    -   A meteor shower or comet
    -   The aurora borealis (the northern lights)
    -   A meteorite crater
    -   A mountain range over 3,000 meters (over 10,000 feet) in elevation
    -   The top of a cloud
2. What three questions about Earth would you like to be able to answer by the end of this course?

that is currently generating research in a wide variety of disciplines relating to all components of the earth system, including geology, ecology, oceanography, and climatology. This work involves thousands of scientists across the globe and has implications for the long-term quality of life for you and your families and is likely to require challenging social decisions within your lifetime. Future economic, cultural, and political choices in all the world's nations will depend on the rate and degree of change. We will follow the theme of global change through many of the chapters of *The Good Earth* and use it to show the links among the components of the earth system. As you will see, there is little that happens on Earth that doesn't involve multiple earth system components.

1.2 The Scope of (Earth) Science

Learning Objectives

-   Describe the principal earth system components.
-   Write a one-sentence definition of the term *science*.
-   Identify examples of the tools that scientists use to learn about Earth.

Earth System Basics

In *The Good Earth,* we introduce you to the study of earth science. Earth is a complex system of interacting rock, water, air, and life where the components and interactions cycle energy and mass throughout the system at a variety of timescales. Changes in one part of this complex

a.

b.

c.

d.

Figure 1.3 The four components of the earth system: atmosphere, hydrosphere, biosphere, and geosphere. All components interact with the solar radiation and other elements from space. How many components are featured in each image?
1.3a: USGS; 1.3b: NOAA; 1.3c: USDA; 1.3d: © Dr. Parvinder Sethi

system can cause ripples that influence multiple related variables and processes. The system can also have balancing or reinforcing feedback mechanisms that can serve to moderate or amplify outcomes or processes. Sometimes complex systems have tipping points that cause the entire system to become unstable when a threshold is broken.

Earth science can be broadly defined as the investigation of interactions among the four parts of the earth system—the atmosphere (air, weather), hydrosphere (water, ice), biosphere (plants, animals), and geosphere (land, rocks) (Figure 1.3). Together, these components form an elegant support system for life. In addition, the sun and assorted features from space, collectively termed the **exosphere**, interact with the earth system and are sometimes considered a fifth earth system component. The historic 2011 tsunami involved three of the components—the hydrosphere (ocean), geosphere (seafloor earthquake), and biosphere (people, plants, animals; Figure 1.4). Throughout this book, we will examine the characteristics of each of the components through the lens of human experience. We will look at how the earth system affects us over a wide range of timescales and how we, in turn, affect different earth system components. We will also be interested in how these components interact with one another and how changes in one component influence processes in the others. Representatives of the earth science community spent some time considering the essential principles or big ideas that everyone should appreciate about the earth system. They ended up with nine "big ideas" that can be divided into a series of secondary concepts. The big ideas were as follows:

- Earth is a complex system of interacting rock, water, air, and life.
- Earth scientists use repeatable observations and testable ideas to understand and explain our planet.
- Earth is 4.6 billion years old.
- Earth is continuously changing.
- Earth is the water planet.

Figure 1.4 Interaction of earth system components involved in the tsunami described in this section. This diagram is a concept map. It includes key terms (in this case, three earth system components) and arrows that describe the interactions between the terms.

✓ **Checkpoint 1.2**

○ basic   ○ advanced
○ intermediate   ○ superior

Make a list of the multiple ways that you interact with each of the four components of the earth system.

- Life evolves on a dynamic Earth and continuously modifies Earth.
- Humans depend on Earth for resources.
- Natural hazards pose risks to humans.
- Humans significantly alter Earth.

We already introduced you to the first idea at the start of this section, and we discuss the second item on the list in the rest of this chapter. The remainder of the points will be discussed in several chapters throughout the book. Take a few moments after you read each chapter to reflect on how many of these big ideas were represented in the material you just read.

Science and Discovery

The second word in the term *earth science* is just as important to us as the first. Much of what you learn in college about science will happen in this and perhaps one other course. Therefore, we want you to have a firm understanding of what science is—and what it is not. Science is not a list of facts to be memorized that have no relevance to your life. Just ask Tilly Smith's parents. The only way to understand how to think like a scientist is to learn to use the skills of scientific reasoning. So in this chapter, and throughout the book, we will give you lots of examples to show that science is a process, a way of thinking about the natural world.

Why do we care what you think about science? The United States is a world leader in scientific research and development. Government and corporate science programs flourish because of substantial investment in innovation and the discovery of new ideas. Even though survey results show that Americans are supportive of scientific research, most people have only a shaky grasp of underlying scientific principles. The National Science Foundation has conducted surveys that reveal that less than one-third of the adult population can define what it means to study something scientifically. Even if you do not make a career in science, it is important that you understand how to use scientific reasoning skills to make wise decisions as an informed citizen to help solve daily problems. **Science is a process of discovery that increases our body of knowledge.** Earth science is like all sciences; some of it is known and can be learned, and much of it is still waiting to be discovered. But science has another less tangible but equally important element—the innate curiosity of the scientists as they search for answers. Increasingly, individual citizens have the opportunity to become involved in a variety of scientific research projects that are too large to complete without the participation of a large number of talented amateur scientists. These projects can involve thousands of observations made by citizens over a limited time interval. They may involve volunteers with little or no training in collecting data or citizen researchers working with scientists in an online community. Examples of some of these projects follow. Which would you most like to be involved with?

- *Stardust:* Volunteers help scientists find a handful of interstellar dust particles brought to Earth by the *Stardust* spacecraft and hiding among a million online images.
- IceWatchUSA: This project by the Nature Abounds organization seeks to compile local climate records by making seasonal observations of ice conditions on water bodies.

- World Water Monitoring Day: Tens of thousands of people in more than 40 nations use simple sample kits to collect information on the characteristics of local streams and lakes. Data are then entered into an online database to be compared with results from subsequent years.
- Project Budburst: This project seeks to track the timing of the leafing and flowering of native vegetation around North America.
- FeederWatch: This data set yields information on the changes in native and alien species of birds using backyard bird feeders and can serve as a proxy indicator of environmental change.
- Weather observers: Eleven thousand volunteers in the National Weather Service's cooperative observer program collect daily temperature and precipitation readings to aid in weather forecasting and to help measure long-term climate change.
- Did You Feel It?: This US Geological Survey (USGS) online program uses reports from people who feel earthquakes and graphs the data to create Community Internet Intensity Maps. For example, a recent 5.8 magnitude earthquake in Virginia generated more than 150,000 citizen reports from people in states up and down the east coast.

Earth scientists combine their basic knowledge of facts and concepts with technical skills to explore Earth and solve its mysteries. It is tempting to view science as a list of facts to be memorized and repeated. But the real essence of science is a detective story in which teams of investigators piece together evidence to generate well-founded explanations of the workings of our planet. Scientists constantly refine or challenge these explanations, causing some to be discarded while others gain wide acceptance. Our imaginations and the physical laws of nature present the only limits to science. Throughout this book, we will strive to give you an inside look at how science is done and to initiate you in the process of discovery. Whenever possible, we will feature real-life situations and pose questions that place you in the role of the scientist.

Tools Used by Earth Scientists

Earth scientists use direct measurements, indirect information, and models to better understand Earth. Direct measurements are collected at field locations by scientists or trained technicians (Figure 1.5). For example, they might determine the type of rocks present (see Chapter 7) to create a geologic map, collect water samples from drinking water wells (see Chapter 12), or

Figure 1.5 This instrument consists of numerous bottles that are submerged to collect water samples from the ocean.
NOAA

gather samples of gases erupting from volcanoes (see Chapter 6). Samples are carefully analyzed and cataloged with information about their original location, the conditions under which they were obtained, and any other data that could affect understanding of the importance of that sample. In these cases, the scientist is directly measuring exactly what she is interested in measuring. The actual measurements may be obtained in the field or in a laboratory.

However, it is often not practical to measure some phenomena directly. In these cases, scientists use indirect measurements. Essentially, they measure something that they can then interpret to get a value for something else. For example, scientists cannot readily examine the features below the world's oceans, but they have been able to use a variety of methods to identify different properties of the rocks of the ocean floor. Measurements of the magnetic patterns of the ocean floor were used to determine that the age of the oceanic crust varied from place to place (see Chapter 4). Satellites measure variations in the height of the ocean surface which is related to the distribution of ridges and trenches on the ocean floor (see Chapter 13). Satellites can also make direct measurements of large regions that would be impossible to map on the surface. For example, scientists have used satellite measurements of Arctic sea ice coverage to show a steady decline over the last few decades (see Chapter 16).

In other cases, scientists use models to better understand Earth. Those models can be physical devices such as wave tanks (Figure 1.6), or they may be theoretical models. The latter are often computer models that simulate complex physical processes to allow investigators to examine the relationships between variables. For example, it has become commonplace for us to see the results of meteorological modeling on our evening news weather

Checkpoint 1.3

| ✓ basic | ✓ advanced |
| ✓ intermediate | ✓ superior |

Three of the big ideas listed near the start of this section detail the interaction of humans and the earth system: (1) Humans depend on Earth for resources; (2) natural hazards pose risks to humans; and (3) humans significantly alter Earth. Take a few minutes and write what you can in support of each of these statements. Consider revising your responses as you progress through the semester to see if you can add more items and/or more information.

Figure 1.6 Wave tank used to understand processes occurring in coastal environments.
© Press Association via AP Images

forecasts (see Chapter 15). Elsewhere atmospheric scientists use an array of models to predict the track of hurricanes, and ever more sophisticated models are being developed to predict future climate trends (see Chapter 17).

1.3 Doing Science

Learning Objectives

- Explain how scientists use observations and predictions to test hypotheses.
- Provide examples of inductive and deductive reasoning.
- Analyze the four basic principles of good science as applied to a real-world scientific investigation.
- Describe examples of poor scientific reasoning.

This information-rich world gives us ready access to all the facts we could ever want. In *The Good Earth*, we have cut down the volume of terms and the amount of new information in favor of addressing the skills needed to process that information—to separate the good from the bad, the significant from the trivial. Most people rate the ability to think critically—to analyze and evaluate information so as to make wise decisions—as more important than knowledge of facts or technical skills that can be easily memorized. Critical thinking is also something science does wonderfully well. The benefits of critical thinking extend beyond science to help us interpret information in a variety of forms, weigh the validity of competing claims, and make judgments on which course of action to pursue. Effective critical-thinking and decision-making skills can be applied to many aspects of life, including buying a house, planning a diet, or successfully managing your time in college.

Science advances by the application of the **scientific method, a systematic approach to answering questions** about the natural world. The scientific method implies that sufficient

✅ **Checkpoint 1.4**

✅ basic	✅ advanced
✅ intermediate	✅ superior

Scientific Analysis

Go to the US Geological Survey site (www.usgs.gov) and find an example of an earth science topic that USGS scientists have investigated.

1. Briefly describe the research, using no more than six sentences.
2. Identify:
   - The types of questions the scientists investigated.
   - The types of tools the scientists used.
   - An example of the data they collected.

observation will reveal patterns that provide clues to the origin and history of Earth. We assume that the components of the universe interact in consistent, predictable ways. Scientists use their observations as an aid in predicting future events in the earth system and, in some cases, the universe.

From Observation to Hypothesis

All of us make observations that we use to mold our personal views of the cultural and physical worlds we inhabit. **Scientists also use observations to shape ideas.** Their ideas are known as hypotheses. **A hypothesis is a testable explanation of facts or observations.** For example, if we owned a Ford Mustang that broke down frequently, we might form a hypothesis that Mustangs or even all Fords were poorly built cars. Through experience, we test the limits of our personal world, allowing those limits to expand to accommodate a positive stimulus or shrink from a negative interaction. Suppose a friend is pleased with the performance of her Mustang. That might require us to modify our original interpretation. Personal observations may vary with the individual, but **valid scientific observations are empirical—that is, they can be measured and confirmed by others.** In that regard, we could collect data on a large number of Ford Mustangs and determine the average number of repairs per car over a specific period of time. We could then compare these data with repair rates for comparable vehicles to support or refute our original hypothesis. Further, others might be prompted to test a similarly large set of Mustangs to confirm our interpretation.

Inductive and Deductive Reasoning

The scientific method is not a single set of steps like a recipe. It can take a variety of forms but includes some or all of the following—making observations, forming and testing hypotheses, developing predictions, planning and conducting experiments, analyzing data, and evaluating results. **A scientific hypothesis is developed to provide a potential explanation of observations.** Hypotheses can be generated and tested using two basic procedures: inductive reasoning and deductive reasoning (Figure 1.7). **Inductive reasoning occurs when scientists draw general conclusions from specific observations.** The success of this method comes

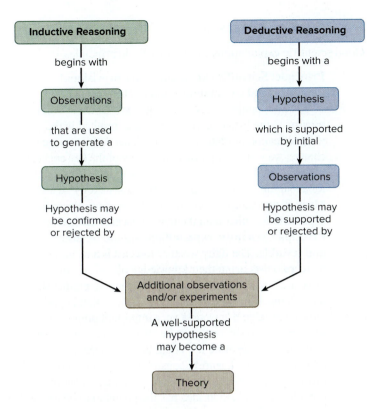

Figure 1.7 Concept map depicting two types of reasoning. Inductive reasoning results when scientists draw general conclusions from specific observations. Deductive reasoning occurs when scientists draw specific conclusions based on general principles.

from recognizing patterns and identifying similarities between comparable systems.

Consider the following situation. Anne, an earth science professor, has students discuss the characteristics of different types of volcanoes with their neighbors during one of her classes. Later, she notices that the students do better on the exam questions about volcanoes than on other questions. She forms a hypothesis that students learn concepts better if they work together. She begins to assign daily exercises for the students to complete in small groups during class. Her average exam scores jump by 4.7 percent over previous classes. Anne is pleased that her students appear to be learning the material more thoroughly, and the students are pleased when she tells them that their grades are higher than expected. Anne has used **inductive reasoning, drawing a general conclusion from specific observations.**

Anne describes her teaching experiment to Don, an instructor in the physics department. He tells her that he too started using group assignments in class. Don tells Anne that he read an article in a physics journal by a colleague at another university that described how the use of groups resulted in improved student learning. Don had done some additional reading about the use of these methods in other science classes and determined that several instructors achieved similar improvements. Consequently, he concluded that if groups could improve learning in other science classes, they would probably also help his students. Don used **deductive reasoning** because he **used a general principle to reach a specific conclusion.**

Students frequently confuse inductive and deductive reasoning, so consider two other examples. What type of reasoning (inductive or deductive) is used in the following pair of statements and why?

1. All hurricanes form as low-atmospheric-pressure systems over oceans. Hurricane Harry is forming in the Atlantic. Hurricane Harry is a low-pressure system.
2. Three massive hurricanes caused large amounts of damage to the United States during the 2005 hurricane season. Hurricane Katrina had a pressure of 902 millibars (mbar); Hurricane Rita, 898 mbar; and Hurricane Wilma, 882 mbar. Therefore, massive hurricanes with air pressures of around 900 mbar or less will cause large amounts of damage if they make landfall.

The first scenario starts with a general statement about hurricanes and concludes with a specific statement about a single hurricane. Therefore, the reasoning is deductive—general to specific. The second scenario starts with specific data (air pressures) and ends with a general conclusion. This is inductive reasoning—specific to general. Most science involves components of both inductive and deductive reasoning.

From Hypothesis to Theory

The best hypotheses are logical and can be readily tested by experiment or by more observations. Continued observations over time will either confirm that a hypothesis is accurate or reveal that it is not quite right and needs to be either further refined or rejected. New information sometimes becomes available with the development of increasingly sophisticated technology and may lead to minor or major changes in existing hypotheses.

An initial hypothesis is a reasonable explanation on the basis of current science and needs further examination. After rigorous testing, bulked up with supporting facts and observations, a hypothesis may become a theory. The US National Academy of Sciences defines a **scientific theory as "a well-substantiated explanation of some aspect of the natural world that can incorporate facts, laws, inferences, and tested hypotheses."** Note that in science, a theory is not just an opinion or a guess; it is a well-supported explanation of a natural phenomenon. (For example, in Chapter 4 we will discuss the theory of plate tectonics.) An even higher standard of scientific scrutiny is reserved for laws. Scientific laws are statements that are so strongly supported by theory and observations that they are considered unchanging in nature. The law of gravity is an example.

✓ **Checkpoint 1.5**

Scientists suggested the dinosaurs became extinct when an asteroid collided with Earth. They noted that *the rare element iridium was present in 65-million-year-old rock layers around the world.* The text in italic is an example of:

a. A hypothesis
b. A prediction
c. An observation
d. A theory

In our constantly changing world, hypotheses or even theories will be modified, and none can ever be completely proved. Widely accepted ideas will be confirmed and strengthened by the work of many scientists, but it is always possible that the next person to test the idea will discover a slightly different result and challenge part, or all, of the original hypothesis. **The willingness to continually question prevailing ideas and to modify or discard them as new information becomes available is the strength of science.** Science is an open book, a perpetual lie detector, limited only by the imagination and abilities of its practitioners. Given the complex nature of Earth, no scientist makes an observation, suggests a hypothesis, or develops a theory alone. The work of every scientist relies on the work of others who have gone before. Even Isaac Newton, whose law of gravity has withstood the test of time, noted, "If I have seen further, it is by standing on the shoulders of giants."

Isaac Newton

✓ Checkpoint 1.6

Observations, Hypotheses, and Mellinarks

Examine the images below. Based on your observations, form a hypothesis as to how many of the images in the bottom row represent Mellinarks.

What was the thought process you went through to arrive at an answer? Try to separate out the "thinking steps" that you took, identifying observations, predictions, and hypotheses. On a separate sheet of paper, briefly describe the steps.

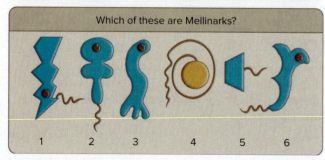

Source: Question adapted from an article by Anton E. Lawson, *Journal of College Science Teaching*, May 1999, pp. 401–11

The Characteristics of Good Science

Good scientific explanations follow four basic principles:

1. **Principle: Scientific explanations are provisional (tentative) and can and do change.** These changes are not always popular. A few years ago, new data on bodies orbiting at the fringe of the solar system resulted in the reclassification of Pluto as a dwarf planet. While science can change, we would emphasize that many of the key concepts discussed in this and other science books have been stable for a long time, some of them well established for centuries. Many of the changes occur in the details of the scientific explanations, rather than the major concepts themselves.

2. **Principle: Scientific explanations should be predictable and testable.** The daily weather forecast is a result of meteorologists using their knowledge of how air and moisture circulate through the atmosphere to predict short-term changes in weather patterns. The idea that hypotheses must be testable brings up a very important point about science. In science, one must be able to test a hypothesis or theory to determine if it could be false. That seems odd, does it not? Scientists construct a hypotheses or a theory to explain nature and then argue that the work is not science unless their idea has the potential to be found to be false. The central idea is this: science deals with the physical world. When we propose a hypothesis, we think it is true. Any additional experiments or observations should either support the hypothesis or show it to be false. Science progresses as hypotheses and theories are tested and shown to be supported (or not). If you cannot test a hypothesis, you are not doing science. Can you think of an idea that cannot be shown to be false (or true for that matter)?

3. **Principle: Scientific explanations are based on observations or experiments and are reproducible.** For example, scientists studying glaciers on different continents have noted that most are decreasing in size (see Chapter 16). These observations can be made routinely by photographing and measuring the dimensions of individual glaciers over several years or decades and are readily confirmed by others willing to visit the same locations. The important characteristic of science lies with the empirical, reproducible data used to support or refute a hypothesis. Scientific results are discussed openly at conferences and published in journals so that all ideas are exposed to review by other scientists (peers). This peer review process ensures that published research is original and adds to the body of valid scientific information. Rogue scientists who publish false data to advance their careers are often discredited by other scientists who are unable to reproduce the original results.

4. **Principle: A valid scientific hypothesis offers a well-defined natural cause or mechanism to explain a natural event.** Science looks for a cause for every effect. A tsunami is caused by an underwater earthquake; an earthquake is caused by movement of tectonic plates; tectonic plates move because Earth is releasing internal heat energy (more on all this in Chapter 4). Sometimes these cause-and-effect relationships are complicated, with multiple causes

✓ Checkpoint 1.7

Identify the application of the four characteristics of good science in the passage that follows.

Science in Full View: The Hutchinson Gas Explosions

Jerry Clark heard the explosion and ran outside in time to see debris fall to the ground. Everyone in the city of Hutchinson, Kansas, heard or felt the blast at 10:45 A.M. on Wednesday, January 17, 2001. Glass covered downtown streets, and two stores were soon in flames. City emergency workers initially assumed that the blast was the result of a gas leak and so shut off gas supplies to the area. But the fire burned on. That evening, jets of gas erupted from the ground on the edge of the city east of the explosion site. A second fiery explosion on Thursday killed two people at a mobile home park near the gas jet site. By this time, no one knew what to expect next. The local police and National Guard quickly evacuated residents living near the second blast site, and the governor's office declared a state of emergency.

Both explosions and the gas jets were linked to abandoned wells, some drilled perhaps as much as a century earlier. It soon became clear that the gas was escaping through the wells from a source far below the city. Large underground caverns in the nearby Yaggy Corporation gas storage facility (see the figure) experienced a significant pressure drop on Wednesday morning as natural gas was being pumped into the storage caverns. Later investigations revealed that the source of the leak was a fist-sized hole in the casing of one of the wells used to pump gas into and out of the caverns. Kansas Gas Service (KGS) operated the Yaggy facility, one of a network of natural gas storage centers across the nation.

The Kansas Geological Survey immediately volunteered to help investigate the explosions. Survey geologists had previously produced a geologic map of the county around Hutchinson. This survey had provided sufficient information about local rocks to determine that it would be possible for gas to rapidly migrate the 10 kilometers (6 miles) from the Yaggy facility to Hutchinson. The scientists could do little more until they were able to analyze the situation more closely. At this early stage in the study, most investigators thought the gas had escaped from a leaking gas well, migrated upward, and then traveled toward Hutchinson in a rubblelike rock layer that capped the salt deposit containing the storage caverns (see the figure). KGS employees drilled numerous wells in an effort to allow any remaining gas to escape safely to the surface. The drilling revealed that the gas was not traveling along the rubblelike layer but was closer to the surface in a thin band of rocks about 100 meters (330 feet) below the city.

On January 30, the governor of Kansas sent a team of geologists, geophysicists, and engineers from the state Geological Survey to Hutchinson to find any remaining gas buildups. Survey representatives coordinated their work with local officials, KGS employees, and other state agencies and soon launched a website to keep residents and interested observers updated with the latest findings. The search for answers was made transparent as information was shared through the website or in public meetings where Geological Survey personnel answered questions and addressed concerns of local residents, officials, and reporters—even when sometimes the answer was "We don't know."

The scientists used ground imaging technology to examine the characteristics of the rock layers underlying Hutchinson and to search for possible locations where gas had collected. KGS drilled wells in two of those locations and vented large volumes of gas. Rock samples recovered from the gas-bearing zone contained a rock type known as dolomite. Geologists hypothesized that a series of connected fractures in the dolomite had served as a pathway for the gas to travel from the Yaggy facility to Hutchinson. Although they never recovered rock samples showing the presence of fractures, these scientists concluded that fractures in the dolomite layer represented the only reasonable passageway through which the gas could move quickly over that distance. Continued investigation revealed no remaining gas deposits, and on March 29, the Geological Survey's representatives told Hutchinson residents: "From a geological viewpoint, the city is safe."

The example of the Hutchinson gas explosions shows that scientific investigations are driven forward by the curiosity and persistence of scientists who systematically rule out potential solutions to arrive at an explanation. This example also illustrates that science doesn't have unlimited resources, personnel, or time. In some cases, it may be necessary to walk away and settle for the best answer available under the circumstances.

influencing the end result. For example, changes in Earth's climate are caused by effects from every component of the earth system, including the composition of the atmosphere, oceanic circulation patterns, the combustion of fossil fuels, plant respiration, and our position relative to the sun. Ongoing research is attempting to determine just how much each of these factors influences climate change.

Geoscience is a discipline based on making observations of the Earth and testing hypotheses about Earth's history. The methods used by geoscientists include comparing features preserved in the rocks of the geologic record with modern processes to interpret how those ancient features formed; finding commonalities and differences; developing converging lines of evidence and testing through prediction. Geoscientists also think about the Earth using a framework that recognizes that the Earth is very old and has a dynamic history that is shaped by a continuum of many long-lived, low-impact processes and fewer short-duration, high-impact processes. These scientists also value collaboration as a strategy for effectively moving forward our understanding of the Earth. Let's see how these principles were applied in one of the most widely publicized scientific discoveries of recent years.

An Example of Good Science: The Alvarez Hypothesis

More than 30 years ago, a team of scientists led by the father-son pair Luis and Walter Alvarez suggested that the extinction of the dinosaurs was caused by a collision between Earth and an asteroid or comet. Walter Alvarez was a young paleontologist, a geologist specializing

Figure 1.8 Badlands near Drumheller, Alberta, Canada, where erosion has exposed the K-T boundary. The boundary is located approximately where the light- and dark-colored rocks meet in the upper part of the outcrop.
© Glenn Larson

in using fossils to decipher the history of Earth. He was investigating the geologic history of the Mediterranean region, and his research took him to a large outcrop of rocks near the town of Gubbio in the Apennine Mountains of central Italy. There, he examined layers of rocks that represented the time in Earth's past that spanned the extinction of the dinosaurs. Dinosaurs (and many other species) died out 65 million years ago (over a period that may have lasted somewhere between 100,000 and 3 million years). Rocks formed before the extinction are classified as Cretaceous in age; those formed after are classified as Tertiary (more on these names in Chapter 8). Most species living on Earth during the Cretaceous became extinct prior to the start of the Tertiary time period. At Gubbio, a thin clay layer marked the Cretaceous-Tertiary boundary (abbreviated as the K-T boundary) between the different ages of rocks. The K-T boundary has subsequently been identified elsewhere (Figure 1.8). You may wonder why it wasn't known as the C-T boundary. Some other periods of geologic time also begin with a C, so, to avoid confusion, it was decided to use the letter K, taken from the German word for chalk 'kreide,' which is characteristic of rocks from that time.

At the time Alvarez was doing his field work, earth scientists had published a variety of different hypotheses seeking to explain the demise of the dinosaurs and other species. For example, some scientists believed the climate got too hot or cold for the dinosaurs, others thought that they were harmed by radiation from a supernova explosion, still others suggested that the evolution of flowering plants affected dinosaur eating habits, while yet others hypothesized that smaller organisms ate their eggs, causing a rapid population decline (provisional hypotheses; principle 1). However, some of these hypotheses could not be readily tested (violated principle 2), and none of the others had been widely accepted, so research continued. Walter Alvarez sought to estimate the rate at which species changed on either side of the K-T boundary by measuring the rate at which space dust had been deposited in the clay layer at the boundary (he used principle 3). Space dust falls to Earth daily and contains rare elements that can be readily measured in the lab, although in low concentrations of parts per billion.

Alvarez returned to the University of California, Berkeley, with samples of the clay layer. His physicist father, Luis Alvarez, suggested that his colleagues Helen Michel and Frank Asaro perform a chemical analysis of the clay material. The analysis revealed the rare metallic element iridium in the clay. Iridium is normally present in concentrations of 0.3 part per billion in rocks of Earth's crust, but Michel and Asaro found concentrations of 9 parts per billion, 30 times the expected amount. They found similar concentrations at other K-T boundary sites in Denmark and New Zealand (supporting data; principle 3). In seeking an explanation for the increase in iridium concentration over such a wide area, the Alvarez team recognized that objects such as asteroids and comets contained elevated levels of iridium and other rare metals. They hypothesized that a relatively large amount of iridium was deposited when an asteroid or comet collided with Earth (natural cause; principle 4; Figure 1.9). They published their hypothesis in 1980 in the journal *Science,* as a paper titled "Extraterrestrial cause for the Cretaceous-Tertiary extinction." The Alvarez hypothesis interpreted the data to suggest that a collision with an approximately 10-kilometer-wide (6-mile-wide) asteroid would have generated so much debris that it blocked incoming sunlight for several years. Vegetation would have died in the absence of light, leading to the deaths of plant-eating dinosaurs and the collapse of the global food chain. The decade following publication of the *Science* article saw a surge in research interest in the extinction event as scientists sought to find data that would support or refute the Alvarez hypothesis (continued study using principles 1–3 above). Within a dozen years, over 2,000 articles and books had been written on the topic.

Soon, several researchers had confirmed the presence of high concentrations of iridium in rocks at the Cretaceous-Tertiary boundary at multiple sites around the world (more data; principle 3). Scientists Alan Hildebrand and William Boynton discovered tsunami deposits in K-T boundary rocks in southern Texas, as well as thick layers of debris in deposits of the same age in Mexico and Haiti; and they predicted that an asteroid impact site should exist somewhere around the Gulf of Mexico (a prediction; principle 2). Soon it was discovered that two petroleum geologists had previously published the results of geophysical exploration in the Yucatan Peninsula, Mexico. They described a buried, near-circular feature (Chicxulub Crater) over 200 kilometers (125 miles) wide that Hildenbrand and Boynton now thought could be the possible impact site (another hypothesis; principle 4). The crater was linked to the K-T collision event as it lies in rocks that are older than the impact event and is covered by rocks that are less than 65 million years old. (For more on impacts, see Chapter 3.)

However, this story is not yet over. The provisional (tentative) nature of science makes it possible that other hypotheses may yet better explain some aspects of the extinction event. In the dinosaur extinction debate, some scientists suggest the source of the iridium was actually a massive series of volcanic eruptions that took place over an interval of half a million years around the same time as the impact event. More-recent research has discovered that iridium can be produced in relatively large quantities by some volcanic eruptions (another observation; principle 3). The span of the volcanic eruptions also better matches the more-gradual die-off of some of the extinct species. Of course, it is possible that both events combined to kill the dinosaurs. Stay tuned: it is likely that the public's fascination with dinosaurs means that we haven't heard the last of this scientific debate.

Limitations of Science

We must keep in mind that scientific explanations may be limited by available technology or other factors. For example, prior to the invention of the telescope, knowledge about Earth's position in space was based on observations made with the naked eye. Astronomers such as Galileo used some of the first telescopes to make observations of planetary orbits that would support the hypothesis that the sun, not Earth, was the center of our solar system.

Although science is a powerful method for examining unresolved questions, *science cannot answer all questions*. Questions that center on ethics or theology often have more to do with cultural or social norms than with scientific concepts. For example, recent concerns about the potential for cloning humans can be separated into two distinct questions; one is scientific, the other ethical. Can we successfully clone humans? is a scientific question, and the current answer is no, although research suggests that a future response could be yes. Should we clone humans? is an ethical question. If the answer is no, we may never clone a person even if the scientific knowledge exists to do so.

Figure 1.9 Asteroid impact with Earth. Luis and Walter Alvarez hypothesized that an event like this occurred 65 million years ago and contributed to the extinction of the dinosaurs.
© Sanford/Agliolo/Corbis

The Characteristics of Bad Science

Poor scientific reasoning rarely reaches a public forum because of the checks and balances inherent in the scientific process. However, sometimes hypotheses are unveiled in the media before they can be rigorously tested. Unfortunately, on further analysis, some of these ideas may be proved wrong, prompting increased skepticism toward the scientific process and scientists in general. Some telltale signs indicate when an argument is not based on sound scientific thinking. Here are some things to look out for when people claim to disagree with a scientific hypothesis or offer unsupported explanations for natural phenomena:

- **An attack on the scientist, not the science.** Science does not advance based on the personalities of the scientists but on verification of facts and observations. (However, it can move backward when a scientist fabricates data to support a hypothesis. This happens only rarely.)
- **People who argue from authority.** Just because a person is important or powerful does not make him or her right. The extremely powerful Roman Catholic Church disagreed with Copernicus when he pointed out that Earth rotated around the sun, but it turned out that he was correct.
- **Confusion over cause and effect.** This type of thinking is often summarized by the Latin phrase *post hoc, ergo propter hoc:* "it happened after, so it was caused by." For example, a student may claim that he did well on an exam because he wore his "lucky" shirt.
- **The use of bad statistics.** Even the weakest scientific arguments may look appealing if supported by statistics that are based on a biased sample or on a sample size that is too small to be representative.

Scientists who don't engage in peer review to evaluate their research will not have their work published and may be discredited if their results cannot be reproduced by others. Alternatively, some hypotheses receive publicity before they have had an opportunity to be critically reviewed by experts. One example of a very public—but failed—hypothesis is described next.

✓ Checkpoint 1.8

Employees at the Ripley's Believe It or Not! Museum in Myrtle Beach, South Carolina, declare that female visitors who come in contact with a pair of African fertility statues are more likely to become pregnant some time later. The statues, from the Boule tribe of the Ivory Coast, stand near the museum's entrance. Some visitors have volunteered the information that they gave birth 9 months after touching the statues and credit the statues. The museum notes that some couples travel from as far away as Texas to rub the statues.

1. What hypothesis is presented in this story?
2. Is the hypothesis supported by sufficient observations? Explain.
3. What prediction could be made to verify or falsify the hypothesis?

Source: Summarized from an article by Isaac J. Bailey, *Houston Chronicle*, October 15, 2000

An Example of Bad Science: Prediction of a Midcontinent Earthquake

Self-proclaimed climatologist and businessman Iben Browning proposed that an earthquake would occur on the New Madrid fault zone in southeastern Missouri on or around December 3, 1990. He based this prediction on the fact that New Madrid had been the site of an extraordinary series of major earthquakes (sometimes called the Mississippi Valley earthquakes) over a 3-month span from December 1811 to February 1812. Browning hypothesized that tidal forces due to the gravitational pull of the sun and moon could trigger another big earthquake on the New Madrid fault zone . . . or maybe one in Japan . . . or in California . . . well, somewhere in the Northern Hemisphere.

The hypothesis generated widespread media interest in the region and raised public anxiety sufficiently that many local schools closed in anticipation of an impending quake. Browning's claims were widely denounced by earthquake specialists, who were also frequently quoted in local newspapers. By the time the fateful day arrived, the hype had taken over, and the area was besieged with reporters who, as it turned out, were able to report that nothing happened.

This was bad science because Browning did not offer an accepted mechanism to explain the occurrence of the potential earthquake. Although tidal forces do exist, they had not been linked to earthquake activity. But this story also illustrates that even a clear, unambiguous message from experts (there's *not* going to be an earthquake!) can get lost in the shuffle. Scientists analyze situations; they do not write the newspaper stories or determine how schools and other public services should respond.

1.4 Science and Society

Learning Objectives

- Identify physical or chemical and social or cultural aspects of the earth system.
- Compare and contrast protection and adjustment procedures related to natural hazards.
- Explain the four principal roles that earth scientists play in society.
- Describe examples of how citizens interact with the natural environment at local, national, and global scales.

Why should you care about science—and earth science in particular? Most of us are removed from the process of science. (How many scientists do you know?) Past experiences have convinced some people that they will never understand science, whereas others may view the study of earth science as irrelevant to our comfortable twenty-first-century lifestyle. Many citizens are understandably bewildered by media reports that portray battling teams of scientists presenting opposing explanations for complex scientific problems. If the experts cannot agree, they reason, how can we be expected to make a decision? Besides, even if we understand environmental problems, we are often frustrated by the apparent inability of those responsible to do anything about them. This can range from simple individual actions (Why doesn't my neighbor recycle?) to issues of corporate responsibility (Why do companies pollute the air?).

How can we become enlightened citizens capable of identifying problems that will affect us all and participate in their solutions? We have to combine the critical thinking we described in Section 1.3 with civic thinking that involves the analysis, planning, and evaluation of actions that may help society to arrive at solutions to these problems. In this context, *society* may refer to anything from a small town up to the global community. Here we suggest a simple three-step process: know, care, act.

- **Know.** We must take responsibility for our world by knowing how it works.
- **Care.** Our society works best when we care about how our actions affect others. But we should also be aware of how we are affected by the actions of others.
- **Act.** Do something. Make your opinion known. Go to a town meeting, blog about it, write a letter to your local paper, contact your congressperson or senator, vote. To quote anthropologist Margaret Mead: "Never doubt that a small group of thoughtful, committed citizens can change the world. Indeed, it is the only thing that ever has."

The Role of Earth Science

We have been interacting with Earth's environment since our human ancestors began to roam Earth's surface. Our demands on the planet have been magnified as technology developed and Earth's population increased. It has become increasingly necessary to monitor fundamental features of the environment around us. **The principal elements of the environment (air, water, soil) have specific chemical and physical characteristics that can be readily measured.** Scientists can measure the volume of dust in the air or the abundance of a chemical in a stream to determine if the air or water quality falls below community standards.

Social or cultural influences on decisions affecting the environment are more difficult to quantify than physical and chemical conditions. Consequently, environmental decisions are complex to evaluate and are often the subject of vigorous debate. Furthermore, the influence of these social and cultural factors changes with time as perceptions change. For example, our view of the role of

wilderness has evolved in the 400 years since the earliest European settlers arrived on the North American continent. The early colonists considered the virgin forests home to unfriendly natives and mythical beasts, so they regarded wilderness with hostility. However, as the population expanded and the number of wilderness areas dwindled, the remaining natural lands began to be considered important cultural assets and were consequently protected by legislation such as the Wilderness Act (1964).

Given the complexities of people's relationships with our planet, earth scientists have several roles to play in modern society. These roles have become more crucial as our global population climbs past 7.1 billion people, with about 77 million more added each year. We are concerned about protecting life and property from the dangers of natural hazards, obtaining sufficient natural resources to maintain or improve our standard of living, and protecting the health of the natural environment. A final, more comprehensive goal, ensuring the future of our own species, has recently received increasing attention as we glimpse a future in which human actions modify the composition of the atmosphere and we recognize the global-scale devastation that may result from an asteroid impact event. All of these roles require a similar scientific approach. Acting as representatives for society at large, scientists seek to:

- Identify and measure natural processes and phenomena.
- Understand the natural processes involved.
- Monitor these variables over time and predict future trends or events.
- Engage citizens and their representatives in using this information to make effective decisions to meet society's needs.

Protecting Against Natural Hazards

Scientists play a vital role in understanding and determining the potential risks from natural phenomena that may harm people and damage property. Natural processes such as earthquakes, landslides, floods, volcanic eruptions, tornadoes, and hurricanes are considered hazards when they occur in populated areas (Figure 1.10). The detailed study of hazards in one area can help predict the potential risks elsewhere. For example, scientists used

Checkpoint 1.9

☑ basic ☑ advanced
☑ intermediate ☑ superior

Read the following summary of the Comprehensive Environmental Response, Compensation, and Liability Act (CERCLA, 1980). Does this law involve the measurement of physical or chemical characteristics of the environment, or did it arise from social or cultural concerns, or were both important? Explain your answer.

CERCLA established prohibitions and requirements concerning closed and abandoned hazardous waste sites, provided for liability of persons responsible for releases of hazardous waste at these sites, and established a trust fund to provide for cleanup when no responsible party could be identified. The law authorized two kinds of response actions: short-term removals, where actions may be taken to address releases or threatened releases requiring prompt response; and long-term remedial response actions that permanently and significantly reduce the dangers associated with releases or threats of releases of hazardous substances that are serious, but not immediately life threatening.

Figure 1.10 Principal natural hazards for some US cities. How does the type of hazard vary by region?

Figure 1.11 Hurricane Katrina approaching the Gulf Coast, August 2005. Katrina devastated coastal communities in Louisiana, Mississippi, and Alabama. The storm was responsible for more than 1,800 deaths and was the costliest natural disaster in US history. NOAA

Checkpoint 1.10

basic · advanced
intermediate · superior

Is evacuation of a city before a hurricane an example of the prevention of a hazard or adjustment to a hazard?

a. Prevention   b. Adjustmeant

What other examples of prevention or adjustment have been described in the chapter so far?

the information they learned from investigations of a 1980 volcanic eruption of Mount St. Helens in Washington state (see Chapter 6) to accurately predict the size and timing of the larger 1991 eruption of Mount Pinatubo in the Philippines. (Was this an example of inductive or deductive reasoning?)

Each decade the cost of property damage from natural hazards more than doubles (when adjusted for the rate of inflation). In some cases, the risks from natural hazards depend more on development decisions made before these events occur than on the natural phenomena themselves. Human beings are unlikely to be able to stop volcanoes from erupting or to banish earthquakes. However, the application of scientific knowledge and appropriate technology can help save lives and protect property. The principal advantage of technology is to provide accurate information to maximize the safety of people living in areas at risk for natural hazards. The effects of some potentially destructive phenomena can be partially offset by the following technological advances:

- Weather satellites are used to track hurricanes and predict landfall sites, allowing timely evacuation of residents (Figure 1.11).
- Doppler radar installations have more than doubled the amount of advance warning time for tornadoes.
- Networks of monitors linked by satellites measure stream flow and predict the magnitude and timing of floods, allowing emergency construction of levees and evacuation of residents.
- Engineering structures and strict building codes are deployed in areas at risk from large earthquakes to ensure that buildings, although damaged, will remain standing when the shaking subsides.
- Arrays of instruments at more than 150 volcanoes are monitored by the USGS National Volcano Early Warning system to predict the timing, location, and magnitude of future eruptions.

Mitigation represents actions that prevent or reduce the probability of a natural disaster or lessen the effects of the event. Although we cannot prevent most natural hazards from occurring, we can make adjustments that will minimize their impact through careful land use planning, the enforcement of building codes, and the purchase of insurance policies. These steps ensure that some areas at risk are not heavily populated, that key structures are built to withstand the hazard, or that funds are available to repair damages following a hazardous event. Floods and landslides are clearly linked to streams and slopes, allowing scientists to make local alterations to the environment in efforts to mitigate future hazards. For example, building levees to contain rising streams or reservoirs to store floodwaters can locally diminish or eliminate flooding. However, we should be aware that any alteration of a natural system has the potential to cause unanticipated changes. For example, building a levee may reduce flooding locally but actually increase the flood risk downstream where the stream is in its natural state. As the devastation of New Orleans following Hurricane Katrina amply illustrates, our confidence in engineered solutions such as levees and flood control structures often prompts increased development, ultimately resulting in greater loss of life and property when disaster inevitably strikes.

In assessing the risks associated with natural hazards, earth scientists must try to answer several questions: How often do such hazards occur? How large an area will be affected? How grave is the risk to people and property? What actions can be taken in both the short and long term to prevent some of these events or lessen their impact? Determining the correct answers to these questions requires knowledge of earth processes, the characteristics of the landscape, the type and distribution of rocks underlying a region, and the rocks' physical and chemical properties. These scientific questions are more than academic—they focus on the very safety and security of human lives. We will explore the science behind natural hazards in Chapters 5, 6, 10, 11, 13, and 15.

Finding and Sustaining Earth's Resources

Life on Earth requires the use of resources. The term *resources* covers everything we use, including such basic assets as air, soil, timber, and water; fuel resources such as coal, oil, and gas; and mineral resources (Figure 1.12), such as sand and gravel. These natural resources may be renewable or nonrenewable. Renewable resources are replenished constantly (wind, soil), on short-term timescales measured in months (crops) or over longer intervals of several years (timber). Nonrenewable resources either are lost following consumption (fossil fuels) or may be recycled to be used in other products (metals). In 1900, renewable resources (agriculture, food, forest materials) accounted for 41 percent of the consumption of US raw materials. Today, they represent less than 10 percent of total materials consumed (by weight).

The United States uses more mineral and energy resources than any other nation, and our economic growth is dependent on continued access to these resources. There are 1,300 billion barrels of oil available in global oil reserves. Each day, the world uses approximately 90 million barrels (33 billion barrels per year). How long before oil reserves are depleted, according to those numbers? Currently, the United States imports a substantial fraction (nearly 50 percent) of its oil, and we rely exclusively on other nations for more than a dozen critical mineral commodities. One-third of the world's people live in the rapidly expanding economies of China and India, placing even greater demands on global mineral and energy supplies. In an ideal world, the human race would develop into a sustainable society, a society that satisfies its desire for resources without completely consuming resources essential for future generations. However, given the pace of global economic development, it is unlikely that we will achieve sustainability in the near future.

Figure 1.12 Strip mining for the metal manganese in South Africa. USGS

Earth scientists work to understand the distribution, abundance, and origin of resources. Will there be sufficient resources to support the growing global population 50 years from now? What steps can we take to preserve and protect the most heavily exploited resources? What alternative energy sources can be utilized to make the nation less dependent on foreign suppliers? Successfully answering these questions requires earth scientists to explore ever more remote parts of Earth's surface, including rain forests, rugged mountain ranges, and the deep ocean floor.

It is theoretically possible that Earth could support many times its current population, but such speculation takes no account of the quality of lives people would be required to lead to ensure sufficient food (and other resources) for all. Individual actions such as turning on a light switch or pouring a glass of water involve relatively modest resource use and require little thought except in the most extreme conditions. However, multiply those actions several billion-fold and divide some resources across international borders, and we can readily imagine situations where resource exploitation can have wide-ranging consequences (Figure 1.13). The distribution and exploitation of natural resources are examined in Chapters 2, 7, 9, 11, 12, 14, 16, and 17.

Protecting the Health of the Environment

The biosphere (plants, animals) has exhibited dramatic changes throughout Earth's history, but recent population growth has contributed to environmental change, albeit over a much shorter timescale. Global population more than quadrupled since 1900, and we will add several billion more people this century. As population has expanded, so has industrialization and consequently pollution of land, air, and water. Pollution is still readily visible in

Figure 1.13 Human-generated lights on Earth. These patterns indicate the distribution of population and serve as a proxy (substitute) indicator of energy consumption for different nations. Densely populated, developed regions (United States, Europe) show brighter lights than heavily populated, developing nations (India, China). Sparsely populated regions are dark (South American rain forest, central Australian desert). NASA

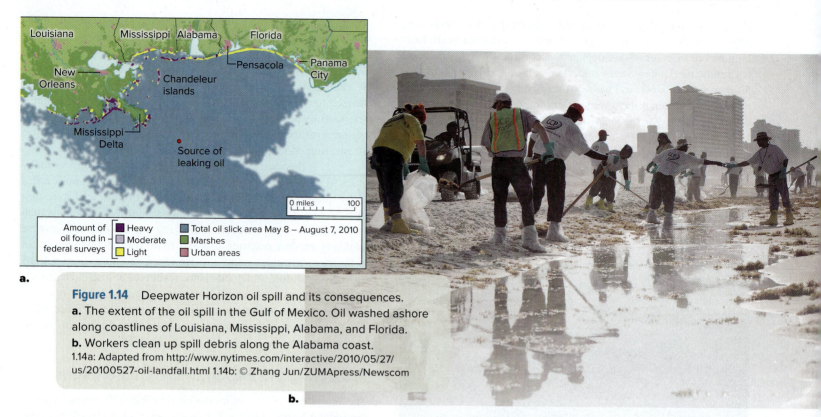

a.

b.

Figure 1.14 Deepwater Horizon oil spill and its consequences.
a. The extent of the oil spill in the Gulf of Mexico. Oil washed ashore along coastlines of Louisiana, Mississippi, Alabama, and Florida.
b. Workers clean up spill debris along the Alabama coast.
1.14a: Adapted from http://www.nytimes.com/interactive/2010/05/27/us/20100527-oil-landfall.html 1.14b: © Zhang Jun/ZUMApress/Newscom

developing countries, but in the United States, its effects are muted and much more subtle, as indicated by reports of respiratory ailments and contaminated drinking water supplies. We can consider these threats as examples of slow-motion hazards. While they don't have the sudden impact of an earthquake or tornado, their long-term consequences have the potential to be much more devastating.

Human activities have the potential to endanger human life and natural ecosystems. For example, we have found a variety of ways to contaminate the hydrosphere. One of the most spectacular was the April 2010 explosion of the *Deepwater Horizon* oil rig, which resulted in the spill of 4.9 million barrels (206 million gallons) of oil into the Gulf of Mexico. The disaster continued for three months as crews struggled to cap the gushing well on the floor of the gulf, 1,600 meters (5,100 feet) below the sea surface. Eleven rig workers lost their lives in the initial explosion, and the spill eventually contaminated 1,064 kilometers (665 miles) of coastline in multiple states (Figure 1.14a). The oil spill itself, and the chemicals used to disperse the oil, combined to decimate marine species in much of the northeastern Gulf. British Petroleum, the company that leased the well, has spent approximately $40 billion cleaning up the region, and legal challenges remain. As we write this, six years after the disaster, it is too early to determine if the biological communities that were devastated by the spill had fully recovered. We will explore how the health of the environment is affected by both human and natural causes in Chapters 2, 6, 8, 9, 12, 13, 14, 15, 16, and 17.

Ensuring the Future of Human Life

The issues discussed so far occur at the local, regional, or national scale and involve events that are significant on timescales measured in hours to years. However, if we take a more global view, we can identify processes that have the potential to affect everyone, everywhere, for decades and perhaps centuries to come: the impact of a large asteroid and the effects of global climate change. Any program that attempts to address either asteroid impacts or global warming would be both complex and expensive, requiring cooperation among many nations and potentially taking decades to complete. Nevertheless, these threats cannot be ignored, and science has the potential to show the way to effective solutions. We examine the science behind these issues in Chapters 2, 3, 4, 6, 8, 13, 14, 16, and 17.

The **impact of a large asteroid with Earth represents a global-scale natural hazard** (Figure 1.9) that has the potential to end all life as we know it or to devastate a continent-sized area of the planet. Concerns about such an impact increased in recent years as we became aware that such events were more commonplace in the geological past than was previously thought. Although scientists have many ideas about how to stop an asteroid on a collision course with Earth, no mechanism yet exists for dealing with such an event. We will address issues related to asteroid impacts in Chapter 3.

An international panel of scientists has concluded that **global warming represents an alteration of global climate patterns resulting from human activity.** Most scientists think that carbon dioxide and other gases of human origin have altered global climates over the last century. Higher concentrations of carbon dioxide are associated with warmer temperatures. Warmer conditions have the potential to cause wholesale changes in natural systems around the world.

Scientific research on global change is an example of "big science" because it involves researchers around the world working on thousands of different projects, each contributing a small piece to a much larger puzzle. The 1990 Global Change Research Act required the federal government to implement a climate change research program. The US government budgets nearly $2 billion for climate change research each year through the Climate Change

✓ **Checkpoint 1.11**

Complete the following concept map to summarize the characteristics of the four principal roles that earth scientists play in society.

Earth Scientists

study Earth to learn more about | study Earth to learn more about | study Earth to learn more about | study Earth to learn more about

natural hazards | [ ] | [ ] | [ ]

such as / such as | such as / such as | impacted by / impacted by | threatened by / threatened by

tornadoes | floods | [ ] | coal | [ ] | hazardous waste | [ ] | asteroid impact

such as | such as

[ ] | [ ]

Science Program, which involves workers from more than a dozen government agencies. The investigations of the program involve hundreds of scientists working in research teams and examining many different topics that will contribute to our understanding of global change. Each team of scientists must make a research plan, collect data, make observations, draw conclusions, present their work at professional meetings, and write technical articles during the term of their research. Researchers seek to piece together a story about past and future global change by reading literally thousands of publications and synthesizing hundreds of ideas to build sophisticated computer models. This process represents a lot of hard work, and the process moves forward slowly in careful increments. This is the nature of science.

The Anthropocene: A New Time on Earth?

Nobel Prize–winning scientist Paul Crutzen suggested that human activity has produced such sweeping changes to the planet that we have entered a new period of Earth history, informally termed the Anthropocene. Crutzen and others point out that global temperature, sea level, and atmospheric chemistry were relatively stable for the last 8,000 years or more. This apparent stablity was disrupted in the last few centuries as the rapid increase in human population was accompanied by economic growth and industrialization, resulting in widespread resource exploitation and environmental change. Geologists differentiate distinct time intervals when they can recognize some characteristic features in the rock layers formed at that time. Some researchers suggest that thousands of years from now, future scientists will be able to recognize several key markers preserved in various parts of the earth system that will be readily interpreted as the result of changes during the Anthropocene time interval. Key changes include:

- Increasing erosion rates: More than one-third of Earth's land surface is exploited by humans. Layers of sediment on the seafloor will record activities such as agriculture and construction that remove more earth materials than natural processes.

- Changing atmospheric chemistry: Gas bubbles trapped in thick ice sheets will show higher concentrations of such gases as carbon dioxide, methane, and sulfur dioxide.

- Decreasing biotic diversity: Fewer species will be represented in fossil sites as many become extinct or less adaptable as single-variety forms used in agriculture replace the variability of natural species.

- Ocean chemistry: Surface ocean waters are becoming more acidic as they absorb higher concentrations of carbon dioxide. This hinders the growth of some species (for example, corals) and dissolves the shells of others, removing them from the rock record.

- Sea level change: Short-term changes will be less than a meter over the next century but may be measured in tens of meters by the end of the millenium.

The concept of the Anthropocene represents significant global changes linked to all parts of the earth system. We will examine different aspects of global change research as we move through *The Good Earth*. By introducing the topic of global change here, we hope to give you an opportunity to look over the shoulders of the researchers to see how our understanding unfolds as scientists try to figure out how our home planet works.

✓ **Checkpoint 1.12**

Read the following quote. Discuss why you agree or disagree with the statement.

This is the first generation in the history of the world that finds that what people do to their natural environment may be more important than what the natural environment does to and for them.

Harlan Cleveland, former US Assistant Secretary of State.

the big picture

Throughout this chapter, we have focused on the characteristics and practices of science. We started with examples of earth processes that challenge us to apply the methods of science to understand how our planet works so that we can avoid, or at least be prepared for, similar events in the future. Japan was reeling from the effects of the disastrous Tohoku earthquake and subsequent tsunami two days after the event. As time passed, fewer survivors were pulled from the rubble, and the vast scale of the destruction had become apparent. One fortunate survivor floated unexpectedly into this grim situation, his remarkable story providing a small moment of consolation for a country that had few glimmers of hope in the wake of a devastating disaster. Minutes after the tsunami warning was issued, Hiromitsu Shinkawa rushed home to collect his wife and gather a few belongings, but he was too late. His home in a coastal Japanese village was destroyed as the mass of water surged inland. He was separated from his wife and swept out to sea. He was fortunate to find part of the roof of his house floating by. After several attempts he was able to climb on board. He floated precariously on his makeshift raft for two days, lost among a dense field of debris in relatively calm seas. Despite the best efforts of rescuers, Hiromitsu wasn't spotted by a boat crew until he was 10 miles (16 kilometers) from the coast and had begun to despair of being found (Figure 1.15).

Scientists will learn lessons from the deadly tsunami to be better prepared for the next one. In the wake of the Indian Ocean tsunami, governments of nations around the Indian Ocean worked to create a warning network to allow for timely evacuations in advance of the arrival of future tsunami. This kind of intergovernment cooperation is necessary to address phenomena that have the potential to impact large regions of the globe. (We will describe the link between earthquakes and tsunami in greater detail in Chapter 5.)

The discussion of the Alvarez hypothesis for the extinction of the dinosaurs illustrated how scientists made predictions, tested (and rejected) hypotheses, and arrived at the conclusions that were most reasonable under the circumstances. We finished the chapter with an indication of how humans are affecting the earth, atmosphere, and oceans. More than 100 years ago, the English mathematician and philosopher William Whewell noted that:

The hypotheses we accept ought to explain phenomena which we have observed. But they ought to do more than this: our hypotheses ought to foretell phenomena which have not yet been observed.

Figure 1.15   Hiromitsu Shinkawa is rescued two days after the March 2011 Tohoku tsunami. He was staying afloat on top of part of his roof 16 kilometers (10 miles) from land.
© STR/AFP/Getty Images

The Climate Change Science Program involves thousands of researchers seeking to explain the processes that control our current climate and to foret`ell Earth's future climate. Think of it as a global weather forecast for the next 100 years. We hope you will pardon us for revisiting this story throughout the book, but it will be one constant theme of science that you will hear, see, and read about in the years ahead so we figured we would get you ready for action.

Introduction to Earth Science: Concept Map

To evaluate your understanding of the interactions between the components of the earth system discussed in this chapter, complete the following concept map exercise.

Examine the following list of interactions between pairs of components in the earth system. Match each interaction with one of the lettered links in the concept map provided.

Interaction	Letter
Plants absorb carbon dioxide.	
Earthquake destruction causes deaths.	
Wind blows sand.	
Spacecraft explore deep space.	
Continents deflect ocean currents.	
Plants release oxygen.	
Fish live in oceans.	
Asteroid impacts Earth.	
Volcano emits toxic gases.	
Animals drink water.	
Water evaporates from the oceans.	
Humans mine coal.	
Winds generate waves.	
A stream carves a canyon.	

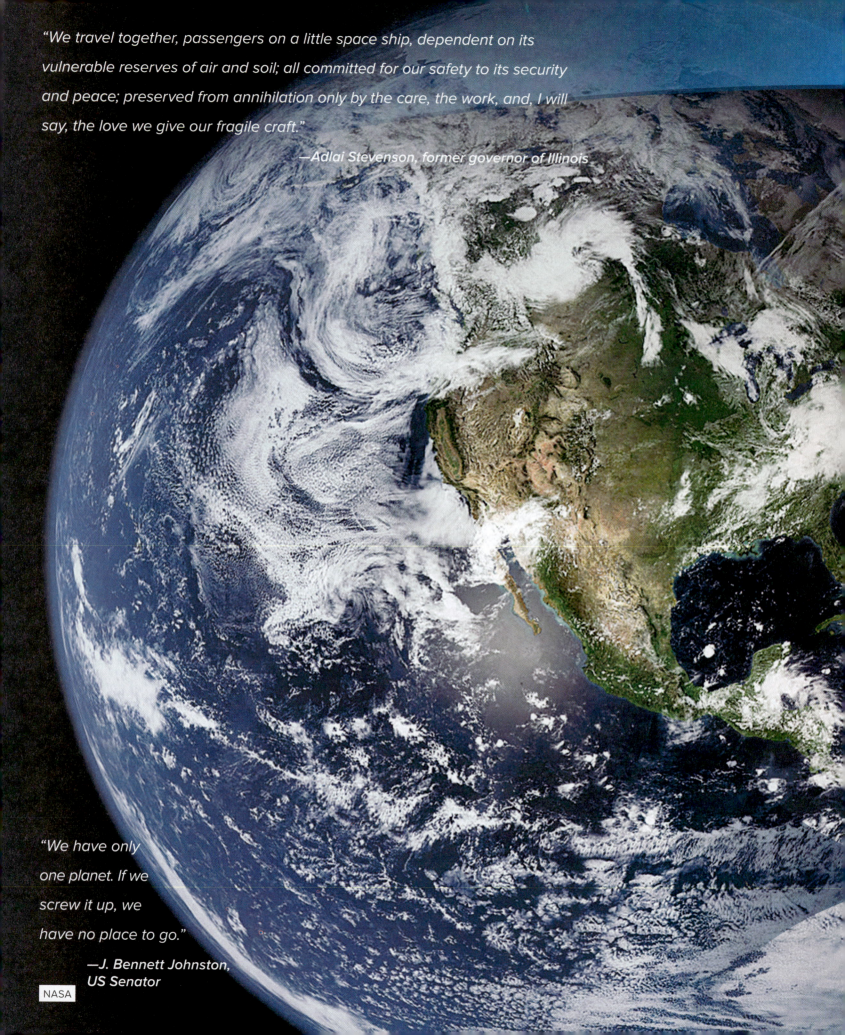

"We travel together, passengers on a little space ship, dependent on its vulnerable reserves of air and soil; all committed for our safety to its security and peace; preserved from annihilation only by the care, the work, and, I will say, the love we give our fragile craft."

—*Adlai Stevenson, former governor of Illinois*

"We have only one planet. If we screw it up, we have no place to go."

—*J. Bennett Johnston, US Senator*

NASA

the big picture

A planet you should know. How many more
of these are there out there?

*See The Big Picture box at the end of this
chapter for the full story on this image.*

We have recently passed through the maximum of sunspot cycle 24 (Figure 2.12), and scientists have enough data to predict that this cycle will hit the next minimum around 2021. Earth receives a slight increase (0.1 percent) in solar energy during a solar maximum and an equivalent decrease during a solar minimum. Solar flares, intense pulses of X-rays and ultraviolet radiation, are often associated with sunspots. These eruptions from the sun's surface can affect telecommunications and other technology systems on Earth.

Actions of Solar Wind. Space has its own weather system, but it doesn't have the moving air and clouds so familiar to us on Earth. Instead, **space weather is dominated by the solar wind, a constant stream of charged particles accelerated from the sun's outer layers by its magnetic field.** These charged particles travel at an average speed of 1.6 million kilometers per hour (1 million miles per hour). **The region of space affected by the solar wind is known as the heliosphere,** and it represents the volume of space in which our sun is the dominant influence. The heliosphere extends far beyond the planets of our solar system and may be likened to a giant bubble that shields us from harmful cosmic rays originating elsewhere in the universe. At several stages of your education, you were shown charts and diagrams of the sun and planets (like the Chapter Snapshot) that show Pluto as the farthest outpost of the solar system, like a lonely "Thanks for visiting" sign on the edge of a small town. In reality, the effect of the solar wind continues for approximately three times the distance to Pluto, all the way to the edge of the heliosphere.

Earth has its own magnetic field that deflects the solar wind around our planet (Figure 2.13), shielding our atmosphere. Even though Earth's magnetic field keeps our protective atmosphere from eroding, we are still vulnerable to the harmful effects of occasional solar eruptions that hurl powerful

pulses of dangerous X-rays, ultraviolet radiation, and charged particles toward Earth. **Intense streams of charged particles can disrupt Earth's magnetic field, generating electrical currents that threaten satellites and result in power surges** with the potential to shut down electrical systems leading to blackout conditions. The economic costs of power outages are measured in billions of dollars. Disruptions in Earth's magnetic field caused by solar wind can also result in spectacular effects such as the dramatic light displays known as *aurora* in the upper atmosphere (Figure 2.14).

Every day, we depend on hundreds of satellites to provide information for a host of needs on Earth, including communications,

Checkpoint 2.11

basic · advanced · intermediate · superior

Use the graph of sunspot numbers (see Figure 2.12) to answer the following questions.

1. When was the most recent solar maximum (month/year)?
2. When was the most recent solar minimum (month/year)?
3. On a slow day in the lab, two graduate students made a bet about how many sunspots would occur two months later in spring semester, 2013. Chad estimated there would be 90 sunspots; Julie thought there would be 30. When the month rolled around, 63 sunspots were recorded. Does this mean that Chad and Julie do not understand the sunspot cycle? Explain your reasoning.

Figure 2.13   Deflection of Earth's magnetic field by the solar wind. Earth's magnetic field is compressed closer to the sun, meaning that the magnetic field lines are closer together on the side of Earth that is facing the sun.
Adapted from John Fix, *Astronomy: Journey to the Cosmic Frontier* 4e, © 2006, Reprinted with permission of McGraw-Hill Education

Figure 2.14 Aurora seen from space, formed by the interaction of the solar wind with Earth's magnetic field. On Earth, these spectacular visual displays can be best observed at high latitudes—that is, closer to the poles.
NASA

defense radar, aircraft navigation, and weather forecasting. For example, Global Positioning Systems (GPS) could be as much as 50 meters (164 feet) off during solar storms. Satellites could be knocked out of action by concentrated streams of solar radiation. NASA has launched several satellites to monitor solar activity and provide notice of potentially damaging bursts of energy heading for Earth (Figure 2.15). Such warnings will be vital to future space exploration that could expose astronauts to the dangerous radiation generated by these solar emissions. People living outside the protection of Earth's magnetic field and atmosphere would be exposed to much higher levels of harmful radiation from the sun than the rest of us on Earth's surface.

Solar Power on Earth. Nonrenewable fossil fuels (coal, oil, natural gas) account for approximately 80 percent of US electric power. Concerns about the availability of imported oil and the potential for global warming have focused attention on alternative energy sources such as solar and geothermal energy (see Section 2.6) that are readily available in North America and don't produce greenhouse gases. Earth receives far more solar radiation than it needs to supply plentiful energy to the global population. Ideally, covering a sunny region about the size of New Mexico with high-efficiency solar cells should be enough to solve the world's energy problems.

Unfortunately, current economic and technological limits prevent more extensive use of this promising resource. Solar energy accounts for less than 1 percent of US electricity generation.

Active (direct) solar energy can be used in two forms. Water or oil may be heated in solar collectors that use mirrors to focus the sunlight onto the liquid. The hot liquid is then used for heating. In contrast, photovoltaic cells convert sunlight directly to electricity. Current solar energy systems are too inefficient for widespread commercial deployment. When coal is burned, approximately one-half the heat generated can be converted to electricity. Commercial solar (and geothermal) energy sources have typical efficiencies of 10–20 percent. Without an increase in efficiency, solar systems will continue to be several times as expensive as using coal or natural gas to produce electricity. However, continuing technological improvements and favorable government policies have resulted in a dramatic increase in the production of photovoltaic cells. Solar cell efficiencies of more than 40% have been achieved and continued research holds promise for a future where inexpensive solar energy systems will supply much of our daily electricity.

Eight, Nine, or Ten Planets?

In contrast to what you may have learned in elementary school, as of 2006, the sun is the centerpoint of a system of eight planets (see Chapter Snapshot). In order, with increasing distance from the sun, the planets are: **Mercury, Venus, Earth, Mars, Jupiter, Saturn, Uranus, and Neptune.** In recent years, there has been considerable debate in the space science community about exactly what constitutes a planet. In the summer of 2006, Pluto was shown the door of the planet club. The planets are now divided into two groups. The smaller planets—Mercury, Venus, Earth,

> ✓ **Checkpoint 2.12**
>
> ⬤ basic ⬤ advanced
> ⬤ intermediate ⬤ superior
>
> Create a concept map that links together the principal characteristics of the sun using the terms listed here, as well as any linking phrases you wish to create.
>
> | | |
> |---|---|
> | The sun | Helium |
> | Solar wind | Nuclear fusion |
> | Solar system | Sunspots |
> | Magnetic field | X-rays |
> | Star | Sunspot cycle |
> | Hydrogen | Differential rotation |

Figure 2.15 Satellite monitoring of solar activity. These three views of the sun were taken by the Solar and Heliospheric Observatory (SOHO) satellite in successive years using the Extreme Ultraviolet Imaging Telescope. The far-right image shows increased solar activity approaching a peak in the sunspot cycle, which could lead to damaging bursts of energy headed toward Earth.
NASA

Figure 2.16   Relative sizes of the eight closest planets to the sun. Mercury, Venus, Earth, and Mars are called the terrestrial planets, while Jupiter, Saturn, Neptune, and Uranus are known as the Jovian planets. The terrestrial planets are closer to the sun—and to one another—than are the Jovian planets, which have much more space between them.
NASA

and Mars—are nearest to the sun, share similar characteristics, and are termed the **terrestrial planets.** The much larger outlying planets—Jupiter, Saturn, Neptune, and Uranus—are called the **Jovian planets** (Jupiter-like; Chapter Snapshot, Figure 2.16). Smaller objects, similar in size to Pluto, are considered members of a newer, less prestigious group of **dwarf planets.**

The average distance from the sun to Earth represents one astronomical unit (AU; 1 AU = 150 million kilometers or 93 million miles). The planets' distances from the sun range from 0.4 AU for Mercury to 30 AU for Neptune (Table 2.1). We can read these values on the page, but these units of distance will make little sense to most of us used to measuring travel in miles or hours. NASA's *New Horizons* space probe, launched in 2006, took nearly a decade to travel the 40 AU to Pluto. In comparison, the terrestrial planets exist in a positively cozy little neighborhood. The four innermost planets all lie within 1.5 AU of the sun, essentially one planet per 0.3 AU. Mars, the farthest terrestrial planet from the sun, is separated from Jupiter, the nearest Jovian planet, by 3.7 AU. This gap contains the asteroid belt, which is composed of thousands of rocky or metallic bodies. The largest asteroids are almost the size of Pluto, and the smallest are little more than space pebbles.

The dwarf planets do not fit with either the terrestrial or Jovian planets. Pluto, smaller than Earth's moon, is composed of ice and rock, like the comets, and has an odd orbit that actually takes it closer to the sun than Neptune for part of its course. Images from the *New Horizons* probe (Figure 2.17) revealed that Pluto has

✓ **Checkpoint 2.13**

○ basic   ○ advanced
○ intermediate   ○ superior

Which one of the following sequences of planets is out of order based on sun-to-planet or planet-to-sun orbital distances?

a.   Mars, Jupiter, Saturn, Uranus, Neptune
b.   Venus, Mars, Earth, Jupiter, Saturn
c.   Uranus, Saturn, Jupiter, Mars, Earth
d.   Jupiter, Mars, Earth, Venus, Mercury

icy mountains and plains but fewer craters than expected, perhaps indicating ongoing processes that renew the surface.

In 2005, Mike Brown, an astronomer of the California Institute of Technology, discovered an object orbiting the sun beyond Pluto. The International Astronomical Union (IAU), the body that coordinates the naming of celestial objects, would eventually name the object *Eris,* after the Greek god of conflict. The name was an ironic nod toward the debate that the discovery of Eris generated among scientists. On the basis of its brightness and orbital characteristics, many scientists argue that Eris is larger and more massive than Pluto and that it is composed of ice and rock, just as Pluto is. Once it became apparent that Eris was bigger than Pluto, science began its slow, inevitable progress toward change. Either Eris, and other objects being found beyond Pluto, had to be

Table 2.1	**Characteristics of the Eight Planets**			
Planet	Size (Radius), km	Orbital Period	Distance from Sun, Million km [AU]	Principal Atmospheric Gases
Mercury	2,440	88 days	58 [0.4]	Helium, sodium
Venus	6,052	225 days	108 [0.7]	Carbon dioxide
Earth	6,378	365 days	150 [1]	Nitrogen, oxygen
Mars	3,397	687 days	228 [1.5]	Carbon dioxide
Jupiter	71,492	11.9 years	778 [5.2]	Hydrogen, helium
Saturn	60,268	29.5 years	1,427 [9.5]	Hydrogen, helium
Uranus	25,559	84 years	2,871 [19]	Hydrogen, helium
Neptune	24,746	165 years	4,497 [30]	Hydrogen, helium

Checkpoint 2.14

Make four generalizations about the eight planets in our solar system, using the information in Table 2.1. For example, planets closer to the sun are smaller than those farther away.

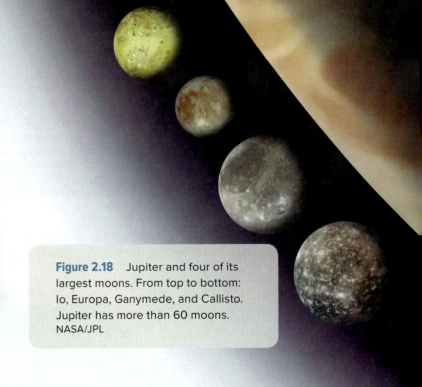

Figure 2.18 Jupiter and four of its largest moons. From top to bottom: Io, Europa, Ganymede, and Callisto. Jupiter has more than 60 moons.
NASA/JPL

added to the club of nine planets, or the term *planet* had to be redefined in a way that would exclude Eris, and that would probably mean the end of the road for Pluto—everyone's favorite eccentric planet. Soon the world of astronomy was full of conflict over the fate of Pluto and all its icy companions.

Oddly, up to this time, there was no generally accepted definition for the term *planet*. The IAU had appointed a panel of astronomers that took 2 years to figure out that they could not agree on a definition. A smaller committee eventually proposed that a planet is *an object that orbits a star and is massive enough for gravity to pull its material into an approximately spherical shape.* By this definition, any roundish object that was about 800 kilometers (500 miles) or more in diameter and orbited the sun would qualify as a planet. This would have let in Eris and at least one asteroid (Ceres) and potentially dozens of other objects that advanced telescopes were beginning to identify beyond the orbit of Pluto.

To counter this generous new "big tent" definition, a more restrictive (little tent) option was proposed that would add a third criterion: that the object would be considered a planet *if it had cleared the neighborhood around its orbit.* By this definition, planets would be the only significant objects lying along the path of their orbits. Because Eris, Ceres, and Pluto all exist in neighborhoods with many other similarly sized objects, they would not be considered as planets but would be assigned to a new class of objects, the *dwarf planets.* While this definition is far from perfect (thousands of asteroids lie in the orbit of Jupiter, and Pluto passes within Neptune's orbit), it was adopted by a majority of scientists attending an IAU conference. So, for the time being, there are officially eight planets and a steadily increasing number of dwarf planets.

Types of Planets

Terrestrial Planets. The terrestrial planets are composed of rock and can be divided into different compositional layers. **For example, the interiors of Earth, its moon, and the other**

Figure 2.17 View of Pluto taken by NASA's New Horizons spacecraft.
NASA/JHUAPL/SwRI

terrestrial planets can each be separated into three layers of contrasting composition and thickness: crust, mantle, and core (Chapter Snapshot). The cores of terrestrial planets are composed of an iron and nickel mixture. Their composition is similar to that of metallic meteorites, which formed from debris that did not coalesce to assemble planets elsewhere in the solar system. As the young, hot Earth cooled, the heavier, metal-rich components sank to form the core, and the lighter, rocky materials made up the planet's outer layers represented by the crust and mantle.

Jovian Planets. The Jovian planets are much larger than the terrestrial planets and are shrouded by dense gases (Chapter Snapshot and Figure 2.16) that create spectacular atmospheric effects. These characteristics led to their nickname, "the gas giants." We generally think of the surface of a terrestrial planet as the interface between the atmosphere and the rocky surface of the continents or the water in the oceans. Such a distinction is more difficult to draw for the **Jovian planets, which all have thick atmospheres over oceans cold enough to be formed of liquid hydrogen, helium, and/or methane.** Any rock present is buried under tremendous pressures at the core of the planet. Consequently, it is difficult to determine where the "surface" of a Jovian planet begins and the gassy atmosphere ends.

The average spacing between the orbits of Jovian planets is over 8 AU, about a 2-year trip if you could hitch a ride on a passing spacecraft. Not only are these planets much larger than terrestrial planets and much farther apart, but also they have much longer solar orbits (see the orbital periods in Table 2.1). The time it takes for a planet to complete a solar orbit increases with its distance from the sun. Mercury orbits the sun in a little less than 3 months, while it takes Neptune 165 years to finish a circuit.

Jovian planets have many moons (Jupiter, 63; Saturn, 62; Uranus, 27; Neptune, 13), and more are discovered almost every year. Figure 2.18 shows four of Jupiter's moons. In contrast, terrestrial planets are moon-poor; only Earth and Mars have moons (Earth, one; Mars, two, named Phobos and Deimos).

THE SOLAR SYSTEM

COMET
Some comets orbit the sun every few years; others take as long as 30 million years to complete an orbit. The nuclei of most comets are less than 10 km (6 miles) across.

Terrestrial planet interiors. Earth and the other three terrestrial planets have three compositional layers: crust, mantle, and core. In all cases, the crust is relatively thin (tens of kilometers) in comparison to the mantle and core. The four planets are drawn to scale in the image here.

EARTH
Without our moon, Earth would wobble on its axis, changing how much solar radiation was received over different parts of its surface, resulting in a much more unstable climate system.

MERCURY
Mercury is just a little bigger than Earth's moon and travels around the sun faster than any other planet, at an average speed of 172,000 km/h (107,000 mph).

MARS
If Earth were the size of a nickel, Mars would be the size of an aspirin tablet.

VENUS
Venus is the hottest planet in the solar system. Its clouds reflect so much sunlight that Venus is the brightest planet in the night sky.

THE SUN
Our sun is one of 70,000,000,000,000,000,000,000 stars in the universe. The sun's interaction with Earth drives our ocean currents, weather patterns, and climate.

 Solid core

 Liquid core

 Mantle

 Crust

MERCURY

VENUS

EARTH

MARS

20,000 km

URANUS

Uranus's axis is tilted at 98 degrees. Its poles are pointed towards the sun. Due to its 84-year long orbit, the planet's poles experience 21 years of night or day depending on their position.

PLUTO

The average temperature on icy "dwarf planet" Pluto is -233° C. Sunlight on Pluto is only 0.0001 times as strong as on Earth.

JUPITER

Jupiter is named for the king of the Roman gods and makes up two-thirds of all the planetary mass in the solar system.

NEPTUNE

Neptune is so far from the sun that it still has not made a complete orbit since it was discovered in 1846.

SATURN

Saturn's density is less than that of water. Saturn's rings cover a distance nearly equal to the distance from Earth to the moon.

Jovian planet interiors. Jovian planets also have compositional layers. On Saturn and Jupiter, the largest Jovian planets, pressures in the interior of the planet are sufficient to generate rare liquid metallic hydrogen. The relatively smaller Uranus and Neptune are still many times larger than Earth and have a mantle dominated by ices composed of methane, water, and ammonia.

ASTEROID BELT

All the asteroids mashed together would make up a body about one-tenth the size of our moon. Asteroids that pass close to Earth are known as near-Earth objects.

Orbits not to scale.
Orbits viewed from an angled-perspective above the orbital

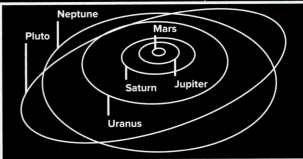

JUPITER

SATURN

URANUS

NEPTUNE

Rock

Water, ammonia, methane ices

Liquid, metallic hydrogen

Liquid hydrogen

Hydrogen, helium, methane gas

50,000 km

Earth's moon plays an important role in generating tides (more about that in Chapter 13). **Most of the Jovian planets' moons revolve in consistent directions around their planets, evidence that the moons formed from the same mass of debris.** There are exceptions, such as Neptune's moon Triton, which orbits opposite the direction of the other moons. How could that happen? Scientists hypothesize that moons orbiting opposite to the norm either were captured by the planet's gravitational fields or are evidence of early collisions between celestial bodies that could have reversed the direction of motion of some of the orbiting matter.

All of the Jovian planets have ring systems, of which Saturn's is the best example (Chapter Snapshot). Rings are composed of individual pieces of rocky debris or chunks of ice that range in size from micrometers to several meters in diameter.

✓ Checkpoint 2.15

| ✔ basic | ✔ advanced |
| ✔ intermediate | ✔ superior |

Venn Diagram: Terrestrial Versus Jovian Planets

Complete the following Venn diagram to compare and contrast the similarities and differences between the two major groups of planets in our solar system. Identify 12 characteristics that either are shared by both terrestrial and Jovian planets or are unique to one of the groups. Then place the number corresponding to each characteristic in the diagram. Characteristics 1 and 2 have been plotted on the diagram for you.

1. Orbits around sun
2. Gas giants
3.
4.
5.
6.
7.
8.
9.
10.
11.
12.

✓ Checkpoint 2.16

| ✔ basic | ✔ advanced |
| ✔ intermediate | ✔ superior |

In 2006, the IAU defined a new class of celestial objects, dwarf planets. Explain why this is consistent with the nature of science described in Chapter 1.

The rings are held in place by the contrasting gravitational pulls of the planets and their surrounding moons. Without the many moons that are characteristic of the Jovian planets, it is likely that the planets' gravity would have gradually pulled the material in the rings down to the planets' surfaces. Each piece is in its own orbit that is positioned such that the material does not fall into the planet or combine with other orbiting materials to form a larger body such as a moon.

2.5 Earth, the Sun, and the Seasons

Learning Objectives

- Describe the how the distribution of incoming solar radiation changes from the equator to the poles.
- Explain how Earth's orbit and rotational tilt angle affect the seasons.

Few scientific questions are more basic than "Why is it colder in winter than in summer?" (Think about it; what is your answer?)

When a small group of seniors graduating from a prestigious university were asked that question, nearly all of them answered incorrectly. The most common wrong explanation offered was that Earth is closer to the sun in summer and farther away in winter. But, in fact, the exact opposite is true for the Northern Hemisphere (where the students lived). **Earth's orbit is almost circular, but the planet comes a little closer to the sun during January and is slightly farther away in July** (Figure 2.19).

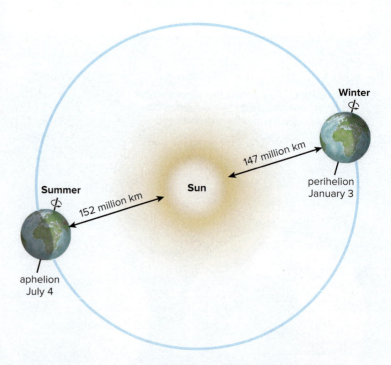

Figure 2.19   The path of Earth's orbit around the sun. In the Northern Hemisphere, Earth is farthest from the sun during early summer and closest during early winter.

Heat from the sun is indeed the factor responsible for colder winters and warmer summers—but not because of Earth's orbital pattern. It is not the amount but rather the distribution of solar radiation on Earth's surface that regulates the order of the seasons. The Northern Hemisphere receives more solar radiation in summer and less in winter. Why?

Distribution of Solar Radiation

To understand why different regions of Earth receive different amounts of solar radiation at different times of year, it is first important to understand that our planet rotates on its axis once every 23 hours, 56 minutes, 4.1 seconds. Further, Earth's rotation axis is tilted relative to its plane of orbit. The axis is tilted 23.5 degrees away from vertical (Figure 2.20a). Because of this tilt, the amount of solar radiation hitting any location varies by season. Think of the axis as the shaft of an open umbrella; imagine the shaft of the umbrella is tilted and resting on your shoulder. You can twirl the umbrella by rotating the shaft. You can adjust how much rain reaches you by altering the angle of the shaft. Similarly, the amount of sunlight received by Earth varies with the tilt of the axis. The direction of tilt of the axis does not change during Earth's annual orbit (Figure 2.20b). Imagine that you are carrying that umbrella as you are walking along a circle in a parking lot but always *facing* in the same direction (you would walk backward for one-half of the circle). The parking lot would represent the orbital plane of Earth, and the circle would be the orbit.

Insolation, the amount of solar radiation received by Earth, is greatest when the sun is directly overhead above a location and decreases as the sun's rays make a lower angle with respect to Earth's surface. When we are discussing insolation, three landmarks on Earth's surface are important to keep in mind: the equator, the Tropic of Cancer, and the Tropic of Capricorn (Figure 2.20a). The equator is an imaginary band encircling Earth at its widest circumference; it passes through northern South America, central Africa, and Indonesia. The Tropic of Cancer is located 23.5 degrees north of the equator, just north of the Hawaiian Islands and Cuba; and the Tropic of Capricorn is 23.5 degrees south of the equator, the same latitude as Rio de Janiero, Brazil, and Alice Springs in the center of Australia.

Approximately one-half of the solar radiation that illuminates Earth actually makes it to the planet's surface. The rest of the incoming sunlight is scattered within the atmosphere or is reflected back into space from clouds and particles. There are two reasons why the sun provides less heat to the poles than to the equator. First, the sun's rays strike Earth at a lower angle nearer

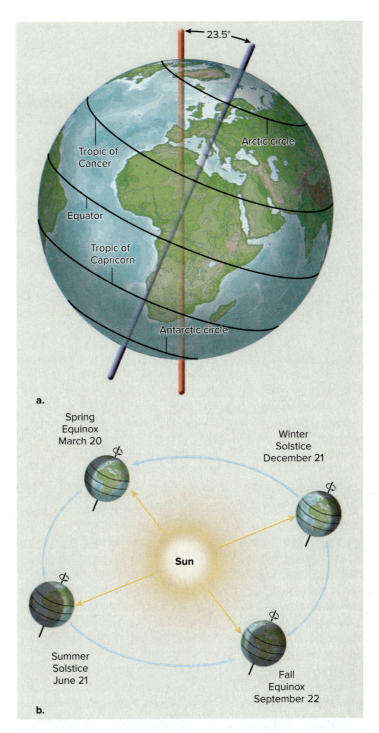

a.

b.

Figure 2.20 The sun's influence on Earth's seasons. **a.** Earth is tilted on an imaginary axis oriented at 23.5 degrees to vertical. The two tropics are 23.5 degrees of latitude north (Cancer) and south (Capricorn) of the equator. **b.** The sun is overhead at the Tropic of Cancer on June 21 and at the Tropic of Capricorn on December 21. It is overhead at the equator during the spring and fall equinoxes. Note that Earth's axis is always tilted in the same direction. Diagram is not to scale.

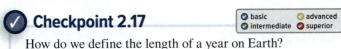

✓ Checkpoint 2.17

☑ basic   ☑ advanced
☑ intermediate   ☑ superior

How do we define the length of a year on Earth?

    a. A year is related to the revolution of Earth around the sun.
    b. A year is related to the rotation of Earth on its axis.
    c. A year is related to the rotation of the sun on its axis.
    d. A year is related to the revolution of the sun around Earth.

the poles (Figure 2.21). Consequently, the sun's rays have a longer path through the atmosphere near the poles than do rays near the equator. **Since a portion of the solar radiation is reflected by the atmosphere, the longer the path in the atmosphere, the less**

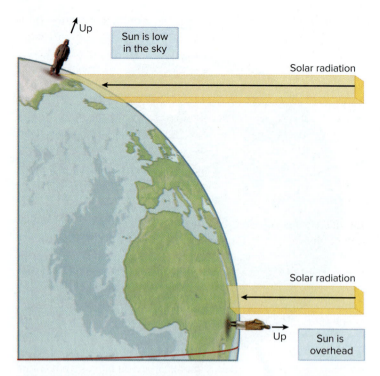

Figure 2.21 Differences in heat distribution. When sunlight strikes Earth's surface at a low angle, as occurs nearer the poles, its energy is diluted over a larger area. The resultant heating is therefore less nearer the poles than in locations where the sun is closer to overhead, such as near the equator.
Adapted from Eric Danielson, *Meteorology* 2e © 2003, reproduced with permission of McGraw-Hill Education

sunlight that reaches the surface. Second, the curvature of the planet's surface results in the same amount of heat energy being distributed over a larger area near the poles than at the equator (Figure 2.21). The equator receives 2.5 times more incoming solar radiation than the poles because the sun is overhead for much of the year and solar radiation therefore takes a shorter path through the atmosphere. As we will discuss in other chapters, **circulation patterns in the atmosphere and oceans transfer heat from the equator toward the poles, moderating Earth's climate.** Without the actions of these heat transfer mechanisms, temperatures at the poles and the equator would be much more extreme.

Why Day Length Changes. What happens to the relative lengths of day and night as we move from summer into winter? The number of hours of daylight changes over the course of a year and by latitude. **The axial tilt places the sun directly overhead on the Tropic of Cancer at noon of the summer solstice** (June 21), our longest day of the year in the Northern Hemisphere. The greater

amount of solar radiation reaching the Northern Hemisphere at this time accounts for the warm temperatures of summer. However, you will have noticed that the longest day is not the hottest time of year, despite receiving the most solar radiation. The maximum average air temperature typically lags behind the date of maximum insolation; this is what we call the **seasonal lag.** Likewise, the coldest part of the year occurs about a month after the shortest day, December 21. On the days leading up to June 21, most locations in the US will gain more heat from solar radiation during the day than they will lose during cooling at night. The same will be true on June 22, June 23, June 24, . . . and each day will have a net gain in heat, and temperatures will gradually increase, reaching a peak in July or August, depending on location. After several weeks, the night time cooling effect becomes greater than the daytime warming and temperatures will begin to decline.

Day and night both last for 12 hours on the first days of spring and fall (equinoxes) when the sun is directly overhead at the equator (Figure 2.20). The hours of daylight increase northward during summer in the Northern Hemisphere and decrease southward in the Southern Hemisphere (where it is winter). At the North Pole, the sun rises above the horizon on the spring equinox and does not set until the fall equinox 6 months later (Figure 2.22).

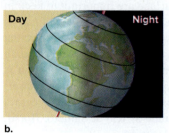

Figure 2.22 Differences in day length. In the Northern Hemisphere, the tilt of Earth's axis results in **a.** 24-hour daylight at the North Pole and almost complete daylight north of the Arctic Circle during summer in the Northern Hemisphere and **b.** perpetual darkness during winter. The situation is reversed south of the Antarctic Circle. Compare the length of time in daylight at the two tropics.

Checkpoint 2.18

basic	advanced
intermediate	superior

How would the amount of incoming solar radiation change at the equator if Earth's axis were vertical instead of tilted?

a. Incoming solar radiation would decrease.
b. Incoming solar radiation would be the same as at present.
c. Incoming solar radiation would increase.

Checkpoint 2.19

basic	advanced
intermediate	superior

Mars has a more asymmetric orbit of the sun than Earth. Mars is 20 percent closer to the sun during its winter than during its summer. How would Earth's climate be affected if Earth had a similarly eccentric orbit, being 20 percent closer to the sun during winter months in the Northern Hemisphere?

Checkpoint 2.20

basic	advanced
intermediate	superior

Imagine that it is your job to explain to a group of middle school students how the distribution of incoming solar radiation varies daily and seasonally on Earth's surface. Assuming you have a basketball and a flashlight to use as props, write a description of how you would have the students use the props in a demonstration.

Imagine daylight 24 hours per day. This pattern is reversed during the winter, when the South Pole is illuminated for 24 hours and the North Pole is dark 24 hours per day (Figure 2.22).

When summer solstice is occurring in the Northern Hemisphere, what is happening in the Southern Hemisphere? The rays are directly overhead in the Northern Hemisphere while the sun's rays strike the Southern Hemisphere at a lower angle and transfer less energy, resulting in lower temperatures. Thus, in the Southern Hemisphere, winter begins in June, and summer (with the sun overhead at the Tropic of Capricorn) begins on our winter solstice (December 21; Figure 2.20), the shortest day of the year in the Northern Hemisphere but the longest day along the Tropic of Capricorn.

2.6 The Unique Composition of Earth

Learning Objectives

- Describe the characteristics of Earth's internal layers.
- Explain the relationship between the planet's interior, the geothermal gradient, and transfer of heat that drives internal Earth processes.
- Discuss how various characteristics of Earth supported the evolution of life on our planet.

Figure 2.23 Compositional layers of Earth. Earth and the other terrestrial planets can be divided into three compositional layers: crust, mantle, and core. Earth's core can be divided further into a liquid outer core and a solid inner core.

Earth's physical characteristics, size, and distance from the sun have contributed to its unique status as the only inhabited planet in our solar system. As noted in Section 2.4, Earth's interior can be separated into three layers of differing composition and thickness: the core, the mantle, and the crust. If we could slice Earth in half like an apple, we would see a bull's-eye pattern of concentric layers, with the core at the center and the crust around the outside (Chapter Snapshot and Figure 2.23). Each layer is made up of different materials and may be further subdivided according to these physical and compositional variations.

Core, Mantle, and Crust

Earth's core is divided into two parts, a solid inner core and a partially melted outer core. The rocks of the core are largely composed of an iron and nickel mixture, metals that can be both molten and solid under the temperatures and pressures of the outer and inner cores, respectively. Scientists concluded that part of the core is liquid because some types of seismic waves (generated by earthquakes) do not travel through liquids and also cannot pass through the outer core. Earth's magnetic field (Figure 2.24) originates from slow-moving convection currents generated as the planet rotates on its axis and as molten material moves from hotter to cooler regions in the liquid outer core.

The mantle consists primarily of rocks composed of the elements oxygen (45 percent), silicon (22 percent), and magnesium (23 percent by weight). The mantle is a solid from the top of the outer core to the base of the crust. Even so, temperatures and pressures about 100 to 200 kilometers (62 to 124 miles) below the top of the mantle are close to the melting point of some minerals. The rocks in that region are hot enough that they can slowly

✓ Checkpoint 2.21

| ✓ basic | ✓ advanced |
| ✓ intermediate | ✓ superior |

What are the three compositional layers in Earth's interior?

a. Asthenosphere, lithosphere, core
b. Oceanic crust, continental crust, asthenosphere
c. Lithosphere, mantle, core
d. Crust, mantle, core

Figure 2.24 Earth's magnetic field. The inclination of Earth's magnetic field varies with latitude. The magnetic field is inclined downward in the Northern Hemisphere and upward (away from Earth's surface) in the Southern Hemisphere.

Figure 2.25 Comparison of physical and compositional layers near Earth's surface. Convection cells in the mantle are associated with oceanic ridges, regions of high heat flow on the ocean floor. Regions of cooler crust are often located along the margins of ocean basins, near oceanic trenches. Oceanic lithosphere forms at the oceanic ridge system and is consumed at trenches. (Features are not to scale.)

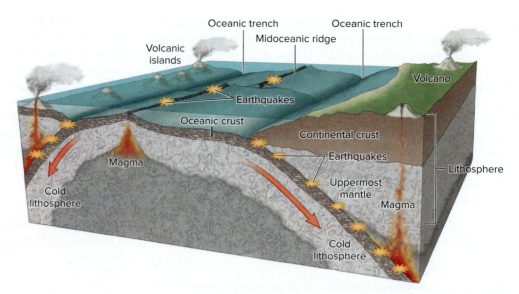

flow (though they are not molten). These weaker rocks form a layer called the **asthenosphere** that is about 200 kilometers (124 miles) thick (Figure 2.23).

The crust is made of lower-density rocks that are composed chiefly of oxygen and silicon (75 percent by weight). **A rigid outer layer called the lithosphere, composed of the crust and uppermost mantle, overlays the asthenosphere** (Figure 2.23). The lithosphere is broken into large slabs termed *tectonic plates*. Although the crust and mantle parts of the lithosphere are composed of different rocks, the two of them together compose a tectonic plate (Figure 2.25; described in detail in Chapter 4).

Geothermal Gradient. Miners working deep underground have long recognized that Earth has a geothermal gradient—that is, its temperature increases with depth. The temperature in the crust increases by approximately 25°C per kilometer (72°F per mile) near the surface. The gradient varies with location and with depth and shows that the interior of the planet is much hotter than the exterior. In areas of volcanic activity where the gradient is higher, the surface is generally hotter than it is elsewhere. In South Africa, ice must be pumped down cooling systems in deep mines to reduce temperatures of over 50°C (122°F). (Imagine working in conditions as hot as the hottest day in Death Valley, California!)

If we can harness the heat from Earth's hot interior, we may be able to provide a source of plentiful, clean energy. Geothermal power uses hot water from underground sources to generate steam utilized to drive turbines and generate electricity. The water is then

recycled back underground to recharge the geothermal reservoir and continue the process. Geothermal power is more commonly utilized in western states because of their underlying geology. States such as California, Utah, Nevada, and Hawaii have sites of recent volcanism fed by shallow magma chambers that provide ready heat sources for circulating groundwater. Recent studies have suggested that technologies used in oil exploration could be adapted for alternative "heat mining" operations that would provide wider access to geothermal energy resources in regions without volcanism. Heat mining requires water to be pumped down a deep well (5–10 kilometers, 3–6 miles), circulated through an underground reservoir where it would be heated, and then brought back to the surface through a second well (Figure 2.26). Earth's geothermal gradient would be sufficient to ensure that water would be heated to high temperatures over large

Figure 2.26 Paired deep wells would be drilled into artificially fractured rocks of a geothermal reservoir. Depending on the local geothermal gradient, the wells would be far enough below Earth's surface to result in heating of water to high temperatures.
Adapted from http://www1.eere.energy.gov/geothermal/pdfs/egs_chapter_1.pdf

✔ **Checkpoint 2.22**

basic ○ advanced ○
intermediate ○ superior ○

Much of our understanding of the character of Earth's interior comes from analyzing seismic waves that travel through Earth. As these waves move through Earth's interior, they may pass through, bounce off (reflect), or bend (refract) at boundaries between different rock types. The time it takes a seismic wave to travel from a source in one location to a recording station at another location can be used to decipher the internal structure of Earth. Identify three similar methods that are commonly used to view the interior of objects in daily life without cutting or breaking them open.

Checkpoint 2.23

basic | advanced
intermediate | superior

The following graphs illustrate four idealized geothermal gradients for Earth.

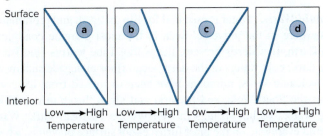

Bearing in mind that Earth is 4.6 billion years old, which plot is most likely to represent the present-day gradient?

Which plot is most likely to represent the gradient 2 billion years ago?

Checkpoint 2.24

basic | advanced
intermediate | superior

Earth's Layers Concept Map

Complete the following concept map by writing the correct terms in the appropriate blank locations as key terms or connecting phrases/terms. A partial list of the terms that you will need is provided here, but some of them may not be applicable to this diagram.

1. Compositional layers
2. Crust
3. One of three
4. Oceans
5. Is the source of Earth's
6. An upper rigid
7. Characteristic of terrestrial planet

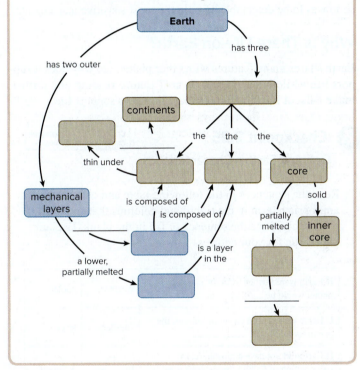

regions of the continent, rather than just at locations associated with volcanism (Figure 2.27). Under this scenario, large western cities such as Tucson and Salt Lake City could be generating much of their electrical needs from geothermal power in the next few decades.

The processes that operate on the surface of Earth and within the planet's interior are driven by energy from different sources. External processes derive energy from solar radiation (see Section 2.5). Most of the heat-driving internal processes come from the radioactive decay of elements in Earth's interior, and a smaller fraction is heat that remains since the time Earth formed. Rocks are poor conductors of heat. Therefore, it takes billions of years to lose the planet's internal heat through the surface. **All terrestrial planets were much hotter when they formed and have cooled with time.** Mercury and Mars, relatively smaller planets, lost their heat hundreds of millions of years ago. In contrast, the greater mass of the larger terrestrial planets acts as insulation, preserving their internal heat. Earth still has a hot interior, as evidenced by the more than 1,500 active volcanoes found around the world. Though no active volcanoes have been observed on Venus, it is thought to have a hot interior based on analyses of recent data. Those data are interpreted to suggest that volcanoes on

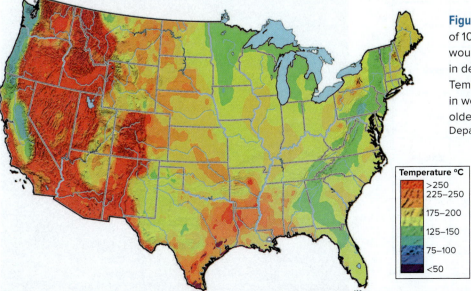

Figure 2.27 Estimated temperatures at the bottom of 10-kilometer-deep (6.2-mile-deep) wells. Waters would be superheated to more than 250°C (480°F) in deep wells throughout the western United States. Temperatures would be about 100°C (212°F) cooler in wells drilled in parts of many eastern states with older rocks exposed in the Appalachian Mountains. Department of Defense

Temperature °C
>250
225–250
175–200
125–150
75–100
<50

Venus are 250,000 to 2.5 million years old. Even a 2.5 million year old Venusian volcano is considered active in a geologic sense, so scientists hypothesize Venus still has a hot interior.

Such geological processes as volcanic activity and the slow movement of tectonic plates are indications that heat is being transferred from the interior toward the surface. Heat transfer occurs in the mantle by conduction and convection. **Conduction is simply the transfer of energy from one location to another through vibrations of molecules.** For example, the uninsulated handle of a metal saucepan becomes hot when left on the stove because heat is transferred from the stove through the pan to the handle by conduction. **Convection is the transfer of heat through movement of hot material from one place to another.** For example, convection occurs in water as it boils on the stove. Convection is thought to occur in the mantle and plays a role in the movement of tectonic plates (see the arrows on Figure 2.25). Heat escapes from all parts of Earth's surface, though at most locations this occurs at rates so low as to be detectable by only the most sensitive instruments.

Why Is There Life on Earth?

Earth shares many features with other planets, so why does it support life while other planets do not? Before reading any farther, make a list of Earth's features that allow it to support life.

✓ Checkpoint 2.25

| ○ basic | ○ advanced |
| ○ intermediate | ○ superior |

Defining Earth's Characteristics

Read the sentences in the following table and circle the appropriate term in the right-hand columns that could be used to complete the sentence or fill in the blank. Explain the reason for your choices.

If Earth were farther from the sun, the planet would be _____.	warmer	cooler
If Earth were a little closer to the sun, the oceans would be _____.	smaller	larger
If Earth did not have a biosphere, the composition of Earth's atmosphere would be _____.	the same	different
If Earth's biosphere had not evolved yet, we would have _____ oxygen in the atmosphere.	less	more
If Earth did not have an atmosphere, its surface would have _____ craters formed by meteor impacts.	fewer	more
If Earth did not have an atmosphere, life on the land surface would be _____ developed than it is today.	less	more
If Earth were larger, its gravity would be _____.	weaker	stronger
If Earth were smaller, its atmosphere would be _____.	thicker	thinner
If Earth were smaller, it would have a _____ interior.	hotter	colder

Does your list include any of the following: the existence of liquid water, gravity strong enough to allow for an atmosphere, a mixture of gases suitable for life, or a strong magnetic field to deflect harmful radiation?

Liquid Water. Water is essential for life on Earth and presumably elsewhere in the universe. Even in the most extreme conditions on Earth, if water is present, life is too. Tube worms flourish in the dark, cold waters of the deep ocean floor alongside superheated volcanic vents; microbes have been recovered from ice cores near the South Pole; and bacteria and algae thrive in the scalding waters of hot springs in Yellowstone National Park. Water would evaporate on the blast furnace–hot surface of Venus, where temperatures average 464°C (867°F), and water would freeze on Mars, where temperatures resemble those of Earth's polar ice caps (−63°C, −81°F). But on Earth, the average temperature is a pleasant 15°C (59°F), making it possible for water to exist as a liquid.

The relatively mild temperatures on Earth are the result of the planet's distance from the sun combined with the presence of an atmosphere and a magnetic field. The atmosphere provides energy redistribution and insulation from the extreme cold of space. The magnetic field deflects solar radiation and keeps it from stripping away the atmosphere.

So why does Earth have liquid water while Venus and Mars do not? Scientists hypothesize that Venus, Earth, and Mars all had water when they originally formed. The closer proximity of Venus to the sun resulted in higher temperatures and, consequently, the

Figure 2.28 Was there once water on Mars? This image shows exposed layers of rocks that may have been deposited by water. Chapter 7 describes more recent findings about the geology of Mars.
Malin Space Science Systems/NASA

Figure 2.29 Earth's fragile atmosphere. In this photo of a sunrise over the Caribbean Sea taken by the crew of space shuttle *Discovery,* the thin blue band is the stratosphere, and the red-orange color is the top of the troposphere, the lower part of the atmosphere where we live.
NASA

conversion of liquid water to water vapor by evaporation. The water molecules were split into their constituent oxygen and hydrogen atoms by intense ultraviolet radiation, with the lighter hydrogen lost to space. The original oceans on Venus likely evaporated over several million years, producing the dry, desertlike conditions that are now characteristic of the planet's surface.

Exploration on the Martian surface and interpretations of satellite data (Figure 2.28) support the hypothesis that Mars once had active rivers, lakes, and perhaps a small ocean. However, as usually occurs in science, there are lingering questions. The amount of water, how long it existed, and whether it still exists in large amounts are hotly debated topics (see www.nasa.gov for the latest). As its internal heat was lost, Mars was too far from the sun to sustain sufficient heat to maintain liquid water. Ongoing exploration missions have discovered substantial ice deposits just below the surface of Mars that have been preserved in the planet's present cold state.

Gravity and a Protective Atmosphere. Meteorites, moons, and planets are all held together by gravity. **Smaller planets have weaker gravity and thus thinner atmospheres; larger planets have stronger gravity fields and much thicker atmospheres.** Earth is large enough to have accumulated an atmosphere that protects the planet from all but the largest incoming space projectiles (comets, meteorites) and absorbs harmful radiation from the sun. Gravity holds Earth's layer of atmospheric gases close to the planet's surface (Figure 2.29). A fast-moving meteorite encounters gas molecules, mainly nitrogen and oxygen, as it plunges through Earth's atmosphere. Heat generated by the compression of the atmospheric gases is sufficient to melt the surface of the meteorite. This process continues until the rocky object is destroyed or until it plows into Earth's surface. The atmosphere thus reduces all but the largest space objects to relatively harmless debris.

Checkpoint 2.26

basic · advanced · intermediate · superior

Venn Diagram Exercise: Earth and Mars

Complete the following Venn diagram to compare and contrast the similarities and differences between Earth and Mars. Identify 12 characteristics that either are shared by both planets or are different for each planet. Then place the numbers corresponding to the characteristics in the most suitable locations on the diagram.

1.
2.
3.
4.
5.
6.

7.
8.
9.
10.
11.
12.

"So haste not, bright meteor; waste not strength,
O fair planet, singing-sister."

Doolittle, Hilda "The Dancer," Life &
Letters Today, December 1935

"Comets are the nearest thing to nothing that
anything can be and still be something."
—The National Geographic Society

© Denis Scott/Corbis

Chapter

3.

Near-Earth Objects

the big picture

What are the chances that an asteroid will land in your neighborhood?

See The Big Picture box at the end of this chapter for the full story on this image.

NEO IMPACT WITH EARTH

HOT FRAGMENTS
Pieces of molten rock blasted out
of the crater would fall back to Earth
to generate wildfires.

TSUNAMI
An impact event in the ocean
would generate a tsunami,
a giant wave that could drown
coastal regions and travel far inland.

AIR BLAST
The collision would send out
a powerful air blast that would
flatten everything for hundreds
of kilometers in every direction.

WILD FIRES
Would add smoke to
the rapidly darkening skies.

EJECTA
There would be sufficient dust in the atmosphere to block sunlight, potentially for several months. Debris would fall to the ground thousands of kilometers from impact site.

BLOCKED SUNLIGHT
Would cause temperatures to decline resulting in a short-term cooling trend.

FIREBALL
Heat from the impact would set trees on fire and burn skin thousands of kilometers from the impact site.

IMPACT
A big NEO (10 km diameter) would smash into Earth at such high speeds that those near the impact site may have just a few seconds between the first sighting of the NEO and its impact. A crater up to 200 km (124 miles) across would be formed.

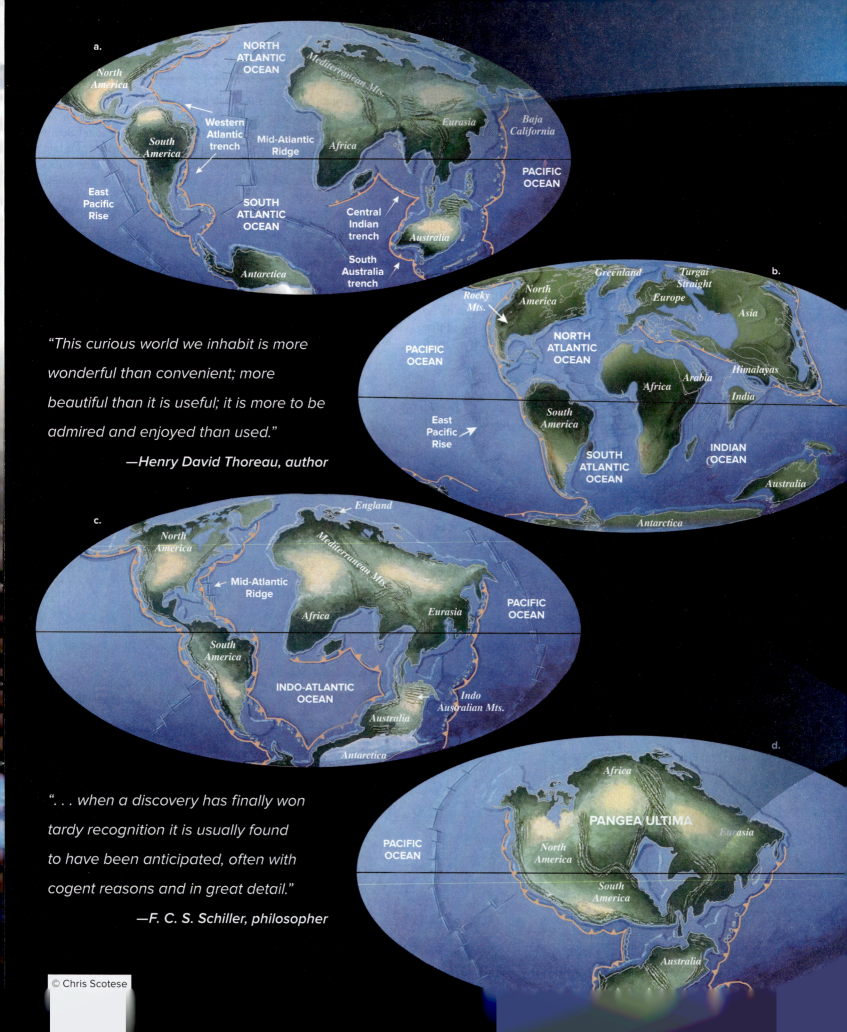

a.

North America
NORTH ATLANTIC OCEAN
Mediterranean Mts.
Eurasia
Baja California
South America
Western Atlantic trench
Mid-Atlantic Ridge
Africa
PACIFIC OCEAN
East Pacific Rise
SOUTH ATLANTIC OCEAN
Central Indian trench
Australia
Antarctica
South Australia trench

b.

Greenland
Turgai Straight
North America
Europe
Asia
Rocky Mts.
PACIFIC OCEAN
NORTH ATLANTIC OCEAN
Africa
Arabia
Himalayas
India
East Pacific Rise
South America
SOUTH ATLANTIC OCEAN
INDIAN OCEAN
Australia
Antarctica

c.

England
North America
Mediterranean Mts.
Mid-Atlantic Ridge
Africa
Eurasia
PACIFIC OCEAN
South America
INDO-ATLANTIC OCEAN
Indo Australian Mts.
Australia
Antarctica

d.

Africa
PANGEA ULTIMA
Eurasia
PACIFIC OCEAN
North America
South America
Australia

"This curious world we inhabit is more wonderful than convenient; more beautiful than it is useful; it is more to be admired and enjoyed than used."

—Henry David Thoreau, author

". . . when a discovery has finally won tardy recognition it is usually found to have been anticipated, often with cogent reasons and in great detail."

—F. C. S. Schiller, philosopher

Chapter

4

Plate Tectonics

Chapter Outline

the big picture

Which of these images best represents what the geography of Earth will look like 50 million years in the future?

See The Big Picture box at the end of this chapter for the full story on this image.

4.1 Science and Santa Claus

Chapter Learning Outcomes

- Students will explain how and why Earth is constantly changing due to plate tectonic processes.
- Students will analyze maps to make predictions about earthquakes, volcanoes, mountains, seafloor topography, and heat flow.
- Students will sketch and label cross sections to illustrate the types of geologic processes and motions that occur in association with plate boundaries.

When I was 6 years old, I asked Santa Claus to bring me a volcano for Christmas. It seemed like an innocent enough request at the time. As I looked around, I couldn't help but notice that the relatively modest landscape around our house could benefit from an interesting landform or two. To my young imagination, no landform was more interesting than a volcano. Of course, I wanted a fully operational volcano, with magma included. Needless to say, I was underwhelmed by the toy trucks and socks that showed up under the tree that year. It was just a few years later that one of my schoolmates informed me that Santa Claus was, in fact, a mythical figure and that I should be taking any requests for volcanoes directly to my parents. As I adapted to this new version of Santa Claus reality, I underwent a personal paradigm shift.

A paradigm is a generally accepted view of how some aspect of the world works. **A paradigm shift occurs when we undergo a fundamental change in our view or rethink our understanding of a basic concept.** We discussed a famous example of a scientific paradigm shift in Chapter 2: the change from the geocentric to the heliocentric explanation of Earth's relationship to the sun. This chapter describes another key paradigm shift in our view of Earth.

Hey, Good Lookin'

What do you look like? Take a moment to think about it. Really! Stop reading and write down a short description of yourself.

Now do the same for Earth.

When you were describing yourself, did you mention your primary features—head, arms, legs—or did you jump ahead to secondary features such as hairstyle, eye color, or height? In terms of only our primary features, we all look pretty much alike. It's the same for the inner planets. They are all spherical, made of rocks, and orbiting the sun. Things get more interesting when we examine the secondary features. On Earth, the surface of the planet is clearly divided into two elevations (Figure 4.1). Most of the land surface lies within a few hundred meters of sea level, while much of the rest of the planet's surface lies at the bottom of the oceans at a depth of several thousand meters.

Looking more closely at the North American continent, we can observe that much of the eastern half of the continent is relatively flat and ringed with sandy beaches. In contrast, the western half is experiencing a geologic party time as characterized by numerous snow-capped mountains, chains of active volcanoes, and stretches of rugged, cliff-lined coast. What is it that makes California so different from the Carolinas? Why does one region experience frequent earthquakes while another does not? Why are the volcanoes of Hawaii relatively safe, while the volcanoes in the Pacific Northwest sometimes explode violently? Why does Earth have a bilevel surface?

Finding the reason why such phenomena occur required earth scientists to abandon accepted views and develop a new theory to explain such observations. That theory holds that Earth's surface can be divided into enormous mobile plates. The slow motion of those plates has opened and closed oceans and has caused continents to migrate across the globe. The most destructive earth hazards, such as volcanic eruptions and most earthquakes, are associated with processes occurring on the edges of these plates.

Dynamic processes that shape the surface topography of other planets and moons share some common elements with plate tectonic processes on Earth (for example, volcanism). However, they differ in many other respects. We need only look to our sister planet Venus to learn that the alternative to plate tectonics would be a cycle of cataclysmic events that would destroy everything on the surface of the planet.

In this chapter, we will discuss the development of the theory of plate tectonics, the most significant paradigm shift in earth science. Like all major paradigm shifts, it did not come easily. Plate tectonics grew out of a previous theory known as continental drift that had been proposed by scientists about 50 years earlier. It took decades of careful observations and data analyses to provide sufficient evidence for plate tectonics to evolve as the dominant paradigm in earth science. The story of the journey from continental drift to plate tectonics provides us with a great opportunity to examine how science is done. In the pages ahead, we will revisit themes from Chapters 1 and 2, and we will set the stage for Chapters 5 and 6, which discuss earthquakes, volcanoes, and mountains. Finally, we will consider how plate tectonics controls long-term global climate patterns.

Figure 4.1 Earth's bilevel surface. The continents and ocean floors represent the two principal elevations of Earth's surface. Africa and the Arabian Peninsula are shown in the upper half of the image, with the Atlantic and Indian oceans to the west and east of Africa.
© PhotoLink/Getty Images

4.2 Continental Drift

Learning Objectives

- Discuss at least four observations that Wegener used to support his theory of continental drift.
- Describe some of the ideas that were used to make a case against continental drift.
- Explain how Wegener applied sound scientific principles to develop the theory of continental drift.

Centuries ago, as explorers drew the first maps of the Atlantic Ocean, the geographer Abraham Ortellus noted the matching shapes of the coastlines of Africa, Europe, and the Americas. In the third edition of his *Thesaurus Geographicus,* published in 1596, Ortellus pointed out that the pieces almost seemed capable of fitting together like a jigsaw puzzle. He suggested that America had been "torn away" from Europe and Africa and that the "projecting parts of Europe and Africa" would fit the "recesses" of America. In subsequent years, others commented on the complementary shapes of the coastlines, but these matching patterns were considered little more than a coincidence.

In the seventeenth and early eighteenth centuries, scientists recognized that Earth had a hot interior and that it was losing heat through its surface. Consequently, many believed that Earth was slowly cooling and contracting and that some features on Earth's surface such as mountain belts could be explained as a result of that contraction (similar to the wrinkles on a drying apple). They further proposed that the collapse of some sections of the surface explained the presence of ocean basins. This hypothesis, known as the "contracting Earth model," assumed that features on Earth's surface were approximately the same age. It also suggested that continents and oceans essentially would be fixed in place once they formed and that vertical crustal movements were responsible for differentiating oceans and continents.

Wegener's Theory

Alfred Wegener, a German meteorologist, was the first person to present a well-reasoned alternative explanation for the origin of continents and oceans. In a series of articles beginning in 1912 and later in a book, Wegener proposed his **continental drift theory.** The theory suggested that at some time in Earth's history, **the continents had come together to form a single supercontinent landmass he named Pangaea** (meaning "all lands"). Wegener suggested that Pangaea began to split apart into individual continents about 200 million years ago and the continents then gradually "drifted" to their current positions (Figure 4.2). The formation and breakup of Pangaea accounts only for the distribution of the continents over the last few hundred

Pangaea, 250 Mya

Laurasia and Gondwana, 210 Mya

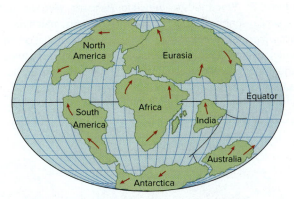
Most modern continents had formed by 65 Mya

Figure 4.2 Continental drift. Wegener hypothesized that the supercontinent Pangaea, made up of a northern group of continents known as Laurasia and a southern group of continents known as Gondwana, broke up about 200 to 250 million years ago (Mya) and that the individual continents eventually drifted to their present positions.
Adapted from Burton Guttman, *Biology* 1e © 1999, reproduced with permission of McGraw-Hill Education

million years. Wegener did not assume that the continents had always been present as a single landmass following the formation of Earth.

Matching Features.   Several geologic features on different continents would line up if the continents were reassembled as Pangaea. Wegener compared it to matching up the lines of print from a torn newspaper page. His observations included these:

- **Plant and animal fossils** found on separate continents form linked patterns if the continents are reassembled (Figure 4.3a). The animal fossils are those of freshwater amphibians and other similar species that would have died in salty ocean water. Therefore, they could not have crossed an open ocean between continents.
- Highlands in South Africa and Argentina (South America) and mountains in western Europe and eastern North America can be aligned to form **continuous mountain belts in a reassembled Pangaea** (Figure 4.3b).
- **Similar unusual sequences of rocks are found in Brazil, West Africa, and elsewhere** throughout the continents of

the Southern Hemisphere. Later researchers were able to identify 2-billion-year-old rocks in Gabon, West Africa, that match up with rocks of the same age on the other side of the Atlantic Ocean in northeast Brazil (Figure 4.3c).

Fit of the Continents.   The opposing coastlines of continents often fit together as do the pieces of a jigsaw puzzle. Later analyses would show that an even better fit occurs if we define the shape of a continent by the edge of the shallow water zone known as the *continental shelf* that surrounds each landmass. The classic fit is seen between the southwest coast of Africa and the east coast of South America (Figure 4.3a, c). Similar fits can be observed elsewhere, especially between the continents of the Southern Hemisphere.

✓ **Checkpoint 4.1**

✓ basic	✓ advanced
✓ intermediate	✓ superior

What type of reasoning did Wegener use in developing his continental drift theory? Explain your reasoning.

   a.   Inductive   b.   Deductive   c.   Both

Figure 4.3   Wegener's evidence for Pangaea. Wegener used the similar shapes of some continental margins, matching rocks, mountains, and fossils, and evidence of glaciation on five continents to infer that the continents must have been joined together as a single supercontinent between 200 and 300 million years ago. It broke apart about 200 million years ago. 4.3a, 4.3b, 4.3c, 4.3d: Adapted from USGS

a. Fossil distribution

b. Match of mountain belts among North America, Europe, and Greenland

c. Fit of continents, matching rock units

d. Glacial features. Arrows illustrate direction of ice movement determined by striations (grooves) cut by glaciers.

Paleoclimates (Ancient Climates). Scientists had long recognized that ice ages had occurred throughout Earth's history. Wegener documented ancient glacial deposits (formed about 300 million years ago) in India, Australia, South America, and southern Africa that all exhibited evidence of thick ice sheets similar to those found in Antarctica today (Figure 4.3d). This glaciation would have occurred as the continents were being assembled to form Pangaea. Such an event would have required a global ice age if the continents were in their present positions. However, the rock record reveals that coal was forming in tropical swamps in the eastern United States and Eurasia at the same time. Clearly, the rest of the world was not in a deep freeze. Such apparently **widespread glaciation could be best explained if certain continents were located close to the South Pole.** Likewise, when Pangaea was reassembled, it was found that areas in North America with coal deposits were located near the equator.

In hindsight, Wegener's theory looks like an open-and-shut case. It may be difficult to believe that his ideas were not championed by every reasonable earth scientist of his time. But, although some scientists embraced the evidence in favor of continental drift, others rejected it, holding fast to the contracting Earth model. Modern research on how our ideas change shows that old concepts (for example, the contracting Earth theory) are rejected when they cannot explain new observations. New theories (for example, continental drift) are accepted when they offer reasonable explanations for these new observations. For the old idea of the contracting Earth theory to be rejected, scientists first had to (1) recognize an inconsistency between the underlying theory and the new observations, (2) believe it necessary to reconcile the inconsistencies with existing concepts, and (3) be unable to make adjustments to incorporate the new observations into their existing theory.

Unfortunately for Wegener, many scientists of his day did not move past the second step in this process. For example, many fossils of the reptile *Mesosaurus* are found in both southern Africa and South America (Figure 4.3a). The fossil is preserved in rocks deposited in streams, so the reptile lived in freshwater environments. *Mesosaurus* could not have traveled across an ocean between the continents. Wegener proposed that the continents must have been located side by side or else *Mesosaurus* would have had to evolve separately on two continents at the same time, an unlikely explanation. Wegener's opponents were aware of this inconsistency with the contracting Earth model but reconciled it by suggesting that reptiles such as *Mesosaurus* migrated across narrow "land bridges" that originally connected the continents and later collapsed into the oceans. Therefore, these scientists did not reach step 3, did not accept Wegener's observations as valid, and therefore saw no significant challenge to their understanding of earth processes.

The opposition to Wegener's ideas was strongest in North America. Some rejected his ideas simply because, as a meteorologist, he was an outsider to the field of geology. Others were critical of the method he used. Wegener applied a deductive approach to formulate his hypothesis. That is, he came up with the idea that continents had drifted and then searched for confirming evidence based on the work of others. While this is a perfectly acceptable way to approach science, it was not the norm for the day. Most scientists studying the earth at that time used a more inductive, observation-based approach. A number of leading scientists contributed to a volume of articles published after a conference on continental drift in New York in 1928. Almost every article was hostile to the continental drift theory. The opposing scientists disputed the significance of many of the correlations that Wegener had relied on, suggesting that the matching features were not really that similar and that his interpretations of ancient climates were faulty.

Perhaps a more significant stumbling block was that Wegener was unable to propose an acceptable mechanism to explain *how the continents had moved.* The idea that huge masses of land could travel over such distances was an enormous intellectual leap. Skeptics could easily argue against continental drift when there was no viable explanation for how it had occurred. Wegener assumed that the continents had pushed through the rocks of the ocean floor, much as a plow cuts through the soil, but he recognized that there was no obvious explanation. To address this

Checkpoint 4.2

Which of these lines of evidence were used to support Wegener's continental drift theory? (Select all that apply.)

a. The distribution of fossils
b. Fit of the continents
c. Match of mountain belts
d. Earthquake locations
e. Paleoclimate data

Checkpoint 4.3

Broken Plate Exercise

Examine the image of the broken (ceramic) plate and answer the questions that follow.

If you were asked to glue the pieces of the plate back together, which pieces would you match up with each other?

a. What features did you use to match the different pieces?
b. What equivalent features did Alfred Wegener use to reassemble the continents to form the supercontinent Pangaea?

✅ Checkpoint 4.4

🔵 basic	🟢 advanced
🟢 intermediate	🟢 superior

Write a paragraph that argues for or against the following statement:

The development of the theory of continental drift was consistent with the characteristics of good science.

weakness in his theory, Wegener speculated that drift resulted from a combination of obscure forces. However, other scientists were able to calculate that these forces were too insignificant to move large landmasses. Consequently, the continental drift theory, although providing a compelling explanation for the distribution of common features on different continents, would have to wait another 50 years before some of its components would return in the theory of plate tectonics.

4.3 Evidence from the Seafloor

Learning Objectives

- Explain how the age of the seafloor varies relative to the locations of oceanic ridges and trenches.

- Compare and contrast global patterns of heat flow and the locations of volcanoes and earthquakes.

- Discuss how evidence from the exploration of oceans and continental margins supports the concept of seafloor spreading and contradicts the contracting Earth model.

- Predict how magnetic properties of rocks would vary in different scenarios.

- Describe how paleomagnetism can be used to provide support for the concept of seafloor spreading.

The amount of information about Earth's surface greatly increased after World War II, as scientists collected data by using new and improved technologies. Much of this technology was intended for national defense but had the added benefit of revealing new truths about Earth. For example, instruments installed to record vibrations that could pinpoint the locations of nuclear explosions were also used to identify the compositional layers in Earth's interior. The features of the seafloor were mapped very precisely to allow submarines to move around the oceans without being detected. Detailed surveys of Earth's magnetic and gravity fields were conducted to accurately predict the flight of long-range missiles carrying nuclear weapons.

Fifty years ago, it could be argued that scientists knew more about the surface of the moon than they did about the floors of Earth's oceans, but the new wave of research soon changed that perception. It became apparent that the old contracting Earth model could not explain the new observations, and a new group of earth scientists soon argued for a paradigm shift in our understanding of how Earth works. By the 1960s, new data about seafloor topography, the age and magnetic properties of the ocean floor rocks, and the global distribution of heat flow, volcanic activity, and earthquakes could no longer be reconciled with the contracting Earth model. A more dynamic explanation was needed.

Eventually, sufficient evidence had accumulated to provide a mechanism for Wegener's continental drift theory; this new mechanism became known as *seafloor spreading*. We divide the description of the evidence in support of seafloor spreading into two parts broadly recreating the chronological order in which the observations were made. Keep in mind that the original scientists did not have all the data and observations laid out before them that we enjoy today.

To understand how the theory of seafloor spreading developed, first we must describe some basic characteristics of the structure and processes of the ocean floor—namely, its topography, its age, and the distribution of heat flow, earthquakes, and volcanoes.

Seafloor Topography

The shapes of the bottoms of modern community swimming pools are in some ways similar to parts of the seafloor. Many community pools start with a "0 depth" area where toddlers can sit without danger of falling in over their heads. The depth then gradually increases up to the point of a flotation line stretched across the pool, where the water suddenly gets deeper. Likewise, the ocean floor gradually deepens from the coast (0 depth), moving seaward across the shallow continental shelf. The edge of the shelf is at approximately a 150-meter (490-foot) depth.

The seafloor gets much deeper (like the transition to the deep part of a swimming pool) beyond the edge of the continental shelf, gradually flattening out to form level areas known as *abyssal plains* at depths of approximately 4 kilometers (2.5 miles) below sea level (Figure 4.4). As we continue to move away from

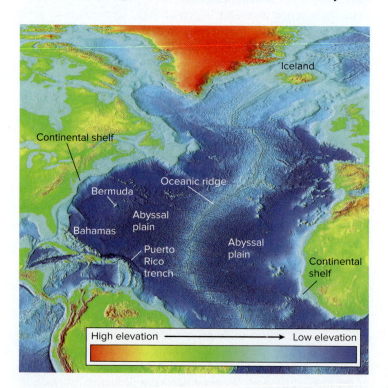

Figure 4.4 Features of the ocean floor. Traveling eastward from the Atlantic coast of North America, we cross a relatively narrow, shallow continental shelf before descending to the abyssal plain. Islands such as Bermuda that are located on the abyssal plain have a volcanic origin, while islands on the continental shelf, such as the Bahamas, are typically formed of sedimentary rocks. NOAA

the continents, the abyssal plains give way to an oceanic ridge system. **The oceanic ridge system consists of broad areas of shallower seafloor** that can be followed continuously through the world's major oceans, like a giant zipper that holds the outer layer of Earth in place (Figure 4.5). There are four principal regions: the Mid-Atlantic Ridge, the East Pacific Rise, the Antarctic Ridge, and the Indian Ridge, and many smaller connecting sections. In comparison to the depth at the abyssal plains, the ocean floor along the ridge system is shallower (less than 3 kilometers, or 1.9 miles, deep). Scientists surveying the ocean floor discovered that the **ridge system is a source of volcanic activity.** Volcanism can be observed firsthand where the oceanic ridge comes to the surface in Iceland. But volcanism does not occur exclusively at the oceanic ridges. In fact, the world's oceans are dotted with hundreds of small islands that represent the tops of volcanoes that grew upward from the ocean floor (for example, Bermuda; see Figure 4.4).

One last set of features found in most ocean basins is **narrow, deep oceanic trenches** (Figure 4.6) along the margins

✓ **Checkpoint 4.5**

⊘ basic   ⊘ advanced
⊘ intermediate   ⊘ superior

After examining the diagram that follows, determine which pair of locations has oceanic trenches.

a. A, C
b. D, F
c. B, E
d. G, H

Figure 4.5 Oceanic ridge system. The purple areas indicate locations of the principal oceanic ridges, which cover approximately one-half of the ocean floor.
Adapted from Charles (Carlos) Plummer, *Physical Geology* 11e © 2007, reproduced with permission of McGraw-Hill Education

Figure 4.6 Distribution of trenches on the seafloor. Trenches are concentrated around the margins of the Pacific Ocean.

of some continents (for example, western South America) or adjacent to volcanic island chains such as the Aleutian Islands, Alaska. These trenches are similar in shape to the valleys we see on land, but ocean trenches are much longer, some may be up to 11 kilometers (6.8 miles) deep, and they form by a completely different process than valleys do.

Notice that the ridges are located near the centers of most ocean basins, while trenches are located almost exclusively near ocean margins (Figures 4.5 and 4.6). An oceanic ridge appears to bisect the entire Atlantic Ocean. Likewise, Africa is surrounded on three sides by oceanic ridges. However, many trench systems are located in the western Pacific and northwest of Australia. In Section 4.4, we will learn that the ridges and trenches play key roles in the process of plate tectonics.

Age of the Ocean Floor

Analysis of rock samples from the ocean floor has revealed that the **oceanic crust is relatively young compared to the age of the continents.** The oldest oceanic rocks are less than 200 million

years (Myr) old. (How these ages are actually determined is discussed in Chapter 8.) In contrast, the oldest known rocks on the continental crust are more than 4 billion years (4,000 million years) old. The **age of the ocean floor is consistently youngest at the oceanic ridges and oldest along the edges of oceans** (Figure 4.7). For example, oceanic rocks along the North American and African coastlines are approximately 180 million years old, whereas rocks adjacent to the ridge are less than 1 million years old. The age of the oceanic crust increases symmetrically moving away from the oceanic ridge system (Figure 4.7). From observation of the relatively young ocean floor, scientists suggested that Earth has a recycling system that constantly creates new oceanic rocks at oceanic ridges and destroys old oceanic floor elsewhere.

Heat Flow, Volcanoes, and Earthquakes

Scientists had long been aware that heat was escaping from Earth's interior; this was a key aspect of the contracting Earth model. Because they assumed a uniformly contracting Earth, early predictions were that heat flow would be relatively uniform around the world. However, scientists were surprised when surveys of the ocean floor revealed that heat flow was greatest along oceanic ridge systems. The ridge system was recognized as a place where **hot, melted rock, called magma,** was forcing its way into the oceanic crust with some erupting onto the seafloor (Figure 4.8a). With the exception of volcanic regions where magma comes to the surface, heat flow at any point on Earth's surface is too low to be noticeable. The region around the Pacific Ocean is called the Ring of Fire because of the many active volcanoes found there (Figure 4.8b). Most active volcanoes above sea level are located adjacent to oceanic trenches. Volcanoes are almost never present along continental margins that are not near a trench (compare Figures 4.6 and 4.8b).

 Checkpoint 4.6

✓ basic	✓ advanced
✓ intermediate	✓ superior

Regarding the relationship between age and the character of the ocean floor, which statement(s) is (are) *true*?

a. Deeper regions of the ocean floor tend to be younger.
b. The Pacific Ocean is larger than the Atlantic Ocean because it contains older oceanic floor.
c. The oldest oceanic crust is present only near trenches.
d. The youngest oceanic crust is near the ridges.

million years

0 20 40 60 80 100 120 140 160 180 200 220 240 260 280

Figure 4.7   Age of oceanic crust. Ages range from young (less than 1 million years old) along the oceanic ridges (red) to old (180 million years old, blue) along the ocean margins (such as the northwest Pacific Ocean). The difference in oldest ages in the northern and southern Atlantic oceans has been interpreted to mean that the northern Atlantic Ocean began to form before the southern Atlantic Ocean. Gray areas represent the extent of the shallow continental shelves around each landmass.
NOAA

When the locations of earthquakes were plotted on world maps, they were seen to occur more frequently in zones on or adjacent to oceanic ridges and trenches. The largest and deepest earthquakes were present in proximity to oceanic trenches in the Pacific and Indian oceans. Near ocean trenches, earthquakes occur from near the Earth's surface down to depths of 750 kilometers (465 miles; Figure 4.8c). The earthquakes are shallowest near the trench and get deeper on one side of the trench, defining an inclined zone called a *Wadati-Benioff zone,* after the researchers who initially recognized this pattern (Figure 4.9).

Checkpoint 4.7

✔ basic	✔ advanced
✔ intermediate	✔ superior

Compare and contrast patterns of (1) topography of the ocean floor, (2) age of the ocean floor, (3) heat flow, (4) volcanic activity, and (5) earthquake activity for oceanic ridges and oceanic trenches, using Figures 4.5, 4.6, 4.7, and 4.8.

Checkpoint 4.8

✔ basic	✔ advanced
✔ intermediate	✔ superior

Explain how the following patterns can be interpreted to contradict the contracting Earth model: (1) topography of the ocean floor, (2) age of the ocean floor, (3) heat flow, (4) volcanic activity, and (5) earthquake activity in the ocean floor.

a.

b.

c.

Figure 4.8 Distribution of global heat flow, volcanoes, and earthquakes. **a.** Global heat flow measured in milliwatts per square meter (mW/m²). **b.** Global distribution of active volcanoes above sea level. **c.** Global distribution of earthquakes for 2005. Large circles indicate the locations of the largest earthquakes. Color represents depth.
4.8a: Adapted from www.geo.lsa.umich.edu; 4.8b: USGS; 4.8c: Adapted from Charles (Carlos) Plummer, *Physical Geology* 11e © 2007, reproduced with permission of McGraw-Hill Education

Figure 4.9
Zone of earthquakes.
Seismologists Kiyoo Wadati and Hugo Benioff discovered that earthquakes become progressively deeper as one moves away from oceanic trenches. The distribution of earthquakes defines the Wadati-Benioff zone. Red X's represent the hypothetical locations of earthquakes. Diagram is not to scale.

Seafloor Spreading Theory

Because the observations just described contradicted predictions of the contracting Earth model, scientists needed a better theory. Along came Harry Hess, a Captain of a US transport ship in World War II. Hess used the new sonar technology onboard his ship to collect seafloor elevation data as he traveled across the Pacific Ocean. He used those data, combined with more collected after the war, to suggest that the seafloor moved away from the oceanic ridges. Hess published his work in 1962, and it soon became one of the most referenced geology publications ever written. **The seafloor spreading theory suggested that new oceanic lithosphere (crust and upper mantle) is being *continuously* formed along the ridge system by magma rising from below, and as this occurs, the existing rocks of the seafloor move away from the ridge** (Figure 4.10). The theory of seafloor spreading led scientists to conclude that the migration of hot magma from below the ridge heats the overlying seafloor, causing it to expand to produce the higher elevations of the oceanic ridge. The ocean floor was interpreted to act as a conveyor belt, gradually moving away from the ridge and creating a gap that was continuously filled with new magma from below.

Because new material is constantly being generated at ridges, old material must be destroyed somewhere else, or Earth would expand. Since Earth is not expanding, there had to be places on Earth's surface where older ocean floor was destroyed. The fact that the deepest earthquakes and some of the older ocean floor are adjacent to trenches led scientists to hypothesize that the **ocean floor descends into the mantle at ocean trenches.**

The concept of seafloor spreading provided a potential mechanism to explain the motion of the adjoining continents that Wegener had described decades earlier in connection with continental drift. A few years later, new evidence related to magnetization of ocean floor rocks strengthened the seafloor spreading theory.

Paleomagnetism

You have probably picked up a compass at some time in your life and know that the needle points north. A compass points north because the tips of the compass needles are magnetized, causing them to line up parallel with Earth's magnetic field (Figure 4.11).

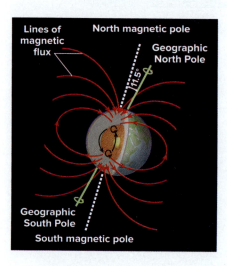

Figure 4.11 Earth's magnetic field. The magnetic field originates from currents in the outer core. A compass needle lines up along the lines of magnetic force (flux) so that it points toward the magnetic poles. The geographic and magnetic poles are not in the same location.

Figure 4.10 Observations of ocean floor features and how they are explained. Young ocean floor forms where magma rises to the surface at the ocean ridge, causing higher heat flow there. This process also pushes existing seafloor away from the ridge. Consequently, seafloor rocks become progressively older with increasing distance from the ridge. Volcanism and earthquakes are common along the margins of oceans where trenches are found. Seafloor rocks sink into the mantle adjacent to trenches, producing deep earthquakes and magma for volcanoes.

Adapted from Charles (Carlos) Plummer, *Physical Geology* 11e © 2007, reproduced with permission of McGraw-Hill Education

You will recall from Chapter 2 that Earth has three compositional layers—crust, mantle, and core—and that the molten outer core is the source for Earth's magnetic field. Planetary magnetic fields require three conditions:

1. A large reservoir of fluid that can conduct electricity—in this case, the iron-rich outer core.
2. Energy to move the fluid. Heat from radioactive decay and planetary formation generates vertical currents in the outer core.
3. Rotation to mix the fluids. The relatively rapid rotation of Earth creates spiraling currents to sustain flow patterns.

The magnetic field has negative (−) and positive (+) poles that are located near the geographic (North and South) poles (Figure 4.11).

Declination and Inclination of the Magnetic Field. The shape of **Earth's magnetic field varies with location and can be defined by its declination (or direction) and inclination** (Figure 4.11). *Declination* simply defines the angle between the direction the north arrow points on a compass at some location and the direction to the geographic North Pole from that location. If a compass needle could rotate in three dimensions, the needle's *inclination* would range from vertical at the magnetic poles to horizontal at the (magnetic) equator. The magnetic field is inclined downward in the Northern Hemisphere and upward in the Southern Hemisphere, and inclination gradually increases between the equator and the poles. In other words, the needle would point straight down at the magnetic north pole and straight up at the magnetic south pole.

Suppose a strong wind blew down some trees in a forest. Most of the trees would be aligned in the direction of the wind. Even without knowing anything about the original wind, you could figure out the direction of the wind by looking at the fallen trees. Geologists use a similar technique to determine where rocks formed relative to Earth's magnetic poles. The inclination of the

magnetic field can be used as an estimate of a past (or "paleo") latitude. When volcanic rocks form on Earth's surface, their magnetic minerals act as miniature compasses. The **magnetic fields of atoms in magnetic minerals, especially those with high iron content, align parallel to Earth's magnetic field.** The inclinations of atoms in magnetized minerals in ancient lava flows can therefore be used to determine the original latitude of the cooling magma when the rock formed. **The study of the magnetic field preserved in these rocks is called *paleomagnetism.***

Magnetic Field Reversals. Studies of continental rocks that form from magma have revealed that the direction of Earth's magnetic field sometimes points toward the geographic north and other times points toward the geographic south. Scientists interpret this contrast to mean that **the positive and negative magnetic poles have switched positions many times in the past** (Figure 4.12). The magnetic poles slowly dance around Earth's surface near the geographic poles and, from time to time, will completely reverse their positions. During these reversals, the poles flip from positive to negative and, thousands to millions of years later, change back again. Earth's magnetic field exhibits "normal" polarity when the negative magnetic pole and the geographic North Pole are adjacent, as they are today. The magnetic field would be considered to have "reverse" polarity if the magnetic polarities were opposite to the present-day pole positions.

Although scientists are not certain of the exact reason for these reversals of polarity, paleomagnetic data from different-aged rocks suggest that magnetic reversals have occurred frequently throughout the history of the planet. Each reversal lasts for an average of about 250,000 years, but some last much longer (37 million years) and others may last for just a few tens of thousands of years. Geophysicists who study the magnetic field have discovered that it can take as little as a few thousand years to change the field's polarity. The most recent reversal occurred 780,000 years ago, and the magnetic field has lost strength in the last few centuries, leading some to wonder if Earth is about to undergo another reversal

Checkpoint 4.9

basic · advanced · intermediate · superior

Which location has the greatest magnetic inclination value (that is, closest to vertical)?

a. Anchorage, Alaska
b. New York, New York
c. Miami, Florida

Checkpoint 4.10

basic · advanced · intermediate · superior

The magnetic north pole has migrated northward over the last century, so how did magnetic inclination readings at Chicago change between 1900 and 2000?

a. Inclination increased.
b. Inclination decreased.
c. Inclination remained constant.

a. Normal polarity **b.** Reversed polarity

Figure 4.12 Polarity of Earth's magnetic field. **a.** The inclination of the magnetic field varies with latitude. The field is horizontal at the magnetic equator, steeper at high latitudes, and vertical at the magnetic poles. The magnetic field is inclined downward in the Northern Hemisphere and upward (away from Earth's surface) in the Southern Hemisphere. **b.** The magnetic field has reversed polarity in the past, meaning that the positive and negative polarities have switched positions.

of polarity. During a change in polarity, Earth may have a weaker than normal magnetic field and could even have multiple magnetic poles. Navigation that relied on compasses would be difficult, and radio communications would be adversely affected. However, such changes occur on geologic timescales and are at least several centuries or millennia in the future.

How Paleomagnetism Supports the Seafloor Spreading Theory. We can use paleomagnetism, this magnetic record in rocks, to deduce how the ocean floor has changed over time. Marine surveys of the ocean floor can measure the intensity (strength) of Earth's magnetic field, and the intensity values reveal the polarity of the rocks. The strength of Earth's magnetic field at any particular location will be slightly increased if magnetic minerals "frozen" in the rocks are aligned with Earth's present-day magnetic field or will be slightly weaker if they are aligned with a reverse-polarity field.

When scientists analyzed very precise magnetic field data recorded over the oceans, the data revealed "stripes" of high and low magnetic intensity corresponding to areas of normal and reverse polarity. These stripes are oriented parallel to adjacent oceanic ridges (Figure 4.13). **Ocean floor rocks reveal a symmetrical pattern of magnetic polarity reversals on either side of oceanic ridges.** For example, a sequence of magnetic reversals (a switch from normal to reverse polarity and back) in

Figure 4.13 The polarity of Earth's magnetic field along the Mid-Atlantic Ridge, south of Iceland. Dark blue stripes indicate normal polarity (N) rocks; purple areas indicate reverse polarity (R). Note symmetrical patterns on either side of the dark band representing the center (axis) of the ridge.

a. Time of normal magnetism

b. Time of reverse magnetism

c. Time of normal magnetism

Figure 4.14 Evidence of seafloor spreading. **a.** Rocks with normal polarity (blue) form when the compass points to the magnetic north pole (as it does today). **b.** Conditions of reverse polarity (tan) represent periods when the compass arrow points to the magnetic south pole. **c.** Alternation of normal and reverse magnetic fields produces a striped pattern of magnetism in the ocean floor rocks (blue, tan, green). This evidence supports the seafloor spreading theory. 4.14a, 4.14b, 4.14c: Adapted from Charles (Carlos) Plummer, *Physical Geology* 11e © 2007, reproduced with permission of McGraw-Hill Education

Inclination is determined for three lava flows preserved in a cliff as shown in the following image. What happened to the continent on which these rocks were formed? (Assume normal polarity throughout. Upper layers are youngest; lower layers are oldest.) The rocks in the cliff moved toward the _____ in the _____ Hemisphere.

a. Equator; Southern
b. Equator; Northern
c. Pole; Southern
d. Pole; Northern

- Sedimentary rocks
- Lava flows

Which magnetic property was more important in providing support for the seafloor spreading theory? Explain the reasoning behind your answer.

a. Magnetic inclination; the inclination of Earth's magnetic field varies with location.
b. Magnetic polarity; the north and south magnetic poles have switched positions throughout Earth's history.

the western Atlantic, offshore from the Carolinas, can be matched with a similar sequence in the eastern Atlantic off the coast of West Africa. These patterns are interpreted to suggest that new crust was divided in half as it formed along the oceanic ridges, and each half moved in opposite directions away from the ridge (Figure 4.14). Scientists found the same sequences of reversals in each of the major oceans. The age and duration of magnetic reversals were initially determined by analyzing lava flows on the continents (Figure 4.15). The seafloor spreading theory predicted that ocean floor rocks would become progressively older with increasing distance from oceanic ridges. This was confirmed when the ages of some continental reversals were compared to those of oceanic reversals. Thus, the record of magnetic polarity in the rocks of the ocean floor provided unequivocal support for the concept of seafloor spreading.

Figure 4.15 The polarity of the oceanic crust adjacent to the oceanic ridge separating the Juan de Fuca and Pacific plates, west of Washington. Dark gray pattern indicates normal polarity, and light gray represents reverse polarity. Note the symmetrical pattern of "stripes" on the ocean floor on either side of the oceanic ridge. These magnetic patterns can be matched to basalt on land to determine when the rocks in each stripe were formed. We can then use the distance of the stripes from the ridge to estimate the rate at which the plates are moving.
Adapted from after http://geomaps.wr.usgs.gov/parks/noca/fg11.gif

4.4 Plate Tectonics

Learning Objectives

- Explain what is meant by the term *plate tectonics*.
- Identify three processes that cause rocks to melt.
- Use a blank map of the world to identify key features associated with plate tectonics.
- Explain how scientists can use data from ocean floor features to estimate the rate of plate motion.
- Draw and label a cross section through a tectonic plate to show two types of crust, the lithosphere and asthenosphere.
- Explain why Earth is the only terrestrial planet with active plate tectonics.

In Sections 4.2 and 4.3, we outlined the concepts of continental drift and seafloor spreading. The names of these concepts imply that both the continents and the ocean floors are in motion, but give little idea about how their motions are related. Our purpose in this section is to draw together all the threads of evidence, to show how the theory of plate tectonics explains many different land- and ocean-based observations. But first we must define the features and processes involved.

Key Layers and Processes

Lithosphere and Asthenosphere. Recall from Chapter 2 that **Earth's interior can be separated into the core (inner and outer),** the **mantle,** and the **crust.** The temperature and pressure for the first few hundred kilometers below Earth's surface are just right to cause most minerals to remain solid while allowing a small fraction to melt. As a consequence, two additional layers can be distinguished: the lithosphere and the asthenosphere (Figure 4.16).

The *lithosphere* is a rigid outer layer composed of the crust and the uppermost mantle. We should consider the lithosphere as the outermost layer of Earth, encompassing both the rocks of the continents (continental lithosphere) and the ocean floor (oceanic lithosphere). As we will see, it is events in the lithosphere that shape the planet's surface features and that generate earthquakes and volcanoes in a predictable pattern.

The *asthenosphere* is a layer beneath the lithosphere. It is about 100–200 kilometers (62–124 miles) thick. Although the asthenosphere is composed almost completely of solid rocks, it flows slowly because of the presence of just a small proportion (about 1 percent) of minerals that experience melting.

Melting of Rocks. Volcanic activity due to the melting of rocks in the lithosphere and asthenosphere is a critical process in plate tectonics. Thus, before going further, we must describe what causes rocks to melt. We know that ice melts to liquid water as its temperature rises above freezing. The higher temperature increases vibration between adjacent atoms or molecules and provides energy to break the bonds that hold the water molecules in the ice together. In the same way, certain processes break the bonds between the atoms and molecules that make up the minerals in rocks. Pressures within Earth's crust and mantle act to reduce vibrations in the atomic structure of minerals and thus to prevent melting. Three basic changes can produce melting of rocks: increasing temperature, decreasing pressure (on hot rocks), and chemical reactions following the addition of water (Figure 4.17).

Note that it is not necessary to completely melt a rock. The three processes break some of the bonds holding atoms or molecules together and lead to *partial melting* of some minerals and the production of magma. All three conditions that result in melting occur as a consequence of plate tectonics. (We will revisit these mechanisms in Chapter 6.)

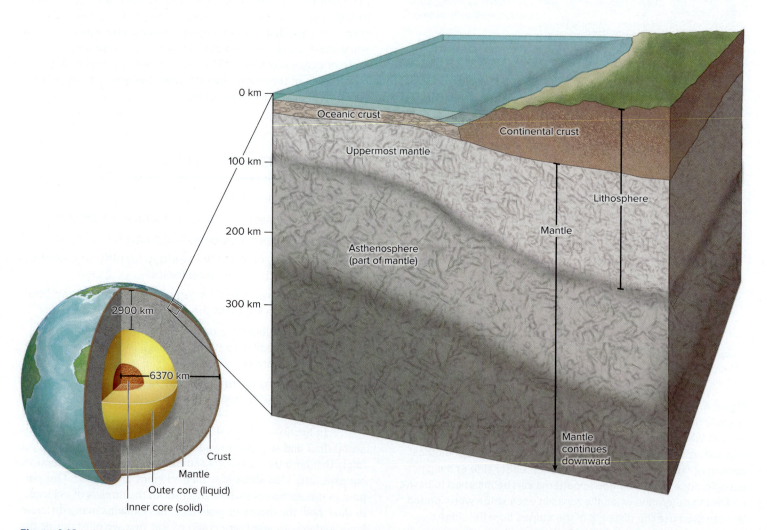

Figure 4.16   Lithosphere and asthenosphere. The outermost part of Earth is divided into two layers, the lithosphere (crust and rigid uppermost mantle) and the asthenosphere (plastic layer in the upper mantle).
Adapted from Charles (Carlos) Plummer, *Physical Geology* 11e © 2007, reproduced with permission of McGraw-Hill Education

The Process of Plate Tectonics

Earth's lithosphere is broken into a series of rigid, mobile **tectonic plates.** There are eight major plates (North American, South American, Pacific, Nazca, Eurasian, African, Antarctic, and Indian-Australian) and several smaller plates (for example, Caribbean, Arabian, Juan de Fuca; see Chapter Snapshot). Reinforcing the tentative nature of science, some scientists argue the Indian-Australian plate is actually two different plates with a boundary starting to form beneath the seafloor of the Indian Ocean. Discussions in this chapter use eight major plates since the location and characteristics of that boundary are not clearly defined. Plates are typically composed of both continental and oceanic lithosphere. For example, the African plate contains the continent of Africa as well as the southeastern Atlantic Ocean and the western Indian Ocean. Oceanic ridges and trenches represent the principal

boundaries of the tectonic plates (compare Figures 4.5 and 4.6 and Chapter Snapshot). **The theory of plate tectonics states that interactions of the plates along these boundaries account for the formation of new lithosphere, mountains, earthquakes, and volcanoes and contribute to the gradual movement of continents and the opening of oceans.**

Formation of New Lithosphere. The oceanic lithosphere is thinnest below oceanic ridges, bringing the underlying asthenosphere closer to the surface. Melting occurs preferentially in the asthenosphere at this location as the thinner lithosphere results in a decrease in pressure. Consequently, partial melting below oceanic ridges is termed **decompression melting** (Figure 4.17b). Melting produces magma that is less dense than surrounding material and rises to the surface (Figure 4.18). The magma cools and solidifies

a. Melting due to increasing temperature
b. Melting due to decreasing pressure
c. No melting at X before the addition of water
d. Melting occurs at X after the addition of water

Figure 4.17 Why rocks melt. A rock (X) located in the crust or mantle may melt because of **a.** an increase in temperature, **b.** a decrease in pressure (decompression melting), or **c.** and **d.** the addition of water. The latter two processes are responsible for most melting. Increasing temperature may occur if a rock is buried more deeply or if it lies adjacent to a heat source such as a magma chamber. Pressure decreases when rocks rise through the mantle or crust. Water may be added to rocks by chemical reactions or seawater infiltration.
4.17a, 4.17b, 4.17c, 4.17d: Adapted from NOAA

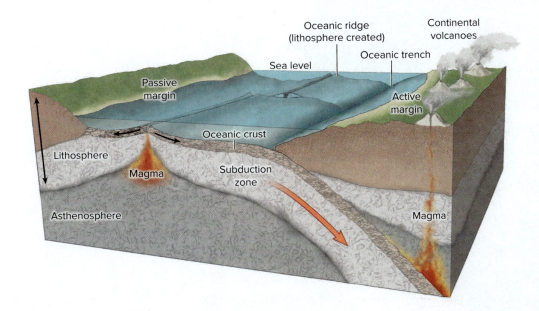

Figure 4.18 A cross section illustrating the plate tectonic cycle. Oceanic lithosphere is created by magma rising from the asthenosphere at the oceanic ridge. Plates move away from the ridge and may be consumed as they descend into the mantle at a subduction zone. Features not to scale.

PLATES
OF THE
WORLD

Eurasian plate

North American plate

African plate

Caribbean plate

Cocos plate

South American plate

Nazca plate

Scotia plate

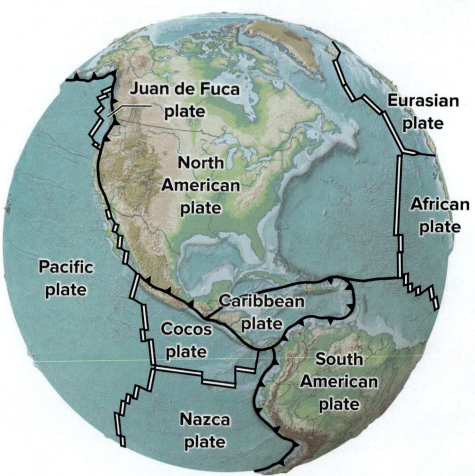

Juan de Fuca plate

North American plate

Eurasian plate

African plate

Pacific plate

Caribbean plate

Cocos plate

South American plate

Nazca plate

TRANSFORM BOUNDARY

CONVERGENT BOUNDARY

DIVERGENT BOUNDARY

Arabian plate

Eurasian plate

African plate

Indian-Australian plate

Antarctic plate

Eurasian plate

Philippine plate

Juan de Fuca plate

Pacific plate

Indian-Australian plate

✓ Checkpoint 4.13

Draw the approximate locations of the plate boundaries on the map to the right. Use different line symbols for the oceanic ridges and oceanic trenches. Name and label as many of the plates as you can.

basic   advanced
intermediate   superior

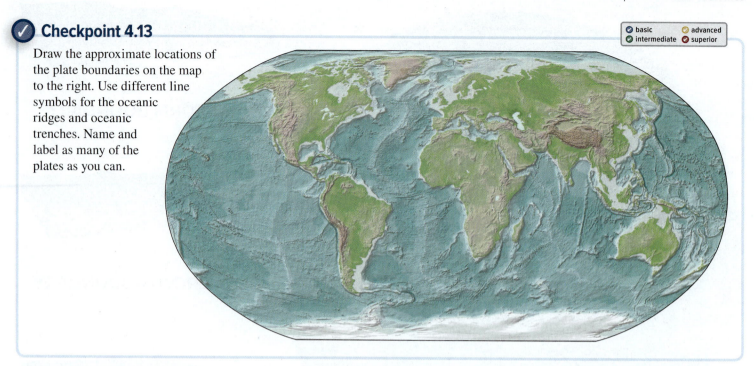

on and below the ocean floor to create new lithosphere that is added to the edge of each of the oceanic plates. This process is similar to the way your fingernail grows as new nail, added at the base, pushes the older part of the nail farther out. This forcing apart of the plates, called *ridge push,* is a process that contributes to the movement of tectonic plates.

Earthquakes, Volcanoes, and the Destruction of Lithosphere at Subduction Zones. To ensure that the size of Earth remains constant, the creation of new lithosphere at the oceanic ridges must be balanced by the destruction of lithosphere elsewhere. The theory of plate tectonics holds that **oceanic lithosphere is consumed as it descends into the mantle adjacent to trenches at regions termed subduction zones** (Figure 4.18). Since the plate is slightly more dense than surrounding material in the uppermost mantle, the plate may actually be pulled downward, a process called *slab pull.* Chemical reactions in the descending plate release water, which lowers the melting point of minerals in the overlying mantle to cause partial melting (Figure 4.17d). The magma generated either rises to the surface to create volcanoes or cools and solidifies within the crust. This addition of material thickens the continental crust adjacent to subduction zones and contributes to the formation of mountain belts.

The descending plate bends and fractures as it is pulled down into the mantle, causing shallow earthquakes to occur. Earthquakes happen only in the cold, brittle lithosphere. The lithosphere eventually loses its brittle nature as it warms up as it descends into the hotter mantle. The deepest earthquakes are thought to be the result of large-scale changes to mineral structures caused by increasing temperature and pressure deeper in the mantle.

Only oceanic lithosphere is consumed at subduction zones. Consequently, the ocean floor is continuously being created

(oceanic ridges) and destroyed (subduction zones), while the continental lithosphere remains at Earth's surface. As we will see later, continents may be broken up into smaller pieces or may combine to form larger landmasses (such as Pangaea), but the mass of continental material does not change significantly.

Most continental margins are not plate boundaries. These continental margins are known as *passive margins* (Figure 4.18) and are generally free of volcanism and earthquake activity. Continental margins that represent plate boundaries are called *active margins* and are characterized by volcanoes and earthquakes. The Atlantic coastlines of North and South America are examples of passive margins, while the Pacific margins of each continent are active margins.

Rate of Plate Movements. Plate movements occur very slowly. Scientists originally determined the rates and directions of plate motions in the same way that you determine your average walking speed. First they determined the age of a seafloor rock, and then they divided that into the distance from an oceanic ridge system (where new material is created). This simple but effective calculation (distance divided by time) was compared to rates of plate motion determined by analyzing the contrasting ages of volcanic rocks on the Hawaiian Islands.

The Hawaiian Islands are not associated with a plate boundary but are formed in the interior of the Pacific plate above a plume of rising hot mantle rock (Figure 4.19). There are many of these rising plumes, and their locations remain essentially fixed relative to the overlying plates. The point where a plume reaches the surface is known as a *hot spot.* The Hawaiian Islands form as the Pacific plate moves over a hot spot, much as a conveyor belt moves above a heat source. The islands become progressively older with increasing distance from the hot spot. The relationship between the distance of the older islands from the hot spot (600 kilometers; 373 miles) and the age of these islands (5 million years) yields the average rate of plate motion:

600 km/5,000,000 yr = 120 km/Myr = 12 cm/yr (5 in/yr)

Today, Global Positioning System satellites allow scientists to record tiny movements of known points on Earth's surface, thus determining the motions of plates (Figure 4.20). Different plates move at different rates, but each has a fairly constant rate of motion:

- **Ultraslow** spreading rates (less than 1 centimeter per year; 0.4 inch per year) are seen at the oceanic ridge in the Arctic Ocean.
- **Slow** rates (1 to 2 centimeters per year; 0.4 to 0.8 inch per year) are found for the North American and Eurasian plates along the ocean ridge in the northern Atlantic Ocean.
- **Rapid** spreading rates of more than 15 centimeters per year (6 inches per year) have been identified for the Pacific and Nazca plates along the East Pacific Rise.

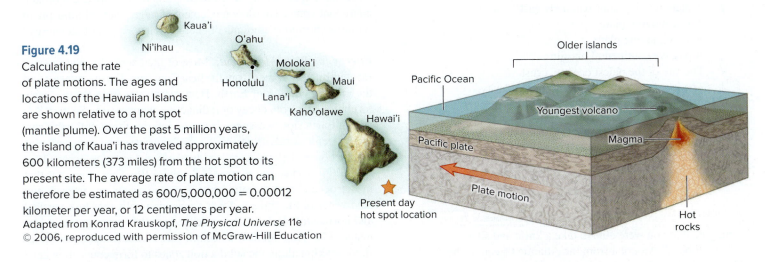

Figure 4.19
Calculating the rate of plate motions. The ages and locations of the Hawaiian Islands are shown relative to a hot spot (mantle plume). Over the past 5 million years, the island of Kaua'i has traveled approximately 600 kilometers (373 miles) from the hot spot to its present site. The average rate of plate motion can therefore be estimated as 600/5,000,000 = 0.00012 kilometer per year, or 12 centimeters per year.
Adapted from Konrad Krauskopf, *The Physical Universe* 11e © 2006, reproduced with permission of McGraw-Hill Education

Figure 4.20 Motions of plates. Arrows show the directions and rates of plate motions determined by Global Positioning System satellites. Plate motions in the Pacific Ocean are nearly five times faster than in the Atlantic Ocean. The arrows point in the direction the plate is moving, and the lengths of the arrows represent rates. (Measurements are in centimeters per year.)
Adapted from Charles (Carlos) Plummer, *Physical Geology* 11e © 2007, reproduced with permission of McGraw-Hill Education

Checkpoint 4.14

| ✓ basic | ✓ advanced |
| ✓ intermediate | ✓ superior |

Using the map from Checkpoint 4.13, identify and label the map with as many of the features on the associated list as you can.

a. A subduction zone
b. A plate that is 99 percent oceanic lithosphere
c. A plate that is over 90 percent continental lithosphere
d. A pair of plates moving toward each other
e. A plate nearly surrounded by oceanic ridges
f. A plate moving approximately north
g. A plate moving approximately east
h. A fast-moving plate
i. A slow-moving plate
j. A passive continental margin
k. The site of the oldest oceanic crust
l. The location of rocks with normal polarity
m. A pair of plates moving away from each other
n. A hot spot
o. A location with very young rocks (0 to 5 million years)

Checkpoint 4.15

| ✓ basic | ✓ advanced |
| ✓ intermediate | ✓ superior |

Imagine that you were able to take a knife and slice through North America from the Atlantic Ocean to the Pacific Ocean. Draw and label a sketch illustrating the locations and positions of the lithosphere, asthenosphere, crust, and mantle.

Current seafloor spreading rates are approximately five times higher for the East Pacific Rise than for the Mid-Atlantic Ridge. The oldest oceanic crust in both the Atlantic and Pacific oceans is the same age (about 180 million years; Figure 4.7), but the Pacific Ocean is much wider than the Atlantic Ocean because of more rapid plate motion. The slow-moving North American and African plates are driven almost exclusively by ridge push forces. In contrast, the more rapid motions for the plates below the Pacific Ocean can be attributed to a combination of ridge push and slab pull.

Although slow in comparison to human activities, plate movements are powerful. Over millions of years, the movements of plates have opened and closed oceans and have formed and broken apart continents. The slow movement of huge plates of lithosphere requires enormous amounts of energy. What supplies that energy?

Energy for Plate Movement. Most of the energy for plate movement ultimately comes from the heat flow from Earth's interior and the gradual cooling of Earth. Heat remaining from Earth's early formation and from decay of radioactive elements (see Section 8.4) within Earth slowly escapes from the interior through Earth's surface. Scientists infer that this heat is transferred by *convection* as hot rock rises below oceanic ridges and colder rock descends at subduction zones (see Figures 4.8a and 4.10). They point out that in the oceans and the atmosphere, warm water or air rises and cold water or air sinks to form convection cells that link vertical and horizontal motions. By analogy, scientists propose that warm material is rising below the oceanic ridges and cold, denser material is sinking along the subduction zones to form convection cells in the upper mantle (Figure 4.21). Since the rate of heat loss likely influences plate tectonic activity, greater heat loss during earlier times in Earth's history may have resulted in faster rates of plate motion and/or more plate boundaries.

Figure 4.21 Mantle circulation system. The flow of mantle rock may occur as two-layer convection cells or as a result of whole-mantle convection. Both models have rock rising from the deep mantle, but subduction zones are interpreted to continue into the lower mantle or end in the upper mantle. Note that the crust, lithosphere, and mantle are not drawn to scale.
Adapted from *Oceanus Magazine,* Woods Hole Oceanographic Institute

Plate movements are also aided by gravity. Recall that oceanic lithosphere is "pushed" away as new lithosphere forms at midocean ridges. At ocean trenches, plates are "pulled" into the subduction zones by the weight of the descending slab of lithosphere. Some plates appear to sink all the way to the core-mantle boundary. Scientists hypothesize that gravity pulling on the plate at subduction zones combines with warm rising rock near ridges to form circulation cells in the mantle, although they debate the details of this circulation (Figure 4.21).

An Ongoing Process. Plate tectonics supplies the big picture of how Earth works. It links the heat flow from Earth's interior and the locations of earthquakes and volcanoes to the distribution of continents and oceans.

Keep in mind that plate tectonics is an ongoing process. The map of plates (see Chapter Snapshot) is just a snapshot in time. Plates and plate boundaries change over time. Movement may stop at some locations, new boundaries will form, and plates will look different in the future. Two hundred million years ago, there was no Atlantic Ocean (see Figure 4.2). Two hundred million years in the future, the world will look much different than it does today.

Do Other Planets Have Plate Tectonics?

Plate tectonics requires not just a layered Earth with lithosphere and asthenosphere but also the excess internal heat necessary to drive the plates. Smaller terrestrial planets such as Mars cooled more rapidly than Earth and lost their internal heat much earlier in their history. There is no evidence that plate tectonics is occurring today on Mars, but surveys reveal that the oldest sections of the Martian crust have paleomagnetic patterns similar to those of Earth's ocean floors. Scientists interpret these observations to mean that Mars experienced plate tectonics very early in its history (4 billion years ago) when it was still hot. When the interior cooled, plate tectonics stopped.

But what about Venus? Venus is almost the same size as Earth and has a similar layered structure. However, Venus does not have plate tectonics. The reason lies in the distribution of lithosphere on Venus. Two-thirds of Earth's lithosphere is oceanic; the rest is continental. Much of Earth's heat is lost in creating new oceanic lithosphere along the oceanic ridge system. This could not occur if older oceanic lithosphere were not being destroyed along subduction zones. The shape and size of volcanoes on Venus suggest it has a thick lithosphere that hinders the motion of plates. Without the destruction of old lithosphere, there is no opportunity to create new lithosphere and thus to allow the planet's internal heat to escape along long-lived stable features such as oceanic ridges. Instead, some scientists hypothesize that heat builds up below the

surface of Venus and is lost in periodic catastrophic volcanic-type events that destroy the existing lithosphere. The resulting magma slowly cools to form new lithosphere. Imagine the mantle boiling up to consume North America every 500 million years. In this scenario, it would be difficult for life to evolve as it has on Earth. This is yet another reason why Earth is unique among the planets of our solar system!

4.5 Plate Boundaries

Learning Objectives

* Explain how relative plate motions result in divergent, convergent, and transform plate boundaries.
* Compare and contrast the features of convergent and divergent plate boundaries.
* Explain the features found in association with a transform plate boundary.
* Interpret a plate configuration to predict the location of key geologic features (for example, mountains, volcanoes, subduction zones), and sketch and label a representative cross section.
* Predict how the current distribution of plates, continents, and oceans will change millions of years in the future.
* Discuss how plate tectonic processes can alter global climate patterns.

Oceanic ridges and trenches, volcanoes, and earthquakes are just some of the phenomena present at the boundaries that separate the tectonic plates. The type and distribution of features are characteristic of the relative motions of the plates on either side of the boundary. We classify plate boundaries into three categories based on their relative plate motions (see Chapter Snapshot).

Divergent Plate Boundaries

Divergent plate boundaries occur where hot, rising mantle rock causes plates to move apart (Figures 4.18 and 4.22a and Chapter Snapshot). The evolution of a divergent plate boundary has three recognizable stages that we can loosely characterize

Figure 4.22 Three types of plate boundaries. **a.** At a divergent boundary, plates move apart. **b.** At a convergent boundary, plates move toward one another. **c.** At a transform boundary, plates move parallel to one another in opposite directions.

✓ **Checkpoint 4.16**

Predict what the map of the world will look like (a) 5 million years in the future and (b) 50 million years in the future. Identify how the distribution of continents and oceans will change. Which plates will exhibit the greatest changes in comparison to today?

as birth, youth, and maturity. The birth of a divergent boundary occurs when an existing piece of continental lithosphere begins to break apart. It may seem counterintuitive that the formation of an ocean begins with the breakup of a continent, but remember that the Atlantic Ocean owes its origin to the breakup of the supercontinent Pangaea. Such a location is characterized by thinning of the continental lithosphere, often accompanied by volcanic activity. This process is happening today on Africa in an area known as the East African Rift zone (Figures 4.23a and 4.24a). As the lithosphere breaks apart, it **forms a wide, steep-walled depression known as a rift valley.** The underlying asthenosphere is close to the surface below the rift valley, and decompression melting generates magma that forms volcanoes. Eventually, the continental crust in a rift valley separates to form a gap where rising magma creates new oceanic floor (Figure 4.24b). Inflow of seawater initially forms a narrow ocean (the youth stage), much like the Red Sea to the north of the East African Rift zone that separates the Arabian peninsula from Africa (Figure 4.24). It takes millions of years for narrow oceans to expand to form a mature ocean like the present-day Atlantic or Pacific Ocean because the rates of plate motions are so slow (Figure 4.23c).

What about a fourth stage—death? Eventually each oceanic ridge system will die; that is, it will stop producing new oceanic lithosphere. Ultimately, the ridge will be lost as the oceanic lithosphere descends into a subduction zone or is added to the margin of a continent by plate collision.

What do you think would happen if the East Pacific Rise, the oceanic ridge in the Pacific Ocean, stopped working?

We would record a slow reduction in the size of the ocean basin. The Pacific and Nazca plates would be consumed by advancing subduction zones along the margins of the South American, North American, Eurasian, and Indian-Australian plates. Plate tectonics regulates Earth's heat loss, so we would expect to see a new divergent boundary start to form or other plate motions to accelerate, if a mature divergent boundary were to shut down.

Convergent Plate Boundaries

Oceanic lithosphere is consumed at subduction zones where it descends into the mantle beneath trenches. The plate junction where this occurs is a convergent boundary (Figure 4.22b). The descending plate has slowly traveled across the ocean floor to the trench for many millions of years. During this time, sediment collected on top of the oceanic lithosphere. Some of the sediment is scraped off the descending plate, and the rest is carried into the subduction zone. Water is abundant in the descending plate. It may be mixed with the sediment, present in fractures at the top of the oceanic crust, or combined in the atomic structure of minerals in the lithosphere. Increased pressures and temperatures in the descending plate compact the sediment, squeezing out water, and result in chemical reactions that flush water from minerals. Even though the addition of water reduces the melting temperatures of the rocks, the temperature of the descending lithosphere is too cold to melt because it has been near Earth's cold surface for millions of years.

In contrast, the temperatures are much higher in the region in the overriding plate immediately above the descending plate.

a. Continent undergoes extension. The crust is thinned and a rift valley forms.

b. Continent tears in two. Continent edges are faulted and uplifted. Basalt eruptions form oceanic crust.

c. Continental sediments blanket the subsiding margins to form continental shelves. The ocean widens and a midoceanic ridge develops, as in the Atlantic Ocean.

Figure 4.23   Cross sections illustrating the development of a divergent plate boundary. **a.** The continental lithosphere breaks up. **b.** A narrow ocean develops. **c.** A mature ocean basin forms. 4.23a, 4.23b, 4.23c: Adapted from Charles (Carlos) Plummer, *Physical Geology* 11e © 2007, reproduced with permission of McGraw-Hill Education

This part of the lower lithosphere is termed the *mantle wedge* (Figure 4.25). The water that is expelled from the descending plate enters the hot mantle rocks of the mantle wedge. Research showing layered structures in and below the mantle wedge suggests that melting may even occur to depths as great as 450 kilometers (280 miles). Water that is forced out of the descending plate reduces the melting temperature of the rocks in the overlying mantle wedge, resulting in partial melting to form magma. The magma rises through the overlying tectonic plate, and some of it reaches the surface to form volcanoes that may occur as a chain of islands or a range of mountains (Figure 4.25). (We will look at plate tectonics and magma generation in greater detail in Chapter 6.)

a.

b.

Figure 4.24
a. The East African Rift Valleys and the Red Sea. The red arrows indicate the relative motions of continental lithosphere on either side of this early-stage divergent boundary. **b.** Satellite photo of the Red Sea.
4.24a: Adapted from Charles (Carlos) Plummer, *Physical Geology* 11e © 2007, reproduced with permission of McGraw-Hill Education; 4.24b: NASA

Figure 4.25 A cross section illustrating magma generation at a subduction zone. Magma is generated where water from the descending plate enters the overlying mantle wedge to cause partial melting of mantle rocks. The magma will rise through the crust, either solidifying or reaching the surface as a volcanic eruption. Reactions in the descending plate that generate the water typically begin to occur at depths of 100 to 150 kilometers (62 to 93 miles). Consequently, the distance of the volcanoes from the trench will depend on the slope of the subduction zone. The volcanic arc is closer to the trench for steeply inclined subduction zones and farther away where the descending plate slopes gently.

In addition to volcanoes, the largest and deepest earthquakes recorded are also concentrated along subduction zones (more on this in Chapter 5).

The density of lithospheric plates that meet across convergent boundaries plays a significant role in determining which plate descends into the mantle and which remains at the surface. When two oceanic plates collide, **the older of the two plates descends into the mantle because it is colder and denser** (heavier). Where oceanic and continental lithosphere collide, the oceanic plate descends because it is composed of denser minerals than the continental plate. **The less dense crust of the continental lithosphere does not descend into subduction zones** but is piled up to form high mountain belts at convergent boundaries. Remember, individual tectonic plates can be mostly oceanic lithosphere,

Figure 4.26 Three types of convergent plate boundaries (features are not to scale). **a.** When two oceanic plates collide, the plate with the older (cooler, denser) lithosphere descends into the subduction zone. A chain of volcanic islands (island arc) forms on the overriding plate. **b.** In an oceanic-continental collision, the oceanic plate descends, and a chain of volcanoes forms on the continental plate. **c.** A continental-continental collision is marked by a high mountain range but little volcanic activity. Crustal thickness is increased, but scientists still debate the fate of the mantle lithosphere in this type of collision. Features are not to scale.

mostly continental lithosphere, or both oceanic and continental lithosphere. When we discuss plate boundaries, we focus on the lithosphere actually involved in the collision. As such, convergent boundaries come in three varieties (Figure 4.26).

Oceanic Plate Versus Oceanic Plate Convergence. The magma produced by partial melting in the mantle wedge of the overriding oceanic plate adjacent to a subduction zone may reach the surface as volcanoes. As the volcanoes grow, they may rise above sea level to form a long chain of islands, known as an island arc (Figure 4.26a). The Aleutian Islands off the tip of Alaska were formed by magma generated when the Pacific plate descended below oceanic lithosphere on the edge of the North American plate (Chapter Snapshot). A 1995 volcanic eruption that devastated the Caribbean island of Montserrat was the result of subduction of the South American plate below an island arc along the edge of the Caribbean plate.

Oceanic Plate Versus Continental Plate Convergence. When oceanic lithosphere collides with continental lithosphere, the thinner, denser oceanic plate will descend into the subduction zone

(Figure 4.26b). For example, on the western edge of South America, the oceanic lithosphere of the Nazca plate descends below the continental lithosphere of South America, forming an oceanic-continental subduction zone. As these plates collide, the Andes Mountains, the second-highest mountain range in the world, are being formed (see Chapter Snapshot).

Continental Plate Versus Continental Plate Convergence. In these collision zones, the crustal part of the continental lithosphere is peeled away from the mantle part of the continental lithosphere, much as soil is scraped up by a bulldozer (Figure 4.26c). The resulting thickened pile of crust creates mountains with no volcanoes or deep earthquakes. The Himalayas form the tallest mountains in the world with several peaks over

Checkpoint 4.17

◎ basic   ◎ advanced
◎ intermediate   ◎ superior

On the following map, the green shaded areas labeled X, Y, and Z represent continents; assume the blue part of the map is ocean.

How many plates are present?

a. 3 c. 5
b. 4 d. 6

▲ Volcano   ▮ Oceanic ridge   — Oceanic trench

Checkpoint 4.18

◎ basic   ◎ advanced
◎ intermediate   ◎ superior

Examine the map in Checkpoint 4.17. Draw arrows on the map showing the relative directions of plate motions. We could add a series of volcanoes associated with one of the plate boundaries. Which boundary would it be and why? Sketch two cross sections through the lithosphere along lines A–B and C–D, illustrating the characteristics of plate boundaries.

Checkpoint 4.19

◎ basic   ◎ advanced
◎ intermediate   ◎ superior

Venn Diagram: Divergent and Convergent Plate Boundaries

Use the following Venn diagram to compare and contrast divergent and convergent plate boundaries. Three characteristics of plate boundaries are provided; identify at least 10 more. Then write the numbers of the features unique to either group in the larger areas of the left and right circles; note features that they share in the overlap area in the center of the image.

1. Rocks on either side of boundary are the same age
2. Example: Nazca and South American plate boundary
3. Associated with oceanic trenches.
4.
5.
6.
7.
8.
9.
10.
11.
12.
13.

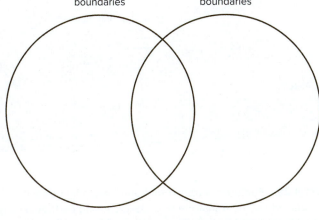

8,000 meters (26,247 feet). These mountains continue to grow to-day as a result of the ongoing continental collision between India and the Eurasian continent (Chapter Snapshot). The collision of the plates began over 40 million years ago, when India began its slow-motion crash into Asia as it finished its long journey north-ward after the breakup of Pangaea.

Mountain belts that formed as a result of collision between two plates of continental lithosphere represent the thickest sections of continental crust on the planet. For example, the crust under the Himalayas is approximately 70 kilometers (43 miles) thick, nearly twice the average crustal thickness under North America (Figure 4.27). Mountains formed along convergent boundaries will continue to rise as long as the plate tectonic forces responsible

The Thickness of the Earth's Crust

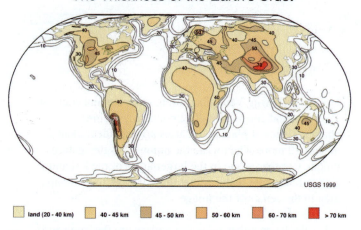

| | land (20 - 40 km) | | 40 - 45 km | | 45 - 50 km | | 50 - 60 km | | 60 - 70 km | | > 70 km |

Figure 4.27 Thickness of Earth's crust. Most of the oceanic crust is less than 10 kilometers (6 miles) thick, while the continents are at least 30 kilometers (19 miles) thick. Less than 10 percent of the continental crust has a thickness of greater than 50 kilometers (31 miles) and is found exclusively below the Himalaya and Andes mountains.
USGS

for the collision continue. The tallest mountains in the world are also among the youngest. The effects of weathering and erosion have worn down older mountain ranges where plate tectonic forces no longer dominate. For example, the 60-million-year-old Rocky Mountains reach altitudes of 4,300 meters (14,000 feet), while the older Appalachian Mountains (250 million years old) top out around 2,000 meters (6,600 feet). Both of these mountain ranges may once have been as tall as the Andes. Given enough time, mountains are worn down to the level of the surrounding land sur-face. (We will discuss the formation of mountains in Chapter 6.)

Transform Plate Boundaries

Imagine trying to gift-wrap a basketball. It would be a difficult job, requiring many small scissor cuts to form the paper to the shape of the ball. Think of Earth as the basketball and the lithosphere as the wrapping paper. The rigid lithosphere does not readily conform to the curved surface. Many small adjustments in the shapes of the plates are needed to make them fit on the spherical Earth. Trans-form boundaries represent those geometric adjustments, much like the small scissor cuts needed to wrap the basketball.

If we are honest, we must admit that the transform bound-ary is a poor relative of the divergent and convergent boundaries. Nothing is created or destroyed. Plates just slide past their neigh-bors like traffic on a two-way street, moving in opposite but paral-lel directions (Figure 4.22c). Since the amount of lithosphere at the boundary does not change significantly over time, these boundar-ies are sometimes termed *conservative plate boundaries*. No vol-canoes form at transform boundaries as there is no mechanism to melt rocks at these sites. However, hazards are possible as plate movement on the transform boundary can generate large earth-quakes, as we will learn in Chapter 5.

Transform boundaries can often be found between sec-tions of divergent boundaries (Figure 4.28 and Chapter Snapshot). These features were discovered because of the earthquakes that occur as the plates slide past one another. The earthquakes only

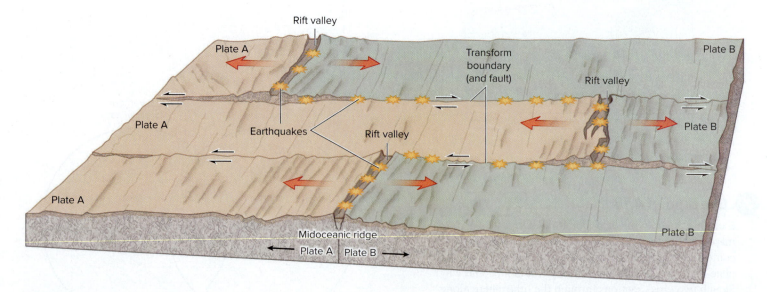

Figure 4.28 Features of transform boundaries that offset divergent boundaries. Plates A (tan) and B (green) move in opposite directions on either side of the transform boundary, causing earthquakes (yellow).
Adapted from Charles (Carlos) Plummer, *Physical Geology* 11e © 2007, reproduced with permission of McGraw-Hill Education

occur on sections of transform boundaries between segments of oceanic ridge (Figure 4.28). The topography of the seafloor can be determined by satellite measurements of the ocean surface. The topography clearly showed that oceanic ridges had offsets along transform boundaries.

Some transform boundaries, such as the San Andreas fault in California, occur on land. Like most transform boundaries, the San Andreas fault joins two other plate boundaries, a divergent boundary in the Gulf of California and the convergent Cascadia subduction zone (Figure 4.29). Land on the west side of the San Andreas fault, including Los Angeles and San Diego, is part of the Pacific plate. San Francisco lies east of the fault and is on the North American plate. Western California is moving slowly to the northwest relative to the rest of the state. Have you heard it said that California is going to fall off into the Pacific Ocean? Well, rest assured that it is not going to happen—but a portion of western California will gradually travel northward along the western edge of the North American plate, eventually colliding with Alaska millions of years from now, if plate motions remain stable that long.

Figure 4.29 San Andreas fault. The San Andreas fault is a transform boundary that separates the North American and Pacific plates. The smaller Juan de Fuca plate lies between these two plates opposite Oregon, Washington, northern California, and part of British Columbia. The Pacific plate moves northwest relative to the North American plate. Over the next several million years, Los Angeles will migrate toward San Francisco.

Plate Tectonics and Climate

Plate tectonics plays both a direct and an indirect role in controlling global climate patterns. Today, the north-south alignment of most of the continents influences the paths of ocean currents that transfer warm waters to cooler, high latitudes and cold waters to warm regions. These patterns reduce the contrast in temperatures between the poles and the tropics. A world with east-west oriented clusters of continents that were located on either side of the equator would contribute to latitudinal oceanic currents that would result in more extreme climates.

Continents located over the poles provide a base for extensive continental ice sheets, increasing the reflectivity of the land surface

✓ Checkpoint 4.20

basic · advanced
intermediate · superior

Continuous Change of Earth's Surface Synthesis Exercise

1. How many plates are present in the central figure?
2. Sketch and label a cross section through the lithosphere along the line X–Y in the central figure.
3. Fill in the upper and lower templates to show what the plate configuration looked like 4 million years earlier and what the plates will look like 15 million years in the future. Draw the configurations relative to the oceanic ridge (that is, assume it stays in the same place and everything else moves).

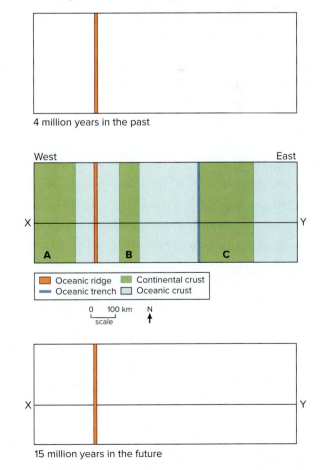

that reduces the amount of solar radiation that would be absorbed at Earth's surface. The southern half of Pangaea, a landmass known as Gondwana, migrated back and forth over the South Pole at least twice during the last 500 million years, resulting in long ice ages each time. The rate at which plates move determines if these events occur at all and how long they last. Processes that occur at plate boundaries are also key contributors to global climate patterns.

Convergent plate boundaries are characterized by the formation of mountain ranges and volcanic activity. High mountains alter regional precipitation patterns and increase erosion rates. Geochemical reactions tied to weathering and erosion can capture atmospheric carbon dioxide. Faster weathering and erosion rates therefore lead to a reduction in greenhouse gas concentrations. In addition, mountains can host alpine glaciers, increasing the amount of sunlight reflected back into space. The haze of dust produced by volcanic eruptions at convergent boundaries can reduce the amount of incoming sunlight that reaches Earth's surface. Consequently, intervals of plate collisions, such as the formation of Pangaea, are often associated with cooling trends. In contrast, the breakup of a supercontinent would have increased the length of

the oceanic ridge, producing more carbon dioxide and reducing erosion, leading to a warming trend.

Faster spreading rates result in the more rapid formation of ocean floor rocks along oceanic ridges. These volcanic processes add carbon dioxide to the ocean and ultimately to the atmosphere, increasing the concentration of greenhouse gases and raising global temperatures. In addition, faster plate motions result in wider ridges and higher sea levels. Higher sea levels, in turn, contribute to lower continental erosion rates. A slowing of erosion can contribute to a climate warming trend by reducing the amount of carbon dioxide extracted from the atmosphere.

Cycles of formation and destruction of supercontinents are thought to have resulted in changes in long-term global climate patterns. Such changes help explain why dinosaurs flourished at high latitudes when Earth was approximately 10 C° (18 F°) hotter 100 million years ago and how the planet was plunged into icehouse conditions of a "snowball Earth" around 600 million years ago. We will explore some of the other factors that contribute to climate change on shorter timescales in other chapters.

the big picture

Alfred Wegener shook up the science community when he proposed that continents drifted across Earth's surface. Wegener's work displayed a key characteristic of the nature of science—imagination. But science is a blend of creativity and logic, so what might start out as a product of imagination must eventually conform to principles of logical reasoning that support conclusions with solid evidence. Developing hypotheses to explain the workings of the world and putting them to the test of logic are creative but disciplined acts.

Many of Wegener's contemporaries met his ideas concerning continental drift with skepticism, which we recognize as a fundamental attitude of a scientific mind. Skepticism keeps scientists from rushing to instant judgment and embracing a new concept without first seeking the evidence behind it. Such verification can take years and is often supported by additional evidence obtained by new technologies. With additional observations and testing of predictions by many other scientists, Wegener's theory grew to become the theory of plate tectonics. Plate tectonics accounts for Earth's structure, numerous geologic phenomena, and processes of change over long intervals of time. Today, global positioning satellites allow scientists to make even more sophisticated observations of plate motion. These observations will inevitably improve as technology evolves and we learn even more about how Earth works. This is the way of science.

The processes and characteristics of plate boundaries are key to understanding the theory of plate tectonics. Remember, divergent boundaries are places where new plates form and move apart. Convergent boundaries are locations where plates collide, and transform boundaries are where plates slide past one another. The boundaries are locations where earthquakes, volcanoes, ocean trenches, and mountains may form, depending on the type of boundary. On the basis of what we know about the current directions and rates of plate motions, it is possible to make some predictions about how the plates will change over millions of years. The chapter-opening image asked you to predict which configuration would most likely resemble the plate configuration on Earth 50 million years from now. Our best prediction is image (a). The Atlantic Ocean will continue to widen, but eventually a subduction zone is expected to form off the eastern coast of North and South America. Once the subduction zone becomes established, the Atlantic Ocean will inevitably begin to shrink as in image (c), and eventually another Pangaea-like landmass will be formed as in image (d). Notice, too, in image (c) that Africa collides with both Europe and southwest Asia in this scenario. For the sake of completeness, image (b) represents the distribution of continents 50 million years ago.

Plate Tectonics: Concept Map

Draw a concept map that illustrates the significant characteristics of the theory of plate tectonics. The majority of features related to plate tectonics are linked to the geosphere component of the earth system. Use the scoring rubric provided here to draw a concept map that can be scored as a 4.

Scoring Rubric	
0	The concept map does not contain any information about plate tectonics.
1	The concept map contains some terms that are significant to plate tectonics, but several key terms are omitted and many linking phrases are either absent or inaccurate.
2	The concept map contains most terms that are significant to plate tectonics, but they are poorly organized and some linking phrases are absent or incorrect.
3	The concept map contains most terms that are significant to plate tectonics, but one or two key term(s) may be absent. The diagram is reasonably well organized, and almost all linking phrases are appropriate.
4	The concept map contains all terms that are significant to plate tectonics in a well-organized display that has appropriate linking phrases for each pair of terms.

"We learn geology the morning after the earthquake, on ghastly diagrams of cloven mountains, upheaved plains, and the dry bed of the sea."
—Ralph Waldo Emerson, US author

Chapter

5

Earthquakes

the big picture

What is this man doing, and what does it have to do with earthquakes?

See The Big Picture box at the end of this chapter for the full story on this image.

5.1 Experiencing an Earthquake Firsthand

Chapter Learning Outcomes

- Students will describe the processes and consequences of earthquakes.
- Students will estimate the relative distance from different earthquakes, using indirect evidence.
- Students will compare and contrast different scales used to measure earthquakes.
- Students will evaluate the risk to society from earthquakes and their effects.

Waiting for an earthquake is a little like watching a well-crafted scary movie. We know what's coming, we just don't know when. As we focus on the obvious villain, we may be taken by surprise by an unexpected character. We can enjoy the thrill of suspense in movies, and even in the scariest moments, we can take comfort in the knowledge that it is an elaborate fiction. Unfortunately, with earthquakes the villain is real and reveals few clues about its identity. This makes preparing for an earthquake a difficult undertaking and predicting the timing of future quakes almost impossible.

During the twentieth century, over 2 million people were killed by earthquakes and associated phenomena. A handful of destructive earthquakes have already claimed nearly one-half that number in the twenty-first century (see Table 1.1). Many of the world's largest cities are located near tectonic plate boundaries that represent zones of frequent earthquake activity. Over 143 million citizens in the continental US live in areas prone to damaging earthquakes. In those areas we also find over 600 fire stations, 800 hospitals and 20,000 schools. If we add Alaska and Hawaii to the total, the number of people subject to earthquake hazards jumps to nearly half of all Americans. Cities such as Los Angeles and San Francisco will inevitably suffer substantial damages and loss of life from a major earthquake in the future; we just don't know when.

Although scientists have identified the sources for many of the earthquakes that occurred in the past century, they can still be surprised by tremors that originate from unexpected locations that have no previous record of earthquake activity. That was the case on January 17, 1994, when the Northridge earthquake shook up much of southern California (Figure 5.1).

On the morning of the earthquake, I had just finished a conference in Los Angeles and was sound asleep in my hotel room. At 4:31 A.M. my bed began to shake violently. As I struggled to wake up, my first reaction was to think that I had overslept and one of my colleagues was shaking me awake to make sure that I had time to get to the airport. Then my science-teacher brain kicked in, and I realized that I was experiencing an earthquake! By now, the bed was rocking back and forth and bouncing up and down. I got up and staggered to the window to see the water in the swimming pool sloshing about, creating surfable waves. I struggled to reach the doorway to the room, but the floor was vibrating so vigorously that I could barely stand up. I tried to stay upright in the doorway and ride out the earthquake. It only lasted for several seconds, but it seemed to go on for much longer. I had just experienced the Northridge earthquake.

There are many large earthquakes each year that result in less destruction, but the location of the Northridge earthquake made it one of the most expensive natural disasters in US history. The earthquake occurred below a densely populated city and caused more than $20 billion in damages. It was felt by 10 million people, of whom 57 were killed and another 9,000 injured. Coincidentally, exactly one year later, a slightly larger earthquake killed

a.

b.

Figure 5.1 Damage from 1994 Northridge, California, earthquake. **a.** Sixteen people were killed when the upper two stories of this apartment building collapsed on top of the lower units. **b.** A parking garage built just 3 years earlier collapsed at California State University, Northridge. 5.1a: NOAA, 5.1b: NOAA more than 5,000 people and destroyed 100,000 buildings in Kobe, Japan. Clearly, earthquakes present a serious threat to humans.

The Hollywood image of the ground fracturing open and people falling into a deep crevasse is not the real danger from an earthquake. Most lives are lost due to building collapse or tsunami (see Chapter 1). Consequently, the death and destruction that follow an earthquake often have as much to do with local building codes and public awareness of earthquake hazards as with the size of the earthquake itself. For example, nations around the rim of the Pacific Ocean have a sophisticated tsunami warning system to protect their citizens. This did not happen by chance. Repeated loss of life from tsunamis coupled with evidence from scientific studies prompted development of the Intergovernmental Oceanographic Commission Tsunami Program with associated systems that provide better early warning.

In contrast, relatively few tsunami have been recorded for the Indian Ocean, and thus, surrounding nations had made little effort to share data on tsunami threats. This lack of a tsunami warning system resulted in massive loss of life from the 2004 Indian Ocean tsunami (see Chapter 1) that followed a giant earthquake off the coast of Sumatra. Though there was opportunity to warn the populace, individual governments did not have adequate disaster plans, failed to educate their people, and failed to communicate with other governments. This failure cost more than 200,000 lives.

Throughout this chapter, we will address several key questions related to earthquakes: Why do such catastrophic events occur? What factors cause earthquakes of similar size to produce very different results? Why do earthquakes occur in some areas but not in others?

We will also discuss the details of the Sumatra earthquake that caused the Indian Ocean tsunami and the more recent Tohoku earthquake in Japan. Apart from their significance as some of the largest earthquakes ever recorded, these events represent potential analogs for earthquakes that could occur along the coast of the Pacific Northwest. Until just a few years ago, no one knew this threat existed, but a combination of historical and geologic detective work brought to light this dangerous hazard and provided us with an excellent example of how science is done. We begin with that story.

5.2 The Science of Ghost Forests and Megathrust Earthquakes

Learning Objectives

- Describe the characteristics of a megathrust earthquake.
- Identify which of the world's largest cities are located relatively close to plate boundaries.
- Explain how researchers used the scientific method to discover that an ancient megathrust earthquake had occurred off the coast of Oregon and Washington.

Historical records indicate that on January 28, 1700, a ship carrying 30 tons of rice ran aground off the coast of Japan. The ship could not reach safe harbor because **a tsunami, a series of high ocean waves generated by vertical movement of the seafloor sometimes caused by an earthquake,** kept the ship at sea. The ship was subsequently caught in a storm and wrecked. The tsunami waves, some 3 meters (about 10 feet) high, damaged buildings in villages all along the Japanese coastline. Documentary evidence of damage reports was collected in a database of historical earthquake records in Japan. Japanese scientists analyzing these ancient records hypothesized that an earthquake had occurred a little before the arrival of the tsunami, but they could find no record of such an earthquake and were consequently puzzled by this apparent orphan tsunami. The consistent height of the tsunami waves at widespread sites was interpreted to indicate that the tsunami had originated from a distant location. Where was the earthquake that had caused it?

Evidence from Trees

All of this might have remained a mystery if scientists working along the coast of the state of Washington had not made an intriguing discovery. They found groves of dead spruce and cedar trees at several locations. Although the trees had died, their stumps and dead trunks were preserved in present-day salt marshes. Scientists recognized that these ghost forests resembled stands of dead trees found on the Kenai Peninsula of Alaska (Figure 5.2). This section of the Alaskan coastline subsided (dropped) during a giant earthquake in 1964, and the trees drowned when their roots were submerged in saltwater. Could the Washington ghost forests be a clue to a past giant earthquake that caused a section of the Pacific Northwest coast to drop below sea level? There are no written records of such a large earthquake, but Native American tribes of the region tell stories of shaking and flooding of the coastal lands. Unfortunately, no specific dates can be linked to these accounts.

In an effort to determine when the trees died, scientists collected samples from the drowned spruce trees at two sites over 50 kilometers (31 miles) apart and determined their ages. These analyses returned a range of dates from 1680 to 1720 for both locations. The 40-year span was the greatest level of accuracy the radiocarbon dating technique they used could provide. (We will discuss how dating methods work in Chapter 8.) Evidently, whatever killed these trees had occurred about three centuries ago and had affected a broad section of the coastline. As evidence started to point toward a large historical earthquake, scientists began to look for a possible earthquake source.

in Further Depth

Geophysicist Ross Stein works for the US Geological Survey Earthquake Hazards program. He has created a relatively simple machine to illustrate how stress builds up prior to movement on faults that produce earthquakes (Figure 1). Stein's device consists of elastic tubing attached to a brick that rests on some sandpaper on a table. Imagine that you are pulling on the elastic. The friction between the sandpaper and the brick initially causes the elastic to stretch without causing the brick to move. But at some point, you pull hard enough so that the brick moves forward and the tubing returns to near its original length. The more friction across the surface between the brick and the table, the more force must be applied to cause the brick to move. Replacing the sandpaper with talcum powder reduces the amount of friction, and thus, less force is needed to move the brick. So, depending on how the model is set up, movements may be slow and gradual, or they may be rapid and catastrophic.

This "brick and bungee" model is similar to what happens during an earthquake. Fault movements may be slow and gradual to generate only small earthquakes, or they may be rapid and catastrophic, causing widespread destruction. The surface of the brick in contact with the table is like the fault surface on which movement takes place. The forces that drive plate movements supply the energy (as when you pull on the tubing). Initially, crustal rocks accommodate some changes as stress builds up (tubing lengthens). However, eventually the stress is sufficient to move the brick (fault moves).

Once the brick has slipped, you must build up additional force by pulling on the tubing again to cause another earthquake. Similarly, once a fault generates an earthquake, plate movements must build up stress again before another earthquake can occur.

Figure 1   Ross Stein's earthquake machine. This simple device, nicknamed the "brick and bungee" model, effectively models the behavior of faults during earthquakes. A nonstretching cord is connected to the elastic surgical tubing. The tubing is connected to the brick that rests on sandpaper. An additional strip of sandpaper is on the base of the brick. As the crank is turned, the tubing stretches until sufficient stress is built up to move the brick. The brick lurches forward before coming to rest again.
Serkan B. Bozkurt/USGS

---

apparent when a map of global earthquakes (Figure 5.13) is compared to a map of plate boundaries (see Chapter Snapshot in Chapter 4).

- Relatively few earthquakes occur in the interior of plates.
- Global oceanic ridge systems and transform boundaries are areas of only shallow earthquakes.
- The largest earthquakes are associated with convergent plate boundaries.
- Most earthquakes occur at shallow depths where the seismic energy is released closer to Earth's surface.
- Deep earthquakes (focal depth greater than 300 kilometers, or 186 miles) are found adjacent to subduction zones along convergent plate boundaries (for example, western South America).
- The only area where deep earthquakes are not present along the Pacific Rim is the San Andreas fault system, which marks a transform plate boundary in the western United States.

Earthquakes in the United States. There are several seismically active areas in the United States. Earthquakes occur most frequently along the western edge of the North American continent (see Figures 5.11 and 5.13). The largest earthquakes in US history have occurred on the southern coast of Alaska, along the convergent boundary between the Pacific and North American plates. However, California has suffered the greatest losses from earthquake activity because of the large amount of development that happens to be near the Pacific and North American transform plate boundary found there. The Juan de Fuca plate is subducting beneath the North American plate off the west coast of northern California, Oregon, and Washington, making that an area prone to earthquakes.

✓ Checkpoint 5.7

If the San Andreas fault moves 2 meters (6.6 feet) per big earthquake, and plate movement is 2.5 centimeters (0.025 meter per year, or 1 inch per year), how many years of plate motion must accumulate to produce one big earthquake? (Assume all plate motion is accommodated by movements on the San Andreas fault.)

a. 4 years   c. 80 years
b. 20 years   d. 200 years

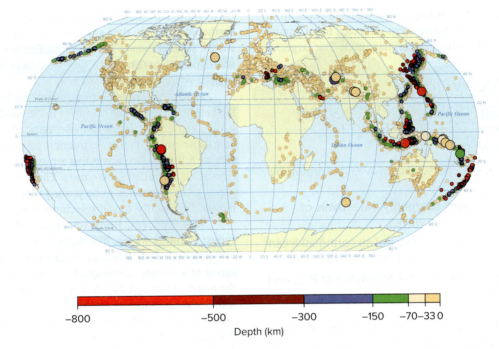

Depth (km)

Figure 5.13 Global distribution of earthquakes for 2015. Large circles indicate locations of the largest earthquakes; colors indicate depth of earthquake foci. Compare and contrast with Figure 4.8c (Earthquakes in 2005) to identify consistent patterns of earthquake activity.

Additionally, some very significant earthquakes have occurred on ancient faults located far from active plate boundaries in the central and southeastern United States in areas such as Memphis, Tennessee, and Charleston, South Carolina.

Recently a new breed of earthquakes, *induced* earthquakes, became increasingly common. These earthquakes are triggered by human actions, mainly the injection of large volumes of waste water generated by oil and gas drilling and production. This water is pumped down deep disposal wells. Induced earthquakes can occur

in cases where the rate of pumping and the volume of wastewater are both relatively high, and where a region is underlain by previously dormant ancient faults. The addition of water can increase the fluid pressure on faults, pushing the sides of the faults apart and causing them to slip and generate small to moderate earthquakes. Approximately seven million people live in locations in the central and eastern US that have the potential for induced earthquakes, with the greatest risk in Oklahoma and Texas (Figure 5.14). Up to the last decade, the central US experienced an average of

✓ **Checkpoint 5.8**

☑ basic ☑ advanced
☑ intermediate ☑ superior

Venn Diagram: Plate Boundary Earthquakes

Use the Venn diagram provided to compare and contrast the characteristics of earthquakes that occur along convergent and divergent plate boundaries.

Convergent plate boundary · Divergent plate boundary

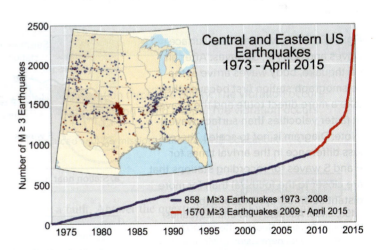

Figure 5.14 Earthquakes in the central and eastern US from 1973–2015. Note the rapid increase in earthquakes beginning in 2009. Almost all of this increase is attributable to induced earthquakes resulting from pumping wastewater deep underground.
Source: http://www.usgs.gov/images/Figure%201.jpg

2004 TSUNAMI

Coastal Wave Amplitudes

- ● >2m
- ▢ 0.5m–2m
- △ <0.5m

Arrival time of first tsunami
(hours after earthquake)

25
24
23
22
21
20
19
18
17
16

29

25

20

16

15

16

STEP 1

Friction along a segment of the plate boundary locks the overriding plate and the subducting plate.

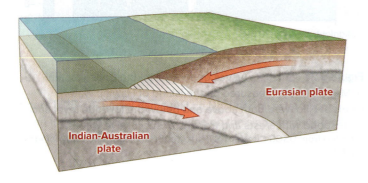

Eurasian plate

Indian-Australian plate

STEP 2

The shape of the overriding plate is distorted as it is pulled toward the subduction zone as the lower plate continues to descend. This can continue for hundreds of years.

Eurasian plate

Indian-Australian plate

Arrival time of first tsunami
(hours after earthquake)

Inset map labels:
BANGLADESH, CHINA, Calcutta, Dhaka, Hanoi, Eurasian plate, INDIA, MYANMAR, LAOS, Rangoon, Vientiane, BAY OF BENGAL, THAILAND, VIETNAM, Madras, Bangkok, CAMBODIA, SRI LANKA, Phnom Penh, Colombo, Burmese microplate, MALAYSIA, Kuala Lumpur, INDIAN OCEAN, Ninety East Ridge, Sumatra, INDONESIA, Jakarta

★ Main earthquake epicenter
○ Aftershock M≥4
— Plate boundary

STEP 3

The stress finally breaks the locked portion of the fault, resulting in an earthquake. When the plate snaps up, it causes the seafloor to move upward pushing the water out of the way and forming a tsunami.

Eurasian plate

Indian-Australian plate

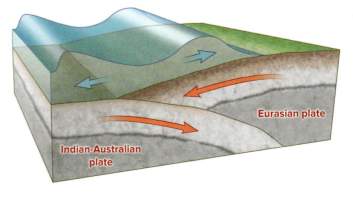

STEP 4

The tsunami moves outward from the source area at speeds of hundreds of kilometers per hour.

Eurasian plate

Indian-Australian plate

the big picture

Earthquakes and their associated effects represent some of the most destructive natural hazards on Earth. In your lifetime, the west coast of the United States will certainly experience more strong earthquakes and, in all probability, at least one major earthquake. Furthermore, in the next few centuries, maybe even tomorrow, another great earthquake will devastate parts of California, Oregon, and/or Washington. Despite years of planning for such events, the human and economic costs will be tremendous. Even if it becomes possible to accurately predict earthquakes to within a specific year, it is unlikely that such an event could be pinpointed to within a few months, let alone weeks or days. Furthermore, we would probably be unable to collect sufficient data to predict earthquakes in areas of infrequent seismic activity. Imagine a devastating earthquake centered near a major city. Given the difficulty in predicting the timing of future earthquakes, we would be well advised to focus instead on technology-based engineering solutions that attempt to earthquake-proof buildings and important landmarks. Such buildings would be able to sustain damage from the largest earthquakes and remain standing, reducing the loss of life due to building collapse.

We began this chapter by comparing the recent large earthquakes and tsunami to a similar pairing of earthquake and tsunami that occurred along the Cascadia subduction zone in 1700. In the chapter-opening image, we asked, "What is this man doing, and what does it have to do with earthquakes?" The man, Guy Gelfenbaum, a USGS scientist, was a member of an

international survey team that traveled to Sumatra just 1 month after the 2004 disaster to identify the characteristics of the tsunami. In the photo, he is measuring how much a section of the northwest coast of Sumatra had submerged as a result of the earthquake. The trees he is standing beside were on dry land before the coastline subsided and they were broken off by the tsunami.

The patterns scientists identified in Sumatra provided further data about a rare geologic event and match other patterns already in the geologic record. For example, by now you have probably realized that the broken-off trees in Sumatra are similar to the ghost forests we described along the coasts of Alaska and Oregon (see Figure 5.2). The sand deposits that were interpreted as tsunami deposits from previous earthquakes (see Figure 5.4) also have their parallels in the Indian Ocean tsunami (Figure 5.28). Scientists analyzing the details of the sand deposits can identify individual layers left behind by separate waves over an interval of a few hours. The survey team documented evidence for wave heights up to 30 meters (100 feet) along the 100-kilometer (60-mile) section of the coast of northwest Sumatra closest to the earthquake epicenter. This region sustained the greatest damage from the tsunami, accounting for about one-half the total casualties.

The comparison of features from several megathrust earthquakes or, more technically, great earthquakes, described in

Figure 5.28   Sand deposits following the Indian Ocean tsunami. In this sediment profile across 410 meters (1,350 feet) of coastline in northwest Sumatra, the dark layers represent original topsoil. The left side (0) of the graph is sea level. The vertical distance between the red and blue lines indicates sand thickness. Sand deposited by the tsunami covers the soil to depths of up to 20 centimeters (8 inches) at this location and as much as 70 centimeters (28 inches) elsewhere. Sand nearest the coast was eroded and washed hundreds of meters inland by the waves.
USGS

this chapter illustrates that scientists build hypotheses by using observations. These hypotheses have predictive power but may not predict every aspect of future events. Incorporating observations of past earthquakes and the theory of plate tectonics assists geologists in predicting the location of future earthquakes but cannot predict the exact time, date, or strength of the next event. Scientists can help us understand the nature of earthquake hazards, especially of tsunami threats, but it is up to us as individuals, and to our social and political institutions, to use these data thoughtfully.

In the self-reflection survey, we asked what action you would take if you lived in a city with a 99 percent probability of a future large earthquake. By now, you have probably realized that the city in question is probably in California. Your hometown may have little risk of experiencing an earthquake but may be at risk for other hazards. What should be the responsible role of a resident of San Francisco (or Los Angeles, Seattle, etc.) in advance of a future earthquake? What should the city do? What about the state and federal governments? Without careful planning, a major or giant earthquake along the west coast has the potential to overwhelm authorities to an even greater degree than Hurricane Katrina affected cities along the Gulf of Mexico.

Earthquakes: Concept Map

Draw a concept map that illustrates the significant characteristics of earthquakes. Use the scoring rubric provided here to draw a concept map that can be scored as a "4."

Scoring Rubric	
0	The concept map does not contain any information about earthquakes.
1	The concept map contains some terms that are significant to the main features of earthquakes, but several key terms are omitted and many linking phrases are either absent or inaccurate.
2	The concept map contains most terms that are significant to the main features of earthquakes, but they are poorly organized and some linking phrases are absent or incorrect.
3	The concept map contains most terms that are significant to the main features of earthquakes, but one or two key terms may be absent. The diagram is reasonably well organized, and almost all linking phrases are appropriate.
4	The concept map contains all terms that are significant to explain the main features of earthquakes in a well-organized display that has appropriate linking phrases for each pair of terms.

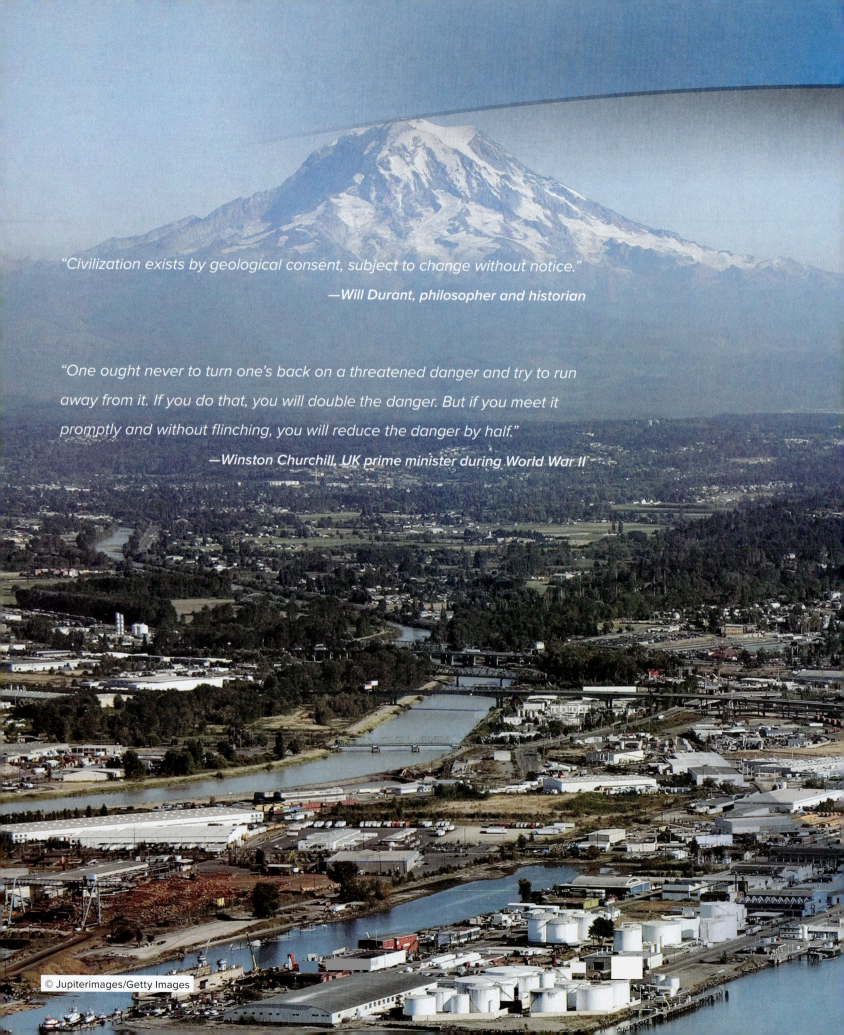

"Civilization exists by geological consent, subject to change without notice."
—Will Durant, philosopher and historian

"One ought never to turn one's back on a threatened danger and try to run away from it. If you do that, you will double the danger. But if you meet it promptly and without flinching, you will reduce the danger by half."
—Winston Churchill, UK prime minister during World War II

Chapter

6

Volcanoes and Mountains

the big picture

What questions would you ask about that mountain if you lived here?

See The Big Picture box at the end of this chapter for the full story on this image.

POTENTIAL FEATURES
OF VOLCANIC ERUPTION

PREVAILING WIND →

TEPHRA AND GASES
Larger volcanic rock fragments fall closer to the volcano vent; smaller particles are carried thousands of kilometers downwind in the eruption cloud. Gases such as water vapor, carbon dioxide, sulfur dioxide, and hydrogen chloride are released from rising magma.

PYROCLASTIC FLOW
A dense avalanche of a mixture of hot volcanic debris and gases that flows rapidly (100 km/h) down the slope of a volcano. The flow loses energy when it reaches the base of the volcano and soon comes to rest.

LANDSLIDE
Steep volcanic slopes may collapse to form fast-moving landslides known as debris avalanches. These materials may combine with rivers to form lahars.

MAGMA
A mixture of melted rock, crystallized minerals, rock fragments, and dissolved gases.

LAVA FLOW
Molten rock that flows over the surface during a volcanic eruption. Depending on the viscosity of the lava, it may flow down the slope of a volcano (low viscosity) or form a volcanic dome (high viscosity).

ERUPTION CLOUD
Tephra and gases form a cloud from an erupting volcano. The eruption cloud may reach as much as 10–20 kilometers into the atmosphere and is carried by prevailing winds thousands of kilometers downwind. Eruption clouds may cause failure of aircraft engines and have been responsible for several near-crash experiences.

ACID RAIN
Sulfur dioxide gas released during the eruption combines with precipitation to increase the acidity of the rainfall.

LAHAR
Lahar comes from an Indonesian word for mudflows of volcanic debris and water. The water may come from rapid melting of snow and ice during an eruption.

Checkpoint 6.16

Volcanic Hazards Associated with Mount Shasta, California

Read the following description of the eruption history of Mount Shasta and answer the questions that follow.

Mount Shasta Volcanic History

Mount Shasta in northern California is a stratovolcano composed of multiple overlapping cones. Eruptions produced andesite lava, pyroclastic flows, and lahars. During the last millennium, Mount Shasta has erupted on the average at least once every 250 years. The most recent eruption occurred in A.D. 1786.

Lava flows issued from the summit and from vents on the slopes of the volcano. Individual lava flows are up to 13 kilometers (8 miles) long. No lava flows extended for more than 20 kilometers (12 miles) from the central summit.

Pyroclastic flows from the summit and at the Shastina vent (see map) have traveled distances of more than 20 kilometers (12 miles). Other vents produced flows that extended up to 10 kilometers (6 miles). Eruptions from the summit crater produced lahars that reached more than 20 kilometers and spread around the base of the volcano. The largest lahars entered the McCloud and Sacramento Rivers.

Tephra deposits cover the ground within 25 kilometers (16 miles) of the summit. A massive debris avalanche occurred around 300,000 years ago. The debris avalanche flowed more than 64 kilometers (40 miles) through the Shasta valley and covered more than 675 square kilometers (260 square miles).

After reviewing the volcanic history of Mount Shasta and examining the map of the vicinity, identify potential volcanic hazards for the area surrounding Mount Shasta.

USGS

Remember that the eruption characteristics and products of future events will be similar to those of historical eruptions.

1. Show the possible extent of selected hazards on the map of Mount Shasta and vicinity.
2. Evaluate whether the cities of Weed, Mount Shasta City, McCloud, and Dunsmuir will face the same types of hazards from a future eruption of Mount Shasta. Which city is at greatest risk? Explain your choice.

6.6 Volcanoes and Volcanic Landforms

Learning Objectives

- Predict the types of volcanoes found at a variety of plate tectonic settings.
- Compare and contrast the characteristics of the three major types of volcanoes.
- Identify and describe at least three other types of volcanic landforms that are not volcanoes.
- Create a summary diagram that synthesizes information about magma composition, the characteristics of volcanic landforms, and plate tectonic setting.

Three Classes of Volcanic Cones

Volcanoes are separated into three basic types: shield volcanoes, stratovolcanoes, and cinder cone volcanoes.

Shield Volcanoes. The Hawaiian Islands are composed of volcanoes that rise from the floor of the Pacific Ocean with **broad, gently sloping sides built up from thousands of fluid, low-viscosity lava flows** (Figure 6.19). The highest mountain outside the Himalayas would be the Mauna Loa shield volcano that makes up much of the island of Hawaii (8,700 meters; 28,000 feet)—if elevation were measured from the seafloor, but it is not. In profile view, these broad, gently sloping volcanoes resemble the shield of a Roman soldier, hence their name—shield volcanoes. The low-silica mobile lavas described in Section 6.5 create shield volcanoes. They are characteristically found above oceanic hot spots (see Figure 4.18) or in association with divergent plate boundaries (see Figure 4.22).

Stratovolcanoes. A majority of the world's active volcanoes are steep-sided, cone-shaped stratovolcanoes (also known as composite cones; Figure 6.20) that form along convergent plate boundaries. **These volcanoes have a violent style of eruption that blasts debris several kilometers into the atmosphere.** They are called composite cones because they are composed of many layers of volcanic debris, pyroclastic flow deposits, and lavas. Most of our familiar images of imposing volcanoes represent stratovolcanoes, but this type is smaller than the Hawaiian shield volcanoes.

Stratovolcanoes develop from more-viscous magmas than shield volcanoes and hence have a narrower base and more steeply sloping sides (see Figure 6.20b).

Cinder Cone Volcanoes. The smallest volcanoes, **cinder cones are composed of coarse tephra produced when gas escapes rapidly from molten lava.** The expanding gases blast blobs of lava into the air. As the blobs cool and solidify, they fall back to Earth around the volcanic vent to build up the cone. Cinder cones are the simplest

Checkpoint 6.17

basic • advanced • intermediate • superior

What type of volcano is Mount St. Helens?

a. Shield volcano
b. Stratovolcano
c. Cinder cone

Figure 6.19 Shield volcanoes. **a.** Mauna Loa in Hawaii exhibits the gentle slopes of a shield volcano. It rises approximately 8,700 meters (28,000 feet) from the ocean floor and is the world's largest active volcano. The peak of a second shield volcano, Mauna Kea, is present in the distance. **b.** Cutaway view of a shield volcano illustrating the layers of lava that build up the gently sloping cone. 6.19a: USGS

Figure 6.20 Stratovolcanoes. **a.** Osorno volcano is a stratovolcano in the Andes Range in Chile. **b.** Cutaway view of a stratovolcano composed of alternating layers of lava (dark gray layers) and tephra (light colors). 6.20a: © Hubert Stadler/Corbis

types of volcano, rarely exceeding 400 meters (1,310 feet) in height (Figure 6.21). They often occur on the slopes of larger shield volcanoes or stratovolcanoes and are easily eroded because they are composed of loose, unconsolidated debris. Lava flows may form from the base of the cone after the eruption's gases have been expended.

Other Volcanic Landforms

Some volcanic landforms are generated when volcanoes collapse or when magma reaches the surface through a linear fissure (a long vertical crack in the crust), rather than through a central vent. Other volcanic features are characteristic of regions where groundwater and magma interact close to Earth's surface.

Calderas. **A caldera is a crater that forms when a stratovolcano or shield volcano collapses into a shallow, empty magma chamber below the volcano** (Figure 6.22). Calderas vary in size

✓ Checkpoint 6.18

basic   ● advanced
● intermediate   ● superior

Decade Volcanoes, Part II

Answer the following questions using the volcanoes illustrated on the map in Checkpoint 6.7. It may be useful to review the map of plate boundaries in Chapter 4 (see Chapter 4 Snapshot) before attempting to answer the questions.

1. Most of the volcanoes on this map are
   a. shield volcanoes.
   b. stratovolcanoes.
   c. cinder cone volcanoes.
2. Name a stratovolcano not associated with the subduction of the Pacific plate.
3. Name a shield volcano.

✓ Checkpoint 6.19

basic   ● advanced
● intermediate   ● superior

Venn Diagram: Shield Volcanoes, Stratovolcanoes, and Cinder Cones

Use the Venn diagram provided here to compare and contrast the three principal types of volcanoes. Place the number corresponding to features unique to each type in the larger areas of the circles; note features they share in the overlap area in the center of the image. Five items are provided; identify at least 12 more.

1. Associated with subduction zones
2. Have a triangular shape in profile
3. Example: Mount Hood, Oregon
4. Mild eruptions
5. Intermediate-silica magma

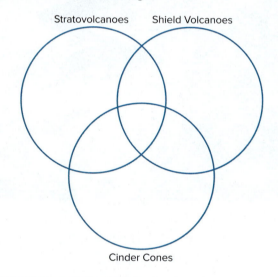

Stratovolcanoes   Shield Volcanoes

Cinder Cones

Figure 6.21 Cinder cone volcano. The 340-meter-tall (1,115-foot-tall) Sunset Crater in northern Arizona was formed less than 1,000 years ago.
USGS

Figure 6.22 A volcanic cone, crater, and caldera. The crater of the Karymsky volcano (Kamchatka, Russia) in the foreground is much smaller than the lake-filled caldera in the background. C.Dan Miller/USGS

a.

b.

Figure 6.23 Lava plateau. **a.** Ancient lava flows are preserved as basalt layers in the Columbia River plateau, Washington. Individual layers are 10 to 20 meters (33 to 66 feet) thick. **b.** Lava associated with the Columbia River plateau covers 164,000 square kilometers (63,300 square miles). 6.23a: USGS;

but can be tens of kilometers across. An oval caldera approximately 80 kilometers (50 miles) long and 50 kilometers (31 miles) wide underlies Yellowstone National Park, Wyoming, and was formed by a supervolcano eruption 640,000 years ago (Figure 6.13). For comparison, the width of the crater in Mount St. Helens is less than 2 kilometers (1.2 miles). Keep in mind that the caldera essentially represents the collapse of the whole volcano above the magma chamber, not just the vent where the eruption occurred.

Lava Plateaus. Low-viscosity lavas that reach the surface through fissures may accumulate to form **lava plateaus,** which are made up of many **layers of solidified basalt lava** (Figure. 6.23). These features represent the greatest volcanic eruptions on Earth. The largest examples, the Siberian Traps in northern Russia and the Ontong-Java plateau in the western Pacific Ocean, each cover an area larger than Alaska with piles of lava layers several thousand meters thick (Figure 6.24). The largest North American example of a lava plateau is the Columbia River plateau, which covers an area approximately equivalent to the size of the state of Washington. On the basis of their size and age, the lava plateaus may have produced lava at a rate equivalent to the total of the global oceanic ridge system. These massive eruptions released huge volumes of volcanic gases, contributing to elevated atmospheric carbon dioxide levels, and may have increased global temperatures, as described in Section 6.5.

Geysers, Hot Springs, Fumeroles, and Mud Volcanoes. Groundwater circulating near hot magma can result in the formation of several different features (Figure 6.25). Where this water is heated under pressure, it may erupt violently to form a *geyser* of steam and hot water. *Hot springs* form when heated waters rise to the surface. *Mud volcanoes* are produced when chemical reactions with heated, acidic waters convert rocks to soupy clays.

Figure 6.24 Lava plateaus of the world. Lava plateaus are equally prevalent on the ocean floor and on the continents. The largest examples are the Ontong-Java plateau and the Siberian Traps.

a.   b.   c.   d.

Figure 6.25   Volcanic features involving groundwater. Yellowstone National Park in Wyoming contains many examples of volcanic features, including **a.** geysers (Castle geyser shown here), **b.** hot springs, and **c.** mud volcanoes. **d.** Geologists sample gases from a fumerole at Mount Baker, Washington.
6.25 all: USGS

✓ **Checkpoint 6.20**   ☑ basic ☑ advanced ☑ intermediate ☑ superior

Create a concept map that relates the types of volcanic landforms with their magma composition and plate tectonic setting.

✓ **Checkpoint 6.21**   ☑ basic ☑ advanced ☑ intermediate ☑ superior

Which mountain range contains the higher peaks?

a.   Appalachian Mountains, North Carolina
b.   Rocky Mountains, Colorado

These hot clays bubble and boil like oatmeal on a kitchen stove, hence the name *mud volcano*. *Fumeroles* form when volcanic gases escape in the absence of water. It is important to note that none of these four features produces lava.

6.7 Mountains: Why Are They There?

Learning Objectives

- Provide multiple explanations for the difference in the heights of mountain ranges on different continents.

- Discuss how you could use wood blocks floating in water as analogs to explain differences in the elevation blocks of continental crust.

- Draw and label a cross section across a continent to illustrate the concept of isostasy and its relationship to topography.

When asked why he wanted to climb Mount Everest (8,850 meters; 29,035 feet), British climber Sir George Mallory famously remarked, "Because it's there." But why is it there? Mount Everest is one of 14 peaks in the world that are over 8,000 meters (26,247 feet) in elevation. All are in the Himalayan Mountains in southern Asia. There are another 39 peaks over 7,000 meters (22,966 feet) high—all in the Himalayas, too. Aconcagua, a mountain in the Andes of western Argentina, is the tallest non-Himalayan

peak in the world at 6,960 meters (22,835 feet). The tallest North American peak is Denali (formerly Mount McKinley) in Alaska at 6,194 meters (20,320 feet).

Why is there such a difference in the elevations of mountains across the world's continents? What is it about the Himalayas that makes them the prominent mountain range in the world? It all has to do with plate tectonics (big surprise!). But it also has to do with the same factors that control how high a marshmallow floats in a mug of hot chocolate or how much of an iceberg sticks out of the water. We will start our discussion by detailing the plate tectonic processes that transfer mass and heat to the crust. Then we will examine two more fundamental properties of materials—density and isostasy—that explain just how high a mountain will rise.

Mountains and Plate Tectonics

As we discussed in Chapter 4, the thickness of the continental crust varies across Earth's surface (see Figure 4.27). **The continental crust is generally thickest below mountain belts and thinnest at the margins of the continents** adjacent to the ocean basins. Mountains form where plates collide at convergent plate boundaries. The highest peaks in Asia (Everest), South America (Aconcagua), and North America (Denali) are all present in young mountain belts that were formed adjacent to convergent plate boundaries (Figure 6.26).

Convergent plate boundaries are characterized by thrust faults that stack up slices of continental crust. This increases crustal thickness where the plates collide (Figure 6.27). Imagine a big table scattered with hundreds of sheets of paper. What would happen if you and a friend began at opposite sides and pushed the loose sheets toward the middle of the table? You would end up with a big stack of paper. That stack of paper is analogous to mountains formed as slices of rock stack up along the edges of continental plates adjacent to convergent boundaries. Each sheet represents a slice of crust carried along by a reverse fault.

The big difference in elevation between the Andes and the Himalayas can be explained as the difference between the styles of plate collisions. The Himalayas are the result of collision between two slabs of continental lithosphere, while the Andes formed as a result of convergence between oceanic lithosphere and continental lithosphere. In addition, the leading edge of the Indian plate has

Figure 6.26 Distribution of mountain belts. The most recently formed mountain belts are adjacent to convergent plate boundaries (red lines) in the Americas and southern Asia (brown indicates higher elevation).
Adapted from Charles (Carlos) Plummer, *Physical Geology* 11e © 2007, reproduced with permission of McGraw-Hill Education

a. Ocean-continent convergence

b. Continent-continent collision

Figure 6.27 Features of a continent-continent collision zone. **a.** India moving north toward Asia with ongoing ocean-continent convergence. **b.** The edges of each continent combine to create a thick wedge of crust that produces high mountains adjacent to the suture zone where the two continental masses collided. The black lines represent faults that stack up and thicken the continental crust.
6.27a, 6.27b: Adapted from Charles (Carlos) Plummer, *Physical Geology* 11e © 2007, reproduced with permission of McGraw-Hill Education

✓ Checkpoint 6.22

✅ basic	✅ advanced
✅ intermediate	✅ superior

Suggest an explanation for why the tallest mountain in Australia (Kosciusko; 2,228 meters, 7,310 feet) is not even one-half as high as the tallest peaks on the other continents.

been wedged below the southern edge of the Eurasian plate. Consequently, the thickness of the continental crust has nearly doubled under the Himalayas.

Not all mountains are formed by adding mass to build up a pile of rocks. In some cases, they are formed by the addition of heat. Oceanic ridges form a submarine mountain range thousands of kilometers long in regions where high heat flow elevates the seafloor. Similar patterns are revealed in continental crust in Africa, where the highest peaks, such as Mount Kilimanjaro in Tanzania, are present in the East African Rift system, an early-stage divergent plate boundary (see Figure 4.24). These are typically areas of anomalously thin crust that are elevated because the rocks beneath expand as they are heated.

Earth's crust is about 40 kilometers (25 miles) thick under much of the continents and is thickest (70 kilometers; 43 miles)

under the Himalaya Mountains. But if the crust is 30 kilometers (19 miles) thicker under the Himalayas, *why are the Himalayas not 30 kilometers higher than the average level of the continents?* To answer that question, we need to understand a little more about two fundamental properties of rocks—density and isostasy.

Density. Imagine that a small block of pine wood is floating in a water-filled container (Figure 6.28). Why does the wood float? The wood floats because it has lower density than the water. *Density* is measured in units of mass per volume. The density of water is 1 g/cm³, and pine has a density of about 0.5 g/cm³. That is, a volume of pine would weigh about one-half as much as an equal volume of water. Consequently, about one-half of the floating pine block would be above the water's surface, and the rest would be below.

Wood blocks float on more-dense water in much the same way that Earth's crust "floats" on a more-dense mantle. Earth's three compositional layers—crust, mantle, and core—are characterized by different densities. When Earth began to cool, the elements that made up the planet became differentiated, with the more-dense materials sinking to the core and the less-dense materials rising to form the crust. The composition of the continental and oceanic crusts differ from one another. The continental crust is less dense (2.7 g/cm³), and the oceanic crust is more dense (3.0 g/cm³). However, both are less dense than the underlying mantle part of the lithosphere (3.4 g/cm³). Note that the density of the continental crust is only 80 percent as dense as the rocks in the upper mantle.

Isostasy. Let us take another look at the floating wood blocks. This time we will use a heavier wood with a density that is 0.8 times the density of water. The wood blocks will represent the crust, and the water will be a stand-in for the mantle. Some varieties of oak have a density of 0.8 g/cm³, which means that a block of oak would float, but most of the block (80 percent) would be submerged. A smaller wood block would not float as high, but neither would it extend as far underwater. The difference in the elevation of the top of each block is much less than the difference in thickness of the blocks (Figure 6.29a). In much the same way, high mountains have a deep crustal "root" that extends much farther down into the mantle. Note the difference in the thickness and

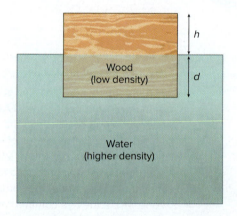

Figure 6.28   Demonstration of density. Wood floats because it has lower density than water. The relative heights of the block above (*h*) and below (*d*) the water level are determined by the relative densities of the wood and the water.

Figure 6.29   Demonstration of isostasy. **a.** Larger blocks of the same wood rise higher above the water but also extend farther below the surface. **b.** In the same way, thicker sections of crust below mountains have a deep crustal root that extends down farther into the mantle than do "normal" sections of continental crust. (Numbers in parentheses are densities.)

depth of each block of continental crust shown in Figure 6.29b; the difference in the depth of each block is greater than the difference in the surface elevation between these two blocks. Compare the crustal root to the foundation under a building. The taller the building, the deeper the foundation needs to be to ensure the stability of the structure. So now let us answer that question: *If the crust is 30 kilometers thicker under the Himalayas, why are the Himalayas not 30 kilometers higher than the average level of the continents?* Much of the 30-kilometer difference (80 percent) is actually in the crustal root. The remainder (6 kilometers) would be added to the average elevation of the continental surface, suggesting that the average height of the Himalayas and the adjacent Tibetan plateau should be around 6 to 7 kilometers (3.7 to 4.3 miles; which is a close approximation of the actual height).

The elevation of the continental crust is dependent on both its thickness and its density when compared with the underlying mantle. The concept that there is **a balance between the topography of Earth's surface and the thickness and density of the underlying rocks is termed** *isostasy.* Higher elevations result from either thicker or less-dense crustal rocks; lower elevations result from either thinner or more-dense crust. The removal or addition of material at Earth's surface can result in isostatic changes in the whole column of underlying crust.

✔ Checkpoint 6.23

| ✓ basic | ✓ advanced |
| ✓ intermediate | ✓ superior |

Different types of wood have different densities. For example, the density of pine (0.5 g/cm^3) is less than that of ebony (0.9 g/cm^3) but more than the density of balsa wood (0.14 g/cm^3). All would float in water but with different proportions of each block lying above and below the surface.

1. Draw a diagram to show what would happen if equal-sized blocks of each type of wood were added to a container of water.
2. What would happen to the blocks if we replaced the water with a liquid with higher density, such as corn syrup?

✔ Checkpoint 6.24

| ✓ basic | ✓ advanced |
| ✓ intermediate | ✓ superior |

The following profile shows the topography of a continent. Draw the relative position of the base of the crust, taking into account the principles of isostasy. Label the continental crust, oceanic crust, and mantle. Explain your drawing.

6.8 The Rise and Fall of Mountains and Temperatures

Learning Objectives

- Explain how density of the crust and age of mountain uplift contribute to the elevation of a mountain range.
- Describe a simple physical model that could serve as an analog for the relationship between erosion and change in elevation of mountains.
- Discuss how the erosion of the Himalayas contributed to a significant change in global temperatures.

Isostasy means that the elevation of a continent will increase as mass is added to the crust or will decrease as mass is removed. The amount of change depends on the relative densities of the crust and the mantle. Let us consider a simple example. Imagine that the wood blocks in Figure 6.30 are composed of several thin slices, each 20 units thick. The central block is made up of five slices, a total of 100 units in thickness. If we were to remove a slice to create an 80-unit pile, the top of the pile would decrease in elevation by only 4 units, while the base of the pile would rise by 16 units (80 percent). In contrast, if we added another 20-unit slice to the central block, it would increase in thickness to 120 units, but the top surface would rise by only 4 units and the base would drop another 16 units. **Isostasy compensates for added material by building a bigger submerged root and for lost material by raising the pile.** In each case, the proportion of the pile that is submerged is always 80 percent because the oak blocks have a density that is 80 percent the density of water.

During collision, the crustal thickening adds mass to the crust. Most of this increased thickness is taken up in an increase in the size of the crustal root. For every kilometer (1,000 meters; 3,280 feet) of material added to the crust by reverse faults, the elevation of the land surface rises by just 200 meters (656 feet). Erosion removes mass and the thickness of the crust decreases, but most of that decrease is accommodated by shrinking of the

Figure 6.30 Elevation changes and isostasy. The central wood block is composed of five wood layers, each 20 units thick. If one wood layer is added or removed, the top of the wood block increases or decreases in elevation by just 4 units (20 percent of the thickness of the wood layer).

Checkpoint 6.25

● basic	● advanced
● intermediate	● superior

How would the elevation of mountains differ if Earth's crust were composed of denser rocks? Mountains would be

a. higher.
b. lower.
c. unchanged in elevation.

Checkpoint 6.26

● basic	● advanced
● intermediate	● superior

Use the information from this section to answer these questions:

1. Why are the 10 tallest US peaks all in Alaska?
2. Why are the Rocky Mountains taller than the Appalachian Mountains?
3. Why can we drive across the site of former mountains in Canada without rising in elevation?

Checkpoint 6.27

● basic	● advanced
● intermediate	● superior

Explain how you could create a wood block-and-water model to demonstrate that erosion can result in an increase in the elevation of the highest mountain peaks.

crustal root. For every kilometer of rock removed by erosion, the underlying crust rises so that the elevation of the land surface lowers by only 200 meters. This process is not instantaneous; rather, it happens over millions of years.

In these simple models, we are assuming that material is added and removed equally across the mountain range. In these cases, erosion always results in a slow decrease in the elevation of the mountains. However, actual conditions may be more complex. For example, erosion in mountainous regions is often concentrated in stream valleys. This results in preferential erosion and deepening of valleys at the foot of the mountains, while erosion is almost nonexistent at the mountain peaks. Isostasy raises the whole mountain range to balance the material lost by erosion. If the high mountain peaks are not being eroded, they can end up actually higher than they were before (Figure 6.31). Consequently, erosion can cause an increase in the elevation of a mountain range! This is especially true in regions characterized by humid climates where stream erosion is concentrated in valleys.

The tallest mountain ranges are still being formed. They are supported partially by isostasy and partly by the forces associated with plate tectonics. India has been pushing into Asia for nearly 50 million years, so it is clear that **mountains must be long-lived features on Earth's surface.** Older mountains, such as the Appalachians in North America, were also formed by continent-continent collision. They were at much higher elevations when they formed but gradually eroded over time. The tallest peak in the Appalachians today is Mount Mitchell, North Carolina, at 2,037 meters (6,684 feet), more than 6 kilometers (3.7 miles) below the maximum elevation of the Himalayas.

Just how long does it take to wear away a mountain range? Scientists who study erosion have learned that erosion rates show considerable variation, but an erosion rate of 0.1 to 0.2 millimeter per year (0.04 to 0.08 inch per year) is a reasonable average. If the rocks of the Appalachians eroded at those rates, it would take 5 to 10 million years to erode 1 kilometer (3,280 feet) of rock. However, isostatic compensation would replace most of this rock, resulting in an overall elevation change of just 200 meters (656 feet). So the elevation of the mountain range would decrease by 1 kilometer every 25 to 50 million years. At that rate, it would take 150 to 300 million years to erode the 6-kilometer difference in elevation between the Himalayas and the Appalachians. This is merely a ballpark figure, but it fits well with what we know about the formation of the Appalachians; the plate collisions that formed the mountains ended around 280 million years ago. The combination of erosion and isostatic compensation can therefore account for the rate of erosion of the Appalachian Mountains. Without the influence of isostasy, the Appalachians would have been as flat as a pancake by now.

Region that erodes into mountains and deep valleys   **Region that erodes almost uniformly downward**

a. Isostatic adjusted highlands before extensive erosion

b. After extensive erosion, without isostatic readjustment

c. Erosion with isostatic readjustment

- - - Original surface
- - - Average surface level

Figure 6.31 How erosion can cause uplift. Right: The loss of material uniformly across a region can lead to an overall decrease in elevation as in a plateau. Left: However, if erosion occurs exclusively in valleys, the original peaks will actually be lifted by isostatic compensation. **a.** Original elevations, prior to erosion. **b.** Erosion removes material from both settings. **c.** Isostatic compensation reduces the change in elevation for the plateau but results in a net loss of elevation. Compensation in the jagged peaks raises the whole range, resulting in higher elevations for the peaks (left). In the real world, erosion and isostatic compensation occur at the same time.

6.31a, 6.31b, 6.31c: Adapted from Charles (Carlos) Plummer, *Physical Geology* 11e © 2007, reproduced with permission of McGraw-Hill Education

Mountains and Climate

Data from the Himalayas show that it takes tens of millions of years to form mountains, and the erosion of the Appalachians illustrates that it takes even longer to wear them away. Mountains therefore represent one of the largest and longest-lived Earth structures other than the continents and oceans. Mountains stretch for thousands of kilometers and reach several kilometers in elevation, forming a significant barrier for atmospheric circulation and influencing regional and perhaps global climate patterns. Nowhere is this more evident than in Asia, where the growth of the Himalayas has been responsible for generating the Asian monsoons, one of the most significant annual weather phenomena on the planet.

Each year, the broad, flat Tibetan plateau, just north of the Himalaya Mountains, is heated by the summer sun. The overlying air warms and rises, pulling in cooler, moist air from the Indian Ocean (Figure 6.32). The moist air is forced to rise as it encounters the Himalayas, and as it rises, the moisture condenses to produce heavy monsoon rains. These rains flow into rivers that erode the base of the mountains, and the eroded sediment is carried back to the ocean by rivers like the Ganges and Brahmaputra, some of the largest rivers in the world. Every year, these rivers transport hundreds of millions of tons of sediment to the Bay of Bengal and the Arabian Sea, where it forms a pair of deep sea features known as the Bengal and Indus fans. These fans represent two of the largest deposits of sediment in the world (Figure 6.32).

This process of erosion and deposition has been operating for about 20 million years, since the mountains became tall enough to cause the monsoon air circulation. One consequence of this process is that rain has been stripping carbon dioxide from the atmosphere, and various chemical reactions have been combining it with minerals that were deposited in the deep sea fans. (We will discuss these reactions in greater detail in Chapter 9.) The steady removal of carbon dioxide contributed to reduction in the global greenhouse effect, cooling the planet by an estimated 5 C° (change of 9 F°) or more. If not for this event, Earth would be much warmer today because of the extra carbon dioxide that was added to the atmosphere by volcanic eruptions around 100 million years ago (see carbon dioxide discussion in Section 6.5). The uplift of the Himalayas helped reduce the atmospheric concentration of the greenhouse gas, returning global temperatures to a more moderate level. This process will continue as long as India continues to drive northward into Asia and as long as isostatic compensation ensures that erosion will bring deeply buried rocks to the surface. Other factors are also important in regulating the global supply of carbon dioxide, but volcanoes and mountains exert a long-term influence on the concentration of this greenhouse gas, which can explain some of the most dramatic shifts in temperature in the global climate record.

Figure 6.32 Monsoon weather patterns. Warm, wet air is drawn in from the ocean and rises along the front of the Himalayas to produce heavy monsoon rains that feed the Ganges and Brahmaputra rivers. The rivers both flow into the Bay of Bengal, producing the largest delta complex on Earth in Bangladesh. Adapted from William Cunningham, *Environmental Science: A Global Concern* 7e © 2003, reproduced with permission of McGraw-Hill Education

 Checkpoint 6.28

Draw a concept map that illustrates how the erosion of the Himalayas resulted in a decrease in global temperatures.

the big picture

When Mount St. Helens began rumbling in 1980, teams of scientists rushed to the mountain with truckloads of instruments to monitor the activity. Still, the May 18 eruption came as a surprise. Despite the experience of the scientists and the sophistication of the devices they deployed, little detailed information on the eruptive history of the volcano had been gathered beforehand and few monitoring instruments had been collecting data. That is no longer the case. In the past quarter-century, scientists have made a concerted effort to place a variety of instruments around the volcano, and even in space, to monitor every rumble and movement. Even with what they know today, it is unlikely that volcanologists would have predicted the precise time of the May 18 eruption. But they would have known enough to have more vigorously encouraged the authorities to move people farther from the volcano itself, dramatically reducing the loss of life.

Educating the public is an important factor in reducing the effects of hazards such as volcanoes. Education should provide a scientifically literate population with the necessary skills to critically respond to scientists' assertions. Deciding what evidence to dismiss and what to pay attention to might mean the difference between life and death for those who live in the shadow of an active volcano. The people living near Mount St. Helens in 1980 weighed the evidence and the accompanying call to action. Some heeded the call to evacuate, while others ignored the evidence provided by the volcanologists, chose to hold their ground, and paid for their decision with their lives.

Mount St. Helens is one of only a few US volcanoes with such a high degree of monitoring. However, the US Geological Survey plans to create a National Volcano Early Warning System that would identify the most threatening volcanic hazards, including the number of people and the extent of property endangered. A preliminary assessment of volcanic threat identified more than 50 volcanoes as high-threat or very-high-threat sites and recommended that each volcano have an extensive network of monitoring equipment to identify the first signs of unrest. Few such networks are currently deployed, and some of these volcanoes have no monitoring systems at all.

One of the volcanoes in the very-high-threat group is Mount Rainier, pictured looming over Tacoma, Washington, at the beginning of this chapter. At 4,392 meters (14,410 feet), Mount Rainier is the tallest and most imposing volcano in Washington. It is located about 70 kilometers (43 miles) southeast of Tacoma. What questions would you ask if you lived in Tacoma?

Historical records indicate that Mount Rainier does not erupt with the frequency of Mount St. Helens. The distance of the peak and the prevailing westerly winds make it unlikely that

tephra would ever reach Tacoma. In addition, lava flows and pyroclastic debris would not extend beyond the foot of the mountain, staying tens of kilometers short of Tacoma. Still, large lahars have the potential to reach the northern suburbs of the city and enter neighboring Puget Sound. Even if Tacoma is safe, many smaller towns lie in stream valleys just a 10-minute trip from the volcano by lahar. It is the residents of towns such as Ashford, Packwood, and Orting (Figure 6.33) who need an early warning system for volcanoes.

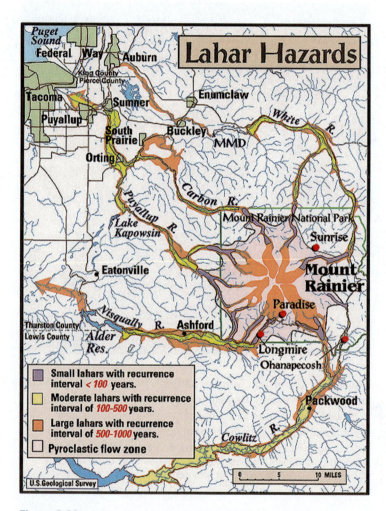

Figure 6.33 Lahar hazards associated with Mount Rainier, Washington.
USGS

Volcanoes and mountains are both part of the geosphere component of the earth system. However, both have the potential to influence, or be influenced by, all other components of the system. While scenic mountains may seem relatively benign, they are formed by movements on faults, movements that generate damaging earthquakes. Building a mountain range like the Himalayas involves thousands of faults that generate millions of earthquakes. Unfortunately, major earthquakes are still common in the Himalaya Mountains and other young mountain belts. A recent magnitude-7.9 quake in south-central China, at the eastern end of the Himalayas, demolished whole towns, killed more than 88,000 people, and left nearly 5 million homeless. The unrest continues; Earth at this very moment is shifting, rumbling, building, and decaying. We must carefully observe and prepare.

Volcanoes and Mountains: Concept Map

Complete the following concept map to evaluate your understanding of the interactions between the earth system and volcanoes and mountains. Match the following interactions with the lettered labels on the figure, using the information from this chapter.

Interaction	Letter
Eruption melted ice on Nevado del Ruiz to cause fatal lahars.	
Sulfur dioxide blocks incoming sunlight.	
Added water causes partial melting of mantle.	
Volcanoes add CO_2 and sulfur dioxide to atmosphere.	
Commercial airlines are at risk from tephra clouds.	
Solar radiation heats Tibetan plateau.	
Rain strips CO_2 from atmosphere.	
Krakatau eruption generated massive tsunami.	
Tephra is carried downwind over cities.	
Some 500 million people are in risk zones for volcanoes; trees are knocked down.	
Industrial materials are swept into rivers and lakes from mudflows.	
Monsoon rains result from air rising over Himalayas.	
Weathering processes break down rocks in mountains.	
Instrumentation of volcanoes.	

"My words are tied in one, with the great mountains, with the great rocks, with the great trees, in one with my body, and my heart."

—*Yokuts Indian prayer*

"*Rocks crumble, make new forms, oceans move continents, mountains rise up and down like ghosts yet all is natural, all is change.*"

Excerpt from "The Wall," from
"*The Awful Rowing Toward Goodbye,*"
Anne Sexton. Copyright © 1975 by
Loring Conant, Jr., Executor of the Estate,
renewed 2003 by Linda G. Sexton.

Chapter

7

Rocks and Minerals

Chapter Outline

the big picture

Are the rocks on Mars the same as rocks on Earth?

See The Big Picture box at the end of this chapter for the full story on this image.

173

7.2 Elements and Atoms: The Basic Building Blocks

Learning Objectives

- Classify an element on the basis of its atomic structure.
- Describe the features of a silicate mineral and explain how they are composed of silica tetrahedra.
- Explain how elements are joined together by bonds to form minerals.
- Summarize the relationships between atoms, elements, minerals and rocks.

Think back on the earth system discussions from Chapter 1. What makes up the geosphere? You might recall that it is mostly composed of rocks. To understand the geosphere and the various interactions that affect it, you need to understand something about rocks—what they are made of and how they form. In this and later sections, we are going to use some foods as analogs for the components of rocks, so you may want to go get a snack before you continue reading. Don't worry, we will wait. . . .

If you could walk outside and pick up a loose rock, what would you notice? (This might be harder to do in some places than others. Look for rocks used in landscaping, building stone, pavements, or gravel driveways.) In general, you are likely to observe that your rock is made of smaller pieces. **Rocks are made up of minerals.** Minerals are composed of elements arranged in characteristic atomic structures.

Geologists have recognized thousands of different minerals (more than 4,000), but only about 20 are commonly found in rocks and we will mention less than a dozen of those here. Individual rocks are typically composed of just a handful of minerals (Figure 7.3). For example, the rock granite is often found in mountain belts and is composed of the minerals quartz, feldspar, mica (such as biotite), and amphibole (also known as hornblende), with small amounts of a few other minerals. If we think of rocks as ice cream, the minerals would be the different ingredients (chocolate, marshmallows, nuts). Different combinations of minerals result in different types of rocks. For instance, the minerals quartz and feldspar show up in a lot of different rocks, and each rock type is characterized by different amounts of quartz and feldspar and different combinations of additional minerals.

Elements

So rocks are made of one or more minerals. What are minerals made of? **Minerals can be categorized by the chemical elements they contain.** Just as we are limiting this discussion of rocks to only a few minerals, we also need to focus on only a select group of elements. There are 92 naturally occurring elements on Earth, but 98 percent of the rocks in the crust below our feet are mainly composed of just eight elements. These eight elements are common ingredients in everyday objects or in the foods we eat (Table 7.1). For example, a bowl of breakfast cereal contains many of the common elements that are naturally extracted from rocks as minerals break down at Earth's surface to form soil (Figure 7.4). These elements are absorbed by crops growing in the soils and often find their way to our bodies via the foods we eat. Many of these elements are essential for good health.

The mineral quartz is composed of just two elements, silicon and oxygen. Other minerals may contain many elements. For example, the mineral amphibole is made up of calcium, magnesium, iron, aluminum, silicon, and oxygen. If we examine each of

Table 7.1	Common Elements in Continental Crust		
Element	Ion	Percent by Weight	Also Found in
Oxygen (O)	O^{2-}	46.6	Air
Silicon (Si)	Si^{4+}	27.7	Window glass, computer chips
Aluminum (Al)	Al^{3+}	8.1	Cans, aircraft
Iron (Fe)	Fe^{2+}, Fe^{3+}	5.0	Meat, cornflakes, your car
Calcium (Ca)	Ca^{2+}	3.6	Milk, cheese, cement, antacids
Sodium (Na)	Na^+	2.8	Salt, bacon, cheese
Potassium (K)	K^+	2.6	Fish, fruit, nuts, fertilizer
Magnesium (Mg)	Mg^{2+}	2.1	Bread, nuts, salt
Others	–	1.5	–

Figure 7.3 A rock composed of different minerals. In this piece of granite, four different minerals are apparent as grains of different sizes and colors. There are two dark minerals, hornblende (amphibole) and biotite; the tan mineral is feldspar, and the white mineral is quartz.
© PhotoLink/Getty Images

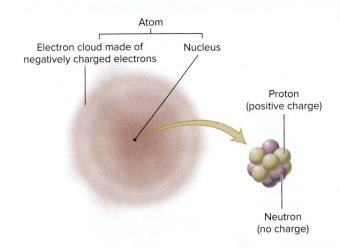

Figure 7.4 Elements in breakfast cereal. Many of the same elements that are common in Earth's crust are present in our daily breakfast cereal. © Nancy R. Cohen/Getty Images

the common minerals in granite (Table 7.2), we notice that some elements are more common than others and are also among the most common in Earth's crust (Table 7.1).

Atoms

Elements are the last stop; they cannot be further divided into other materials, but they can be separated into individual atoms. **An atom is the smallest particle that retains the characteristics of an element.** Atoms are composed of one or more of the same three basic components—**neutrons, protons, and electrons.** Just as two bricks from different buildings may look alike, individual protons look the same, whether they come from oxygen or silicon or potassium. The same is true for neutrons and electrons. The protons and neutrons are present in the atom's nucleus, and the nucleus is surrounded by a swirling cloud of electrons (Figure 7.5).

In much the same way as a building would look different if different numbers of bricks were used, elements look different because the number of protons in each differs. **The number of protons in an atom is unique for each element and is called the element's *atomic number.*** For example, oxygen has 8 protons, and silicon has 14; consequently, their atomic numbers are 8 and 14, respectively. If we were to subtract a proton from or add a proton to oxygen, we would create a different element (nitrogen or fluorine, respectively).

Protons have a positive electrical charge, while electrons are negatively charged. Neutrons are neutral; they have no charge. Electrons are in constant motion around the central nucleus. Usually, electrons are located in specific concentric zones, or energy levels. If the number of electrons (negative) equals the number of protons (positive), the atom has no net charge. For example,

Table 7.2	Most Common Elements in Granite Minerals
	Elements
Quartz	Oxygen, silicon
Feldspar	Oxygen, silicon, aluminum, calcium, sodium, potassium
Mica	Oxygen, silicon, aluminum, iron, potassium, magnesium, hydrogen
Amphibole	Oxygen, silicon, aluminum, iron, calcium, magnesium, hydrogen

Figure 7.5 Model of an atom. The tiny, dense nucleus consists of positively charged protons and uncharged neutrons. Most of the volume of an atom is occupied by rapidly moving, negatively charged electrons, which can be represented as an electron cloud.

✓ **Checkpoint 7.1**   ☑ basic   ☑ advanced   ☑ intermediate   ☑ superior

Examine these atomic models and answer the question that follows. The filled black circles represent electrons. (Note: Electrons and nucleus are not drawn to scale.)

A   B   C   D

Assuming that the number of electrons is the same as the number of protons (which is not always the case), which picture represents the atom of carbon that has an atomic number of 6?

the elements helium and neon have full energy levels and equal numbers of protons and electrons (Figure 7.6). However, the outer energy levels of many elements are incomplete and may gain electrons from other elements to fill gaps or may lose electrons from a sparsely populated energy level (Figure 7.7).

Bonds. Atoms that lose or gain electrons are known as *ions* and have a positive or negative electrical charge. This is so because the number of electrons does not necessarily equal the number of protons. Ions of individual elements can become bonded together to form minerals. In the most common minerals, bonds form in two principal ways—**by balancing the electrical charges of different ions or by sharing electrons.** For instance, the salt that people shake onto their dinner is made up of the mineral halite. Halite is composed of the ions of two elements, sodium and chlorine. A sodium ion is missing an electron, so it has a positive charge (Na^+), making it a *cation,* while chlorine has an extra electron and has a negative charge (Cl^-), making it an *anion* (Figure 7.8a,b). Essentially, we can consider that a spare electron was transferred from sodium to chlorine. **The positive and negative charges balance to create an ionic bond due to electrical attraction between the oppositely charged ions** (proving again that opposites attract). These ions pack themselves together in a gridlike pattern of alternating positively and negatively charged ions (Figure 7.8c).

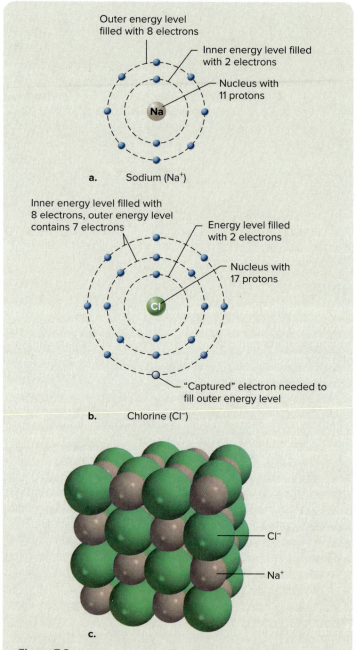

a.   Sodium (Na^+)

b.   Chlorine (Cl^-)

c.

Figure 7.8   Atomic models of sodium and chlorine. **a.** Eleven electrons fill the energy levels of sodium, but the eleventh electron is often absent, giving the ion a positive charge (Na^+). **b.** Seventeen electrons fill the energy levels of chlorine. This results in a configuration that readily accepts an electron in the outer energy level, imparting a net negative charge to the ion (Cl^-). **c.** The two ions will bond together to form the atomic structure of the mineral halite (formula NaCl). Ionic bonds hold oppositely charged sodium ions to adjacent chlorine ions in a grid or cubic pattern.
7.8 a & b: Adapted from Charles (Carlos) Plummer, *Physical Geology* 11e © 2007, reproduced with permission of McGraw-Hill Education; 7.8 c: Adapted from Francis Carey, *Organic Chemistry* 5e © 2003, reproduced with permission of McGraw-Hill Education

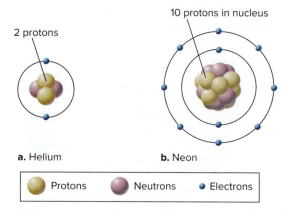

a. Helium   **b.** Neon

Protons   Neutrons   Electrons

Figure 7.6   Atomic models of the distribution of protons, neutrons, and electrons for **a.** Helium and **b.** Neon. In all atoms, electrons lie in energy levels around the nucleus. There are a maximum of 2 electrons in the first energy level and a maximum of 8 electrons in the second energy level. In helium and neon, the number of protons equals the number of electrons, so these elements do not form ions and consequently do not bond with other elements.
7.6a, 7.6b: Adapted from Charles (Carlos) Plummer, *Physical Geology* 11e © 2007, reproduced with permission of McGraw-Hill Education

Figure 7.7   Oxygen and silicon atoms have gaps in their outermost energy levels. If electrons are gained to fill these gaps, the atoms will be transformed to a negatively charged ion. Alternatively, electrons may be lost. For example, if silicon lost its outer four electrons, it would become a positively charged ion (Si^{4+}).

Atoms and ions tend to gain, lose, or share electrons to get eight electrons in their outer shell. (In some cases, no electrons transfer from one atom to another.) For example, water (H_2O), one of the most common substances on the planet, is composed of two atoms of hydrogen (H) and one of oxygen (O). Each hydrogen atom has an electron that is shared with oxygen to give it a full energy level (Figure 7.9). Likewise, oxygen shares an electron with each of the hydrogen atoms. This type of sharing is called **covalent bonding** and typically occurs where atoms share electrons to achieve a stable structure.

A Special Bond: The Silicates. The two most common crustal elements, oxygen and silicon, combine to form a relatively unusual type of bond that is one-half ionic and one-half covalent. Four oxygen atoms bond covalently to a single silicon ion. **The bonding of silicon and oxygen atoms forms a pyramidlike shape known as the silica tetrahedron (or more accurately,** **the silica-oxygen tetrahedron),** in which the smaller silicon atom is surrounded by the four oxygen atoms (Figure 7.10). The resulting multiatom anion has a negative charge (SiO_4^{4-}) and will form ionic bonds with positively charged ions from most of the other common elements (see Table 7.1). Silicon and oxygen make up over 70 percent of the continental crust by weight, so we deduce that these elements would be present in many minerals. Minerals that contain both silicon and oxygen are known as **silicates** and feature atomic structures that are characterized by different arrangements of the silica tetrahedron. These basic building blocks, the tetrahedra, can be joined together (Figure 7.11)

Oxygen is 2 electrons short of 8 electrons in its outer energy level.

The hydrogen and oxygen atoms share electrons to give oxygen an outer energy level of 8, and the hydrogens each have 2.

Figure 7.9 Covalent bonding of hydrogen and oxygen in water. Two hydrogen atoms each share an outer electron with an oxygen atom to fill each atom's outer energy level by covalent bonding. Adapted from Francis Carey, *Organic Chemistry* 5e © 2003, reproduced with permission of McGraw-Hill Education

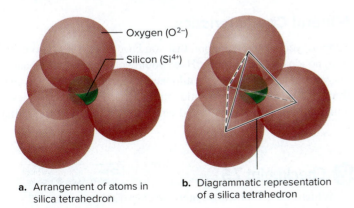

a. Arrangement of atoms in silica tetrahedron

b. Diagrammatic representation of a silica tetrahedron

Figure 7.10 Silica tetrahedron. Silicon and oxygen join to form a pyramid (tetrahedron) by covalent bonding. **a.** Four oxygen atoms surround a smaller silicon atom. **b.** The combination of atoms is illustrated on diagrams by a pyramid (tetrahedron) shape (see Figure 7.11). The corners of the pyramid coincide with the centers of the oxygen atoms.
7.10a, 7.10b: Adapted from Charles (Carlos) Plummer, *Physical Geology* 11e © 2007, reproduced with permission of McGraw-Hill Education

Structure		Mineral examples
Isolated silicate		Olivine
Single-chain		Pyroxene group
Double-chain		Amphibole group
Sheet silicate		Mica group Clay group
Framework silicate		Quartz Feldspar group

Figure 7.11 Common silicate structures. These are formed by combinations of silica tetrahedra and other atoms. Arrows indicate directions in which the structures repeat indefinitely. Adapted from Charles (Carlos) Plummer, *Physical Geology* 11e © 2007, reproduced with permission of McGraw-Hill Education

Checkpoint 7.2

basic · advanced · intermediate · superior

Which of the following mineral formulas represents a silicate?

a. FeS_2
b. $KAlSi_3O_8$
c. Fe_2O_3
d. $CaSO_4 \cdot 2H_2O$

Checkpoint 7.3

basic · advanced · intermediate · superior

The total electrical charges of the ions of the elements in the mineral olivine must balance. From the data in Table 7.1, which is the most reasonable formula for the mineral?

a. $MgSiO_2$
b. $MgSiO_4$
c. Mg_2SiO_4
d. Mg_4SiO_2

in combination with other elements (using covalent or ionic bonds or both) to form different silicate minerals, such as all those minerals found in granite (see Table 7.2).

Recall from Chapter 6 that the amount of silica present in magma is especially significant in controlling magma viscosity. Silica combines with other elements in the magma (for example, iron, magnesium, potassium). In magmas with low silica content, atoms combine to form minerals such as olivine and pyroxene with simple forms such as pairs and single chains (Figure 7.11). With more silica, the tetrahedra form complex double chains, sheets, and three-dimensional frameworks. These larger, more complex forms get tangled together in the magma, making flow difficult and resulting in higher viscosity.

Strengths of Bonds in Minerals. Imagine attaching yourself to another person with a strip of Velcro or tying yourself to someone with a piece of rope. Which connection would be harder to break? It should be easier to separate the Velcro strip. In the same way, some bonds between atoms are stronger than others. Bonds formed by electrical attraction (ionic bonds) are typically weaker than bonds in which electrons are shared (covalent bonds). Therefore, **minerals formed with covalent bonds are typically stronger, and thus more likely to be preserved on Earth's surface, than those with only ionic bonds.** For example, a diamond, the hardest mineral, is formed of carbon atoms that share covalent bonds with four neighboring carbon atoms.

The character of the rocks we will discuss in this chapter depends to a large degree on the atomic structures of their minerals and the way the atoms are bonded together. Some minerals such as quartz are strong and show up in a variety of roles in different rock types. Other minerals such as halite will dissolve readily in water and consequently are often indicative of specific

Checkpoint 7.4

basic · advanced · intermediate · superior

Construct a concept map that illustrates the relationships among atoms, elements, minerals, and rocks. Use the following six terms and add at least three more of your own choosing.

Atom   Bond   Electron
Mineral   Quartz   Ion

marine conditions. Finally, the combination of elements used to form minerals will determine the melting temperature of the rocks and the behavior of the magma produced.

7.3 Minerals

Learning Objectives

- Describe the principal features of a mineral.
- Explain how minerals can be identified on the basis of their physical properties such as crystal form, cleavage, hardness, color, luster, and streak.

Minerals are inorganic solids that occur naturally in Earth. (In this discussion, we will neglect some minerals that are made by humans in a lab.) Each mineral is made up of one or more elements that combine their atoms in a unique structure characteristic of that mineral. The consistent arrangement of atoms means that the mineral has a uniform chemical composition no matter where it is found. Stringing all that together, **minerals are naturally occurring inorganic solids of one or more elements that have a definite chemical composition with an orderly internal arrangement of atoms.** The most common minerals in Earth's crust are composed of the most common elements.

Mineral Characteristics

Suppose you encounter a food that you have never seen before and have no idea how it is made. How would you describe it? Chances are you might describe what it looked like, its texture, and something about its taste. You would remember those characteristics if you were served the same food again in the future. In a similar

Checkpoint 7.5

basic · advanced · intermediate · superior

Which of the following cannot be classified as a mineral?

a. Salt
b. Ice
c. Diamond
d. Glass

Figure 7.12 Diversity of minerals. **a.** Quartz; **b.** muscovite mica; **c.** orthoclase feldspar; **d.** galena; **e.** pyrite; **f.** halite. Galena, pyrite, and halite all exhibit a cubic crystal form. Galena and pyrite have a metallic luster.
7.12a & b © David Steer, 7.12c, d & e: © McGraw-Hill Education/Doug Sherman, photographer; 7.12f: © McGraw-Hill Education/Bob Coyle, photographer

way, geologists use various physical features to identify minerals. Although thousands of different minerals exist on Earth, not that many of them are frequently found in rocks. The same mineral found in different parts of the world has characteristic features and a consistent chemical composition. The most common features used to identify minerals are crystal form, cleavage, hardness, and color, as well as other secondary properties, such as luster and streak.

Crystal Form. If you take a magnifying glass and compare common table salt (a mineral) to pepper (a ground seed), you will notice some differences. (Try it. What do you see?) The salt will have a consistent cubic shape, while the pepper does not. When mineral crystals have lots of room to grow, they form specific shapes (Figure 7.12). For example, we often find well-formed crystals in caves where the mineral is precipitated out of solution and can grow outward from the cave wall into the open interior of the cave. **Crystal form is the shape of the crystal that a mineral forms when it is free to grow unimpeded.** The shape of a mineral crystal is related to the way the chemical bonds form. Atoms or molecules of the mineral align in a way to minimize the energy needed to form the crystal. Common shapes are prisms, pyramids, needles, cubes, and sheets (Figure 7.13). We saw in Section 7.2 that chlorine and sodium ions in halite are organized in a gridlike pattern; this is revealed by the cubic form of halite crystals (see Figure 7.12f).

a.

b.

Figure 7.13 Crystal shapes include **a.** pyramid (calcite) and **b.** needlelike (tourmaline).
7.13a: © medaacek/iStock/Getty Images; 7.13b: © RF Company/Alamy

Cleavage. **The propensity of a mineral to break along one or more planes is called** *cleavage.* Depending on how their constituent atoms are arranged, minerals may break along specific surfaces of weakness called cleavage planes. Cleavage planes mark those parts of the mineral where ions are connected by relatively weak ionic bonds. In contrast, covalent bonds are strong and are less likely to form cleavage planes. Muscovite and biotite, two forms of the mineral mica, have a single series of parallel cleavage planes that causes them to separate into sheets, much like a stack of paper (see Figures 7.12b and 7.14a). In contrast, other minerals may have multiple cleavage planes that define sheets, cubes, or other shapes (Figure 7.14). Some minerals, such as quartz, have no cleavage planes and consequently break with an irregular fracture pattern.

Hardness. Which is harder, glass or a copper penny? Suppose that you do not already know and have a piece of each. What experiment could you do to determine which is harder? Minerals can be ranked by their relative hardness based on their ability to scratch one another. As it turns out, glass can scratch a penny, but a penny cannot scratch glass. **The resistance of a mineral to scratching is measured by its hardness.** Ten minerals make up the Mohs hardness scale (Table 7.3). Minerals not in the table are ranked relative to these. For example, a mineral that could scratch feldspar but not quartz would have a hardness of approximately 6.5. Quartz can scratch feldspar or any mineral with a lower Mohs scale number but cannot scratch topaz, corundum, or any mineral with a higher Mohs scale number. Softer minerals with low rankings on the scale are more likely to break down on Earth's surface, while harder minerals are more likely to survive. That is one reason why the sand on a beach is almost always quartz; most of the other minerals with lower rankings on the Mohs scale are broken down so that only quartz is left. For comparison, a copper penny would rank around 3 and glass would be a little harder than 6 on the Mohs scale. The scale is relative, meaning that it is based

Figure 7.14   Types of cleavage. **a.** One direction; **b.** two directions that intersect at 90° angles; **c.** two directions that do not intersect at 90° angles; **d.** three directions, intersecting at 90° angles; **e.** three directions, not intersecting at 90° angles; **f.** four directions (for example, diamond); **g.** six directions. Some minerals have no cleavage planes (for example, quartz), while others may have several. 7.14a, 7.14b, 7.14c, 7.14d, 7.14e, 7.14f, 7.14g: Adapted from Charles (Carlos) Plummer, *Physical Geology* 11e © 2007, reproduced with permission of McGraw-Hill Education

Table 7.3	Hardness Scales		
	Mohs Hardness Scale	Mineral	Absolute Hardness Scale
Softest	1	Talc	1
	2	Gypsum	2
	3	Calcite	9
	4	Fluorite	21
	5	Apatite	48
	6	Feldspar	72
	7	Quartz	100
	8	Topaz	200
	9	Corundum	400
Hardest	10	Diamond	1,500

Checkpoint 7.6

Finish the partially completed concept map for minerals provided here. How could you add additional levels to the concept map?

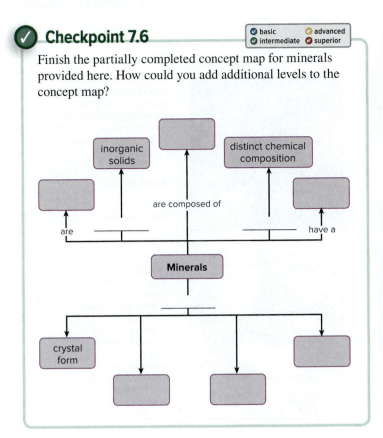

Checkpoint 7.7

Examine the following images of the mineral halite, and identify how many cleavage planes are present.

© McGraw-Hill Education/Bob Coyle, photographer

a.

b.

Figure 7.15 Other properties of minerals that are useful in distinguishing one from another. **a.** A brown streak on white porcelain identifies this mineral as the metallic iron hematite. **b.** When weak hydrochloric acid is added to the mineral calcite, a characteristic reaction becomes visible.
7.15 a & b: © David Steer

on one mineral's ability to scratch another, not on their intrinsic hardness. Therefore, another system exists, called the *absolute hardness scale* (Table 7.3). That scale denotes hardness based on how much pressure must be applied to a diamond point moved across a mineral before a scratch appears.

Color. Just as color might be used to describe a food (for example, a green apple), it is used to describe minerals, which come in a variety of colors (see Figures 7.3 and 7.12). Examples of common dark-colored minerals (black, dark brown, dark green) are amphibole, olivine, pyroxene, and biotite mica. The light-colored minerals (white, pink, gray, translucent) found in rocks that you might pick up around your house are typically quartz, feldspar, muscovite mica, gypsum, halite, and calcite. However, we must be careful when using color to identify minerals because some minerals exist in a wide range of colors. For instance, quartz occurs in over a dozen different colors. Minerals can also change color when they are exposed to changing natural conditions (rain, heat) on or near the surface of Earth.

Other Properties. Other mineral characteristics that geologists may use to distinguish between minerals include **luster (how light is reflected from a mineral;** see Figure 7.12) and **streak (the mark formed when a mineral is scratched across an unglazed piece of porcelain;** Figure 7.15a). For example, gold creates a yellow streak, while iron sulfide (also known as pyrite or fool's gold) makes a black streak. Some minerals have specific traits. For example, the common mineral calcite creates bubbles of carbon dioxide gas when exposed to acids (Figure 7.15b); magnetite is magnetic; sulfur has a bad smell; and halite (salt) has a salty taste (if you are willing to lick a rock).

A *rock* is an *aggregate* of *minerals*. The structure, texture, and types of minerals depend upon the conditions of formation of the rock.

Granite

Quartz (clear)

A mineral is a physically and chemically distinct part of a rock. It has properties you can see with an unaided eye, such as color and cleavage.

Biotite (black)

Feldspar (pink)

Each type of mineral also has its own orderly internal (crystalline) structure at the atomic level. This explains the physical properties seen at a larger scale.

In this example, showing quartz, large oxygen atoms enclose smaller silicon atoms.

We construct models of some crystal structures linking the centers of oxygen atoms together to form silica tetrahedra. This helps us envision the geometric beauty of a mineral's atomic arrangement.

An atom consists of a cloud of electrons at different energy levels enclosing a tiny nucleus.

Nucleus

An atomic nucleus is a swarm of protons and neutrons at the core of an atom.

Figure 7.16   Relationships among rocks, minerals, elements, and atoms.
Adapted from Charles (Carlos) Plummer, *Physical Geology* 11e © 2007, reproduced with permission of McGraw-Hill Education
7.16: © Charles C. Plummer; 7.16 (inset): © Charles C. Plummer

Figure 7.16 summarizes the relationships among rocks, minerals, elements, and atoms described in this section.

Checkpoint 7.8

✓ basic   ✓ advanced
✓ intermediate   ✓ superior

Which mineral characteristics discussed in this section are most closely tied to the bonding of atoms described in Section 7.2?

7.4 Igneous Rocks

Learning Objectives

- Explain the relationship between the cooling rate of magma and the texture of igneous rocks.
- Classify and name images of common igneous rocks.

- Compare and contrast the characteristics of volcanic and plutonic igneous rocks.
- Discuss which igneous rocks would be found in different plate tectonic settings.

Igneous rocks form from the cooling of melted rock (magma), a concept that follows from the old idea of Plutonism. Igneous rocks can be classified as volcanic or plutonic according to whether they form on Earth's surface or within Earth's interior. **Volcanic igneous rocks form when magma in Earth's interior rises to the surface through pipes or fractures in the crust.** Many of the gases trapped in magma (for example, CO_2, SO_2, and water vapor) escape as it nears the surface. At the surface, this slightly altered magma is termed *lava* (Figure 7.17). Landforms such as volcanoes and lava flows, discussed in Chapter 6, are the most easily recognized formations of igneous rocks.

Igneous rocks that cool below Earth's surface are termed *plutonic igneous rocks*. The features they form, termed

Figure 7.17 Lava at the surface of a Hawaiian volcano. Lava is issuing from the vent in the right background and flowing toward the viewer. The black surface is where the rock is cooling and solidifying to form a crust. The crust breaks when moving over obstructions to show the extremely hot orange lava below.
D. A. Swanson/USGS

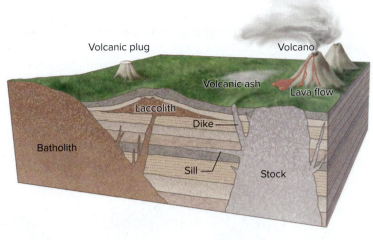

Figure 7.18 Igneous landforms. Volcanic processes form volcanoes, lava flows, and ash falls, while plutonic processes lead to batholiths, laccoliths, dikes, sills, and crystallized magma chambers (called stock).
Adapted from James Zumberge, *Laboratory Manual for Physical Geology* 12e © 2005, reproduced with permission of McGraw-Hill Education

plutons, remain hidden from view until erosion or tectonic uplift exposes them at the surface (Figure 7.18). Plutons come in a variety of shapes and sizes. They may represent a large reservoir of magma that cools to form a large mass (a batholith) or a thin vein of magma that solidifies as it flows through a near-vertical fracture (a dike) or along a mostly horizontal contact between adjacent layers of rock (a sill).

The Classification of Igneous Rocks

Examine the objects in Figure 7.19. Suppose you were asked to organize these objects in some logical way. How would you classify them? We will pause here for a snack; look at the figure. . . .

Most people organize the objects into classes labeled fasteners or paper clips and divide them by size and color. In a similar way, igneous rocks are classified on the basis of the size and arrangement of their mineral grains (known as texture) and their color, which is an indicator of chemical composition.

Texture. Recall that igneous activity occurs when magma rises toward the surface through pipes and fissures (fractures) in Earth's crust. **The size of the mineral crystals that form depends on how fast the magma cools and solidifies.** Plutonic rocks cool slowly far below the surface where the geothermal gradient ensures that it is relatively hot. Therefore, minerals that form in plutonic rocks have plenty of time to grow into large crystals as the magma cools slowly. In contrast, hot magmas undergo a rapid temperature drop of at least 700°C (1,292°F) when they reach the surface. Consequently, **volcanic rocks cool rapidly at the surface.** Individual mineral grains (crystals) do not have much time to grow when magma cools rapidly. Consequently,

the grain sizes in all volcanic rocks are small, too small to be seen without a magnifying glass. The larger grains of plutonic igneous rocks make them easier to distinguish from volcanic

Figure 7.19 Classification exercise. On the basis of your observations only, what features could you use to classify these objects into different groups? Be careful not to use your existing knowledge about properties that you cannot observe (for example, do not base your classification on the strength of material used to make them).
© David McConnell

rocks (Figure 7.20). Individual grains can be readily seen in specimens of plutonic rocks. Keep in mind that, despite the difference in mineral grain sizes, the same minerals are present in both volcanic and plutonic igneous rocks that can form from the same magma.

Color. The color of an igneous rock is representative of its composition—in particular, of its silica content (Table 7.4). Light-colored minerals such as quartz and feldspar contain more of the element silicon. Consequently, their abundant presence signals a silica-rich magma source. **Light-colored igneous rocks such as rhyolite and granite are formed from silica-rich magmas** and contain abundant white, pink, or translucent silica-rich minerals. In contrast, **dark-colored silica-poor igneous rocks** such as basalt and gabbro are dominated by olivine and pyroxene, dark-colored minerals (black, brown, dark green) having a smaller proportion of silica. Rocks of an intermediate silica composition such as andesite and diorite lie somewhere between the light and dark rocks, depending on the minerals present. (Compare Table 7.4 with Table 6.1.)

Figure 7.20   Comparison of volcanic and plutonic rocks. Plutonic rocks have larger grain sizes than volcanic rocks. Rocks with high silica content contain lighter-colored minerals than low-silica rocks. Ferromagnesian minerals are rich in iron and magnesium.
7.20 (granite) & 7.20 (diorite): © Charles C. Plummer; 7.20 (gabbro): © Larry Davis; 7.20 (basalt): © Charles C. Plummer; 7.20 (andesite): © David Steer; 7.20 (rhyolite): © Charles C. Plummer

Correlation Between Rock Types and Magma Types. In Chapter 6, we discussed the variations in magma compositions and plate tectonic settings, but only in terms of the type of magma, not the specific igneous rocks. We labeled the magmas as basaltic, andesitic, and rhyolitic. The lavas they produced are named for the volcanic rocks—basalt, andesite, rhyolite. Where would we find their plutonic equivalents? The answer is simple—under the volcanic rocks (see Chapter Snapshot). If we were to drill through the basalt rocks that make up the new seafloor along oceanic ridges, we would find the plutonic rock gabbro and eventually peridotite, the mantle rock that undergoes partial melting to form basaltic magma. On rare occasions, we can observe the same association on the continents in places where thin slices of oceanic lithosphere have sometimes been trapped and preserved along convergent plate boundaries (Figure 7.21).

Andesite is a product of volcanic eruptions at convergent boundaries; the plutonic equivalent, diorite, would be found where ancient magma cooled and solidified below mountain ranges such as the Cascades and the Andes. Granite is a more common plutonic rock than diorite, and the (rhyolitic) magma that forms granite is more viscous and less likely to reach the surface. Rhyolitic magma forms from the partial melting of the continental crust in locations above subduction zones and over hot spots (see Chapter Snapshot). Granite plutons known

Table 7.4	Silica Content of Igneous Rocks		
Silica Content	Volcanic Rocks	Plutonic Rocks	Common Minerals
High	Rhyolite	Granite	Quartz, feldspar
Intermediate	Andesite	Diorite	Feldspar, amphibole, pyroxene
Low	Basalt	Gabbro	Pyroxene, feldspar, olivine

✓ Checkpoint 7.9

✔ basic   ✔ advanced
✔ intermediate   ✔ superior

Geologists sometimes find a type of igneous rock known as porphyry, which contains both large and small crystals. Which is the best explanation for the formation of this rock?

a. The rock experienced a two-stage cooling process, with initial slow cooling at depth followed by rapid cooling at the surface.

b. The rock experienced a two-stage cooling process, with initial rapid cooling at depth followed by slow cooling at the surface.

c. The rock experienced a two-stage cooling process, with initial rapid cooling near the surface followed by slow cooling at depth.

d. The rock experienced a two-stage cooling process, with initial slow cooling near the surface followed by rapid cooling at depth.

Figure 7.21 Oceanic lithosphere on land. The uppermost part of mantle (peridotite) is visible at Table Mountain, western Newfoundland.
© Bill Church/University of Western Ontario

✓ Checkpoint 7.10

✔ basic   ✔ advanced
✔ intermediate   ✔ superior

Name each of the four igneous rocks pictured here. Describe how each of these rocks formed.

Checkpoint 7.10a-7.10d: © David Steer

ORIGIN OF ROCKS

Glacier

Mountain stream
(conglomerate)

Alluvial fan
(sandstone,
conglomerate)

Meandering stream
(sand, shale)

Reef
(limestone)

Swamp
(shale, coal)

Slope/Rise

Delta
(shale with
sandstone)

Beach
(sandstone)

Lagoon
(rock salt)

Dunes
(sandstone)

Shelf
(limestone)

CONTINENTAL CRUST

OCEANIC CRUST   Basalt
Gabbro

LITHOSPHERE

UPPER MANTLE

ASTHENOSPHERE

Ocean
(chalk, limestone, shale)

Sea level

DIVERGENT
Plate Boundary

CONVERGENT
Plate Boundary

Andesite
Tephra
and Lava

Diorite Plutons

Contact
Metamorphism

Granite Plutons

Regional
Metamorphism

Basaltic Lava

Slate

Rhyolitic magma
from partial melting
of continental crust

Basalt
Gabbro

Basaltic
Magma

Schist

Gneiss

Andesitic magma
from mixing of low-
and high-silica magmas

MAGMA

SUBDUCTION ZONE

PERIDOTITE
Partial melting of mantle

Partial melting
of mantle

Dense low-silica magmas
collect at the base of crust

Checkpoint 7.11

Complete the following table by identifying which of the characteristics in the left-hand column are present in volcanic or plutonic igneous rocks or both and placing a check mark in the appropriate column(s). Do not place a check mark in either column if the characteristic is not present. One characteristic has been completed as an example.

Characteristic	Volcanic Igneous Rocks	Plutonic Igneous Rocks
May form from basaltic magma		
Form at Earth's surface		
Have texture	X	X
Made of small grains		
Granite is an example		
Form as a result of melting		
May form from rhyolitic magma		
May form in the presence of water		
Exposed at Earth's surface only after erosion		
Contain minerals		
Andesite is an example		
Classified based on density of minerals		
Dark-colored examples have low silica content		
Contain visible grains		

Checkpoint 7.12

Finish the partially completed concept map for igneous rocks provided here by filling in the blanks with appropriate terms. Three of the appropriate terms are *magma*, *basalt*, and *plutonic* rocks.

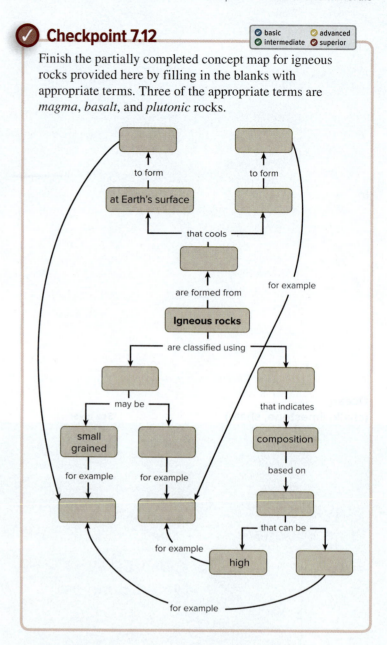

Figure 7.22 Batholiths. The peaks of much of the Sierra Nevada range in California are formed from granite batholiths that have been exposed by erosion.
© McGraw-Hill Education/Doug Sherman, photographer

as batholiths (see Figure 7.18) can account for huge volumes of continental crust and are often exposed by erosion, especially where the crust is thickened below mountains such as the Sierra Nevada range in California (Figure 7.22).

in Further Depth

In Chapter 6, we told you that when a rock is heated, some minerals will melt while others remain relatively unchanged. Silica-rich minerals such as quartz melt first as they have a lower melting temperature than silica-poor minerals. But what happens in the reverse process—when magma cools?

As temperatures begin to drop below 1,200°C (2,190°F) in the magma chamber, some minerals with a *higher* melting temperature are the first to form crystals. What happens next was figured out by N. L. Bowen in a series of experiments about 80 years ago and is illustrated in Figure 1. Bowen observed a sequence of crystallization, Bowen's reaction series, which has two branches.

- The discontinuous branch represents the ferromagnesian minerals, so called because they are rich in iron (*ferro−*) and magnesium (*−magnesian*).
- The continuous branch represents a single type of feldspar that has sodium- and calcium-rich varieties.

Minerals crystallize from both branches at the same time. For example, the same magma will contain crystals of both olivine and plagioclase feldspar.

In an idealized magma, minerals with low silica content such as olivine and the calcium-rich variety of feldspar would be the first to form as the magma cooled (Figure 1). If temperatures remained high, these minerals would continue to form until they used up all the iron, magnesium, and calcium in the magma. However, it is more likely that the temperature of the magma would continue to decrease. In that case, two events are possible:

- The early-formed crystals could settle out and sink to the bottom of the magma chamber (crystal settling). The next mineral in Bowen's reaction series would then form (pyroxene and more calcium-rich feldspar; Figure 1).
- The early-formed ferromagnesian minerals would react with the

remaining magma. These reactions would convert the early-formed minerals to the next mineral in the discontinuous branch of the series. For example, olivine would recrystallize to pyroxene. The early calcium-rich plagioclase feldspars would get steadily bigger but would begin to add more sodium into the crystal structure.

The reactions continue until the magma is used up. Early-formed magmas are rich in pyroxene, olivine, and plagioclase feldspar and have relatively low silica content. These form rocks such as basalt and gabbro (see Table 7.4). Any leftover magma forms silica-rich minerals such as quartz and potassium feldspar, which crystallize to form granite and rhyolite.

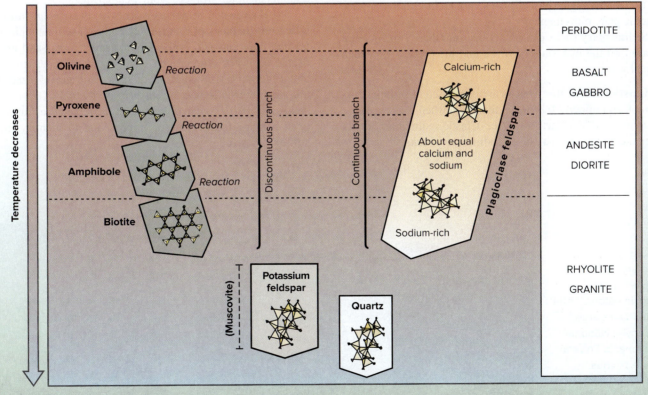

Figure 1 Bowen's reaction series. Silica-poor rocks such as basalt and gabbro form early in the series and are rich in olivine and calcium-rich plagioclase feldspar. Intermediate-silica rocks such as andesite and diorite form if magma solidifies in the middle of the series. Silica-rich magmas remain at the end of the series and are characterized by rocks such as rhyolite and granite, which feature minerals such as potassium feldspar and quartz. This is the unifying concept of how igneous rocks correlate with silica-based igneous minerals and cooling temperature. Compare this diagram with Figure 7.20.

Adapted from Charles (Carlos) Plummer, *Physical Geology* 11e © 2007, reproduced with permission of McGraw-Hill Education

7.5 Sedimentary Rocks

Learning Objectives

- Explain how clastic sedimentary rocks are formed.
- Describe how you could use the properties of a clastic sedimentary rock to make predictions about the conditions under which it formed.
- Compare and contrast the characteristics of chemical and biochemical sedimentary rocks.
- Discuss the geological conditions necessary for the formation of oil and gas deposits.

Take a minute and examine the rock formations in Figure 7.23. What do you notice?

Sedimentary rocks often form in a series of layers called **beds** or **strata.** These layers can be readily identified in nature on the basis of different colors, thicknesses, and resistance to erosion, properties that are linked to the composition and origin of the rocks. This layering is apparent in each of the images in Figure 7.23. Note that some layers stand out; this is so because they are more resistant to erosion than the others.

Sedimentary rocks form in a range of environments, from stream channels to the floor of the deep ocean (see Chapter Snapshot). Geologists study modern environments to understand how rocks form today and use these data to interpret the conditions under which rocks formed millions of years ago. Scientists can use this information to investigate past climates, mountain-building events, and tectonic plate movement.

Sedimentary rocks can be divided into three fundamental types based on the materials involved and the process by which each type forms:

- **Clastic sedimentary rocks** are composed of sediments—rock and mineral fragments that form when rocks break apart at or near Earth's surface. Clastic sedimentary rocks make up the majority of all sedimentary rocks.
- **Chemical sedimentary rocks** are crystallized from a solution (for example, seawater) as a result of changing conditions.
- **Biochemical sedimentary rocks** form by the actions of living organisms or are composed of the remains of dead organisms.

The processes involved in the formation of each of these rock types reveal echoes of the concept of Neptunism described at the beginning of the chapter.

Clastic Sedimentary Rocks

Clastic sedimentary rocks are formed from rock and mineral fragments (clasts). This process occurs in a series of steps: (1) generation of clasts due to the breakdown of an original rock by weathering; (2) transportation of the eroded material from the source area; and (3) lithification, the deposition and subsequent conversion of the material to rock.

Generation. Rocks on or near Earth's surface are physically broken into smaller pieces, and minerals undergo chemical changes to form weaker minerals. This process of disintegration and decomposition is termed *weathering* (see Chapter 9.3 for more details) and is influenced by the original rock type and by climatic conditions. Weathering is more rapid for rocks that contain minerals with well-defined cleavage planes or minerals that can dissolve in water. Climatic conditions characterized by high temperatures and lots of water promote rapid weathering.

Weathered material is known as *sediment* and is classified on the basis of its grain size. Small sediment particles are termed *clay* and *silt,* while larger grains form sand and gravel (Table 7.5). For comparison, sand-sized particles are typically about the same size as grains of sugar, and gravel-sized particles are about the same size as pieces of hard candy. Gravel and most sand-sized particles can be seen with the naked eye, but you would need a magnifying glass or microscope to identify individual silt- or clay-sized particles.

Figure 7.23
Sedimentary rocks from western parks. **a.** Kaibab limestone, Little Colorado River gorge, Grand Canyon National Park, Arizona; **b.** Entrada sandstone, Arches National Park, Utah; **c.** Cliff House sandstone, Cliff Canyon, Mesa Verde National Park, Colorado; **d.** White Rim sandstone, Canyonlands National Park, Utah. 7.23 all: USGS

a.

b.

b.

a.

Table 7.5	Clastic Sediments and Sedimentary Rocks		
Sediment	Grain Size (diameter)	Rock	Grain Size Comparisons
Clay	Less than 0.0039 mm (less than 0.00015 in)	Shale, mudstone	Smaller than granulated sugar
Silt	0.0039 to 0.0625 mm (0.00015 to 0.0025 in)	Siltstone	
Sand	0.0625 to 2 mm (0.0025 to 0.079 in)	Sandstone	Ranges from sugar to coarse salt
Gravel	More than 2 mm (more than 0.079 in)	Conglomerate	Ranges from rice grains to oranges

Transportation. The process by which **sediment is removed from its place of origin by running water, winds, or glaciers is known as** *erosion.* Erosion rates vary over several orders of magnitude depending on the material. For instance, granite erodes at rates as little as 0.002 millimeter per year (0.00008 inch per year) while unconsolidated sediment can be eroded from a stream bank at rates of meters per hour! A muddy river is an indication that sediment is being transported by water, while a cloud of dust is a sign that sediment is being carried by wind (Figure 7.24). The size of the sediment that can be transported depends on how fast the water flows or how hard the wind blows. If the velocity is high, a stream can move a lot of material; for example, a fast-moving stream in a flood can sometimes transport a boulder as large as an elephant.

c.

d.

The process of erosion shapes the landscape and contributes to the formation of many distinctive landforms, such as valleys, canyons, and mountains. As mentioned in Chapter 6, some of the most rapid erosion today is taking place in the Himalayan Mountains, filling the adjacent rivers with sediment. That sediment must eventually be deposited somewhere, which takes us to the next stage in the formation of clastic rocks—deposition and eventual lithification.

Lithification. Imagine leaves blowing in the wind on a late fall day. As the wind slows, the leaves fall. In a similar fashion, when the speed of a stream slows, the sediment it carries drops out. **Clastic sediments are deposited when the velocity of the transporting medium drops** (Figure 7.25). Large material is deposited first, followed by smaller and smaller particles as the velocity steadily decreases. Rivers dump much of their sediment where they enter the relatively quiet waters of an ocean or lake, creating the landform known as a delta (see Figure 7.24a). This material is often transported and redistributed along the coastline by shoreline currents to form beaches. The finest particles remain in suspension in the ocean before eventually being deposited offshore. The largest sediment deposits in the world are the deep sea fans in the Bay of Bengal and the Arabian Sea that formed over millions of years from sediment eroded from the Himalayas (see Figure 6.32). The Bengal fan is more than 2,500 kilometers (1,550 miles) long, hundreds of kilometers across, and several kilometers thick.

The sand, clay, and silt particles at the bottom of a pile of sediment become compacted, squeezing out water and forcing the grains closer together. Slowly, over hundreds of thousands of years, fluids circulating through the pile of sediment precipitate minerals to cement the grains together, converting the sediment into a rock. **The processes of compaction and cementation that convert sediment into a sedimentary rock are together termed** *lithification.* Clastic sedimentary rocks such as sandstone and siltstone have names that reflect the size of their grains (see Table 7.5; Figure 7.26). Different speeds of stream flow or wind serve to sort and concentrate sediment by weight (grain size). Consequently, we often find that the sand grains on a beach are the same size and that clastic rocks have a dominant grain size instead of being a random mix of sand, clay, and gravel-sized particles. The rate of clastic sediment formation is more rapid adjacent to active plate margins characterized by high elevations and faster weathering and erosion rates.

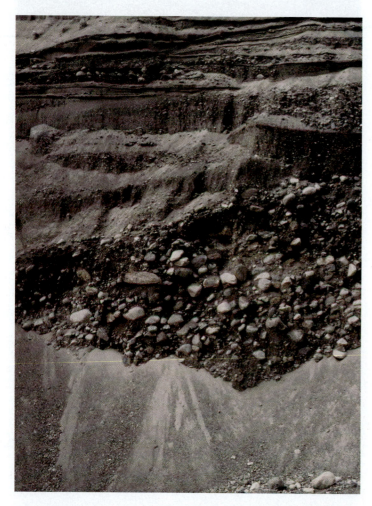

Figure 7.25 Clastic sediment deposit. A lens of gravel separates layers of sand. Pebbles can be seen in thin layers toward the top of the image. This is a typical sand and gravel deposit indicative of a sedimentary environment with alternating intervals of fast- and slow-flowing streams.
U.S. Park Service

✔ **Checkpoint 7.13**

Suppose you were given a plastic jar that is about one-third full of a mixture of dry gravel, sand, and silt. The jar is filled with water, sealed, and violently shaken. Draw a picture of what it would look like after the shaking stops and the jar is placed on the counter for a few minutes. How might the picture change after 10 hours?

Figure 7.26 Examples of clastic sedimentary rocks defined by grain size. **a.** Fine-grained sandstone has small sand grains cemented together that are visible to the naked eye. **b.** Conglomerate composed of gravel- and pebble-sized rounded grains.
7.26a & b: © David Steer

a.

b.

Checkpoint 7.14

✓ basic ✓ advanced ✓ intermediate ✓ superior

The following photo shows a typical section of clastic sediments that geologists might study. What observations can you make about the grain size and arrangement of these sediments that would help determine their origin?

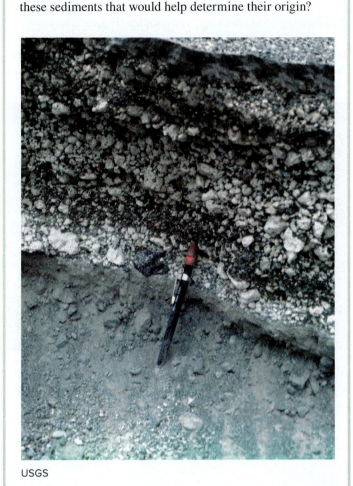

USGS

Note that the formation of clastic sedimentary rocks partially mirrors the brick manufacturing process. First, rocks are broken down to smaller sizes; these grains are sorted during transport to ensure that a rock is made up of consistent grain sizes. The mixture of water and sediment is deposited and compacted to expel excess water and air. With time and slightly increased temperature, the mixture hardens to form a rock (brick). The process of lithification applies to all types of sedimentary rocks formed by the processes we discuss in the next sections.

Chemical Sedimentary Rocks

Have you ever noticed the salty taste of your perspiration after completing an outdoor activity on a hot day? That taste is still there even after the perspiration is gone because the salt crystallizes as the water evaporates. In much the same way, **chemical sedimentary rocks form when minerals precipitate (crystallize out) from a solution as a result of changing physical conditions.** The most common solution is seawater, but minerals are also dissolved in lesser concentrations in groundwater and in freshwater lakes. Some minerals are dissolved from rocks on the continents by weathering and are carried in solution by steams to lakes and oceans. Millions of years ago, vast shallow tropical oceans covered much of what is now the North American continent. Those seas evaporated and left behind deposits that are the source for the bulk of the chemical sedimentary rocks mined in North America today.

The common table salt you sprinkle on your food is actually a mineral, halite. Salt forms when seawater evaporates. As more water evaporates, the salt concentration in the remaining water gradually rises to the point where the solution can no longer hold the salt. The salt begins to precipitate from the water and sink to the floor of the lake or ocean, forming a layer of halite (Figure 7.27). Evaporation typically occurs in restricted basins in hot, dry climates. The salt is later buried under other sedimentary rocks (perhaps clastic rocks). Scientists interpret thick salt deposits as an indication that there was a constant supply of seawater and

Figure 7.27 A chemical sedimentary deposit. A large lake, Lake Bonneville, covered much of Utah less than 10,000 years ago during the last ice age. Most of the lake dried up, and the water evaporated, leaving behind a flat lake bed covered in salt. Today, the Great Salt Lake near Salt Lake City is the last remnant of Lake Bonneville.
© Dennis Flaherty/Science Source

that hot climate conditions were stable at that location for millions of years. Different chemical compounds may form depending on the concentrations of elements present in solution. For example, gypsum (used in drywall) is another mineral dissolved in seawater that is precipitated when the water evaporates. These types of chemical sedimentary rocks are sometimes termed *evaporites* in recognition of the process that formed them.

Biochemical Sedimentary Rocks

Biochemical sedimentary rocks result from the actions of living organisms that cause minerals to be extracted from a solution or are composed of the remains of dead organisms. As such, these rocks represent a link between the biosphere and the geosphere. The actions of many organisms in seawater change the composition of the water, causing the mineral calcite (calcium carbonate, the principal ingredient in limestone) to be precipitated from solution. Calcite is made up of one atom of calcium, one atom of carbon, and three atoms of oxygen. Thus, the atomic structure of calcite traps carbon atoms, preventing the carbon from contributing to global warming.

Biochemical sedimentary rocks come from a wide variety of sources. For example, massive limestone coral reefs in the shallow tropical oceans around the world have formed because of the actions of the coral organisms (Figure 7.28). Other biochemical rocks are composed of the remains of dead organisms. Chalk, a type of limestone, forms from the skeletons of marine microorganisms known as coccolithophores. These tiny creatures are about the size of grains of clay or silt and live in the shallow, sunlit waters of the ocean (Figure 7.29). Unlike coral, coccolithophores are concentrated in colder, subpolar waters, sometimes in such great numbers as to change the color of the ocean and reflect back more incoming

sunlight. The coccolithophore is covered by a mineral armor made up of rounded plates of calcite known as coccoliths. Scientists estimate that 1.4 billion kilograms (1.5 million tons) of coccoliths collect on the seafloor each year to form chalk. The shells of larger organisms, the types you might pick up on a beach, may be broken down, sorted by wave action, and cemented together to form a shell-bearing limestone known as coquina (Figure 7.30a).

The most recognizable example of a biochemical rock is coal (Figure 7.30b). **Coal forms from the compacted remains of dead plants,** such as those that grew rapidly in a tropical swamp

Figure 7.28   Sources of biochemical sedimentary rocks. A coral reef in clear, shallow marine waters. Limestones and coquinas are rocks that commonly form in this type of environment.
© Purestock/SuperStock

Figure 7.29   Coccoliths and coccolithophores. **a.** Coccoliths are the tiny plates that surround a type of plankton known as a coccolithophore that lives in the shallow ocean. **b.** Billions of coccoliths combine to form layers of chalk. **c.** The presence of coccolithophores in the oceans can reflect more sunlight and change the color of the ocean water.
7.29a & 7.29c: NASA; 7.29b: © Bob Krist/ Corbis

environment. Such rocks are sometimes termed *organic sedimentary rocks.* What does the fact that large deposits of coal are found in West Virginia, Pennsylvania, and Kentucky tell you about the past environment there?

Figure 7.30 Two biochemical sedimentary rocks. **a.** Coquina is a form of limestone composed almost entirely of shell fragments. **b.** A massive coal seam in the Powder River basin, northern Wyoming. The seam is up to 60 meters (200 feet) thick in places. Note the large bulldozers for scale. 7.30a: © David Steer; 7.30b: © AdShooter/iStock/Getty Images

Sedimentary Rocks and Fossil Fuels

Current US energy use is weighted heavily toward fossil fuels (oil, natural gas, and coal), which account for approximately 90 percent of all energy used in the nation. Fossil fuels form from decayed organic material. Oil, coal, and natural gas are the most common products of this process. Oil and gas form from organic material in microscopic marine organisms, whereas coal forms from the decayed remains of land plants. The formation of these fossil fuels extracts carbon from the environment, but their consumption rapidly releases carbon dioxide, elevating global greenhouse gas concentrations (see Chapter 17).

Oil and Gas. The principal requirements in the development of oil and gas reserves are time, temperature, and rock type. The steps in the process are:

1. Organic-rich sediments are deposited and gradually buried to greater depths and converted to sedimentary rock (for example, shale).
2. Chemical reactions occur during burial under conditions of increasing temperature and pressure. The reactions occur at temperatures of 50 to 100°C (122 to 212°F).
3. The reactions change organic molecules to hydrocarbon molecules (chemical compounds of carbon and hydrogen). Higher temperatures "boil off" the hydrocarbons; lower temperatures are not sufficient to drive the chemical reactions. With increasing time (millions of years), the hydrocarbons become more mature, changing from heavy oils to lighter oils to natural gas.
4. Oil and gas migrate upward through fractures and spaces in sedimentary rocks. Some hydrocarbons escape at Earth's surface, but others are trapped and collect in sedimentary rocks below the surface to form commercial oil and gas deposits (Figure 7.31).

Global oil reserves will last for several decades at current consumption rates. The United States uses approximately 21 percent of the world's oil, more than any other nation (for comparison,

Figure 7.31 Oil deposits. Oil and gas are trapped in rocks below Earth's surface when they are juxtaposed with overlying or adjacent rocks that they cannot pass through. The hydrocarbons migrate (arrows) to the highest part of the reservoir rock until they meet an obstruction that prevents further movement.

✓ Checkpoint 7.15

☑ basic ☑ advanced
☑ intermediate ☑ superior

Venn Diagram: Chemical and Biochemical Sedimentary Rocks

Use the Venn diagram provided here to compare and contrast chemical and biochemical sedimentary rocks. Identify at least seven characteristics.

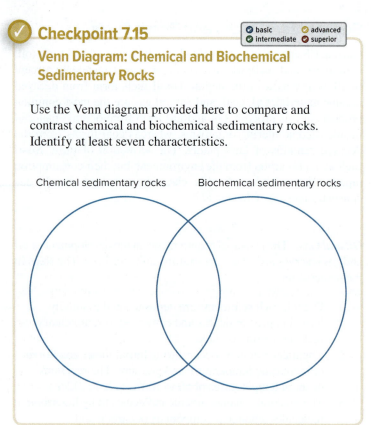

Chemical sedimentary rocks   Biochemical sedimentary rocks

✓ Checkpoint 7.16

☑ basic ☑ advanced
☑ intermediate ☑ superior

Similar organic-rich source rocks are present in rocks below two locations, Oiltown and Dryville. Oil and gas deposits formed in the overlying sedimentary rocks below Oiltown but did not form in rocks at Dryville. Suggest at least four potential explanations for this difference.

produced from both surface and underground mines. Unfortunately, some of the bituminous coals have a high sulfur content and therefore contribute to air pollution. Given the stringent regulations on pollutants, some companies prefer to use lower-grade subbituminous coals from surface mines in Great Plains and Rocky Mountain states (Montana, Wyoming, North Dakota, South Dakota, Colorado).

Over 80 percent of the world's recoverable coal is found in just seven nations. The United States has the greatest reserves, accounting for 25 percent of the world's total, enough to last for more than 200 years at current consumption rates. This suggests that we will have a plentiful supply of electricity into the distant future, but it is of little help as a replacement fuel for refined oil products (gasoline) unless we can assume that automobiles of the future will run on electricity. (Bring on the electric car!) One drawback to this scenario is that burning coal is a major source of carbon dioxide, contributing to the warming of global climates. The continued exploitation of coal may require the widespread adoption of a means of trapping carbon dioxide emissions during power generation (for more on these options, see Chapter 17).

China uses 11 percent). Recent oil discoveries have made the United States the world's third largest producer of oil and reduced imports to less than one-half of the oil we consume. The imports mostly come from other North American nations (Canada, Mexico). However, as more than one-half of the world's oil reserves are located in the Middle East, countries in that region (for example, Saudi Arabia, Kuwait, Iran) may play an increasingly important role in global oil supplies in the decades ahead unless we learn to drive vehicles using energy sources such as ethanol or electricity.

Coal. Coal, the carbon-rich residue of plants, can be classified by rank (carbon content). Coal matures by increasing rank with increasing heat and compaction with progressive burial that drives off water and other chemical constituents. Carbon content increases from low-grade peat (less than 25 percent carbon) to lignite (25 to 35 percent) to subbituminous (35 to 45 percent) and bituminous coal (45 to 86 percent) to anthracite (more than 86 percent). Peat is the least-mature form of coal, containing a large volume of fibrous plant matter. The higher the carbon content, the more heat that is released when the coal is burned. Small amounts of high-carbon coals produce the same heat as larger volumes of low-carbon coal. The volume of ash that remains after burning decreases with increasing rank.

There are three principal coal-producing regions in the United States (Figure 7.32). The first two, Appalachian basin states (Ohio, eastern Kentucky, West Virginia, Pennsylvania) and interior states (Illinois, Indiana, western Kentucky), produce high-rank bituminous coals and anthracite. These coals are

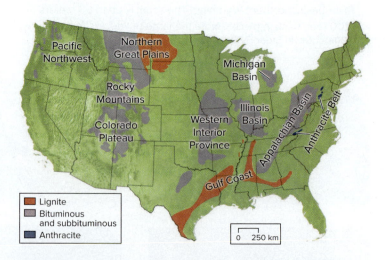

Figure 7.32 Coal-bearing areas of the United States. Eastern coal fields are mainly underground and typically produce bituminous coals. Western coal fields are mostly at the surface and produce subbituminous coal. The eastern coals have a higher carbon content but produce more pollution as a result of their higher sulfur content.

7.6 Metamorphic Rocks

Learning Objectives

- Define the terms *metamorphism* and *foliation*.
- Identify some everyday analogies for the formation of metamorphic rocks.
- Compare and contrast the features of rocks formed by contact metamorphism and regional metamorphism.

The word *metamorphosis* means "change" and was originally used to describe how vegetation changes from a seed to a full-grown plant. In earth science, **metamorphism relates to changes in mineral composition or texture that occur in solid rocks as a result of increasing pressure or temperature.** For example, limestone can become marble when heated near an igneous pluton. Notice that we referred to heating, not melting. If melting occurs, the end result is an igneous rock. Metamorphic rocks require a temperature range that is high enough to trigger the chemical reactions necessary to change minerals but not so high that the minerals will melt. Depending on the combination of pressure, temperature, and the chemical composition of circulating fluids, one rock has the potential to be converted into several different metamorphic rock types. Temperatures of at least 200°C (390°F) are generally needed before some of the minerals in a rock will undergo chemical reactions to cause compositional changes. Depending on their composition, most minerals in rocks near Earth's surface melt at temperatures ranging from 600 to 1,100°C (1,110 to 2,010°F). Consequently, the temperature "window" for metamorphism is from 200 to 1,100°C (390 to 2,010°F), depending on the minerals present in the original rock. Such temperatures are common near magma chambers or at deeper levels in the crust.

There are two main types of metamorphism: contact metamorphism and regional metamorphism. These types of metamorphism occur by different processes and create rocks with different characteristics.

Contact Metamorphism

Contact metamorphism occurs when rocks come in contact with a heat source (usually a magma body). Essentially, the rocks are cooked. A comparable change occurs in roasting meat or baking bread—that is, the composition does not change but the texture of the material does. Rocks do not conduct heat well (they are good insulators), so the zone of contact metamorphism is usually relatively narrow and occurs in the rock immediately surrounding the heat source. Marble is an example of a rock that may be formed by contact metamorphism (Figure 7.33a). Marble forms when limestone is heated to high temperatures. Both marble and

Figure 7.33 Rocks that may be formed by contact metamorphism. **a.** Marble shows large crystals of calcite when viewed through a microscope. Crystals are 2 millimeters (0.08 inch) across. **b.** Quartzite shows quartz crystals approximately 0.5 millimeter (0.02 inch) in width. These are examples of rocks that may form as a result of contact metamorphism of limestone and sandstone, respectively.
7.33a (left and right), 7.33b (left and right): © David Steer

Table 7.6	Metamorphic Rocks Based on Foliation and Texture (Grain Size)		
	Grain size		
Foliation	Fine (<0.1 mm)	Medium (~0.1–4 mm)*	Coarse (>2 mm)*
No	Hornfels	Marble, quartzite	Marble, quartzite
Yes	Slate, phyllite	Schist	Gneiss

*Approximate sizes for comparison purposes only.

limestone may have the same composition, but marble typically has larger grains (Table 7.6).

Back to the bricks (Section 7.1): The final stage in the brick-making process requires the addition of heat. The green brick is heated to high temperatures comparable to those of metamorphism as it makes its way through the kiln. The heat is not sufficient to melt the materials in the brick, but it does cause them to fuse to produce a stronger product than the raw brick that went into the kiln. Similarly, metamorphic rocks are often stronger and more resistant to erosion and weathering than their sedimentary cousins.

Regional Metamorphism

Regional metamorphism occurs when rocks undergo increased temperatures and pressures typically associated with the plate tectonic processes that form mountains (see Chapter Snapshot). In these areas, sedimentary rocks may be buried to great depths (10 to 20 kilometers; 6 to 12 miles), and igneous rocks in the crust may end up 50 kilometers (31 miles) below the surface as a result of crustal thickening. Recall from Chapter 2 that the temperature in Earth's crust increases by approximately 25°C per kilometer (72°F per mile) near the surface. Assuming a surface temperature of 15°C (59°F), metamorphism would begin to occur in rocks at depths of approximately 7 to 8 kilometers (4 to 5 miles). The additional pressure causes sheetlike minerals (for example, mica) in the rock to physically rotate or to grow with a preferred alignment. The combination of both higher temperatures and pressures and the presence of fluids can cause some specific

Figure 7.34 Foliation. Tabular or sheetlike minerals grow perpendicular to the direction of pressure to form a foliation. **a.** Randomly oriented minerals prior to metamorphism; **b.** directed pressure causes minerals to form a foliation.

types of minerals to grow. Those minerals are aligned parallel to one another (like a stack of papers on a table) and perpendicular to the direction of pressure (like pressing down on the stack of papers). **This geometric orientation generates an alignment of minerals into sheets termed *foliations*** (Figure 7.34).

Checkpoint 7.17

basic · advanced · intermediate · superior

The conversion of bread to toast can be seen as an analog for the formation of a metamorphic rock by

   a. Contact metamorphism.
   b. Regional metamorphism.

Checkpoint 7.18

basic · advanced · intermediate · superior

Metamorphic Rocks Defining Features Matrix

Identify which of the characteristics in the left-hand column of the following table are present in rocks formed by contact metamorphism, regional metamorphism, or both. Place an "X" in the two right-hand columns of the table where appropriate. Do not place a mark in either column if the characteristic is not present. One characteristic has been completed as an example.

	Rocks formed by	
Characteristic	Contact metamorphism	Regional metamorphism
Form at temperatures above 200°C		
May originally have been an igneous rock	X	X
Form as a result of increasing pressures		
May surround plutonic igneous rocks		
Slate is an example		
Form as a result of melting		
May underlie several adjacent states		
Found in mountain belts		
May originally have been a sedimentary rock		
May contain a foliation		
Marble is a possible example		
Form on Earth's surface		
Limestone is an example		
May have originally been a metamorphic rock		

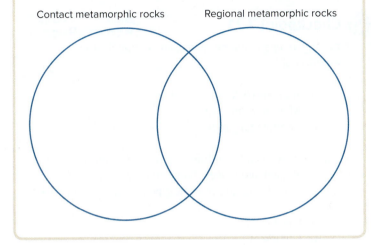
change from the original rock to the metamorphic rock (see Table 7.6). Foliated metamorphic rocks, in order of increasing metamorphic grade (low to high temperature) and grain size (small to large), are slate, phyllite, schist, and gneiss. The grain size in slate is much too small to be seen with the naked eye, and the presence of millions of tiny crystals of sheetlike minerals means that slate splits easily into thin sheets (Figure 7.35a). If you had taken this class a century ago, your teacher would have been using a slate chalkboard. Mica, a common mineral in metamorphic rocks, can give some rocks (for example, phyllite, schist; Figure 7.35b) a shiny, silky appearance when present in larger grains. Gneiss (pronounced "nice") has the largest grain size, with quartz and feldspar often forming much of the rock. Light- and dark-colored minerals typically separate into darker and lighter bands in the rock (Figure 7.35c).

Not all rocks that undergo regional metamorphism form a foliation. For example, rocks that lack sheetlike minerals do not generate the parallel alignment of minerals necessary to create a foliation. Sandstone, composed mainly of quartz grains, is converted to quartzite by regional metamorphism; limestone forms marble. Neither metamorphic rock contains foliations (Figure 7.33).

The higher the temperatures and pressures, the more intense the metamorphism. Rocks formed by regional metamorphism are classified by the size of minerals that form (a temperature- and pressure-dependent process) when they

a.   b.   c.

Figure 7.35 Foliated metamorphic rocks. **a.** Foliations in an outcrop of slate in Antarctica. The foliations are oriented from top right to bottom left. **b.** Phyllite (top) and schist (bottom) both derive their shiny appearance from abundant mica. Schist has a larger grain size. **c.** Gneiss (bottom) formed from the metamorphism of granite with a similar composition (top).

7.35a: P.D. Rowley, USGS Photo Library, Denver. CO; 7.35b (top): © Joyce Photographics/Science Source; 7.35b (bottom): © David Steer; 7.35c (top): © Joyce Photographics/Science Source; 7.35c (bottom): © Andrew J. Martinez/Science Source

7.7 The Rock Cycle and Mineral Resources

Learning Objectives

- Identify some common analogies for the formation of igneous, sedimentary, and metamorphic rocks.

- Describe how an element might move through the rock cycle, explaining the changes in conditions that it would experience.

- Explain how the rock cycle can be interpreted as an example of Earth operating as a constantly changing system.

- Identify at least four different processes that form mineral deposits, and give an example of one metal or commondity formed by each process.

As a society, we have recognized the need to protect our natural resources and have created extensive recycling programs. We collect papers, cans, and bottles, and we convert them to other products used in our daily lives. The materials of Earth are also "recycled" through a process known as the *rock cycle* (Figure 7.36). The rock cycle links together different materials and processes on Earth to form the components of the three principal groups of rocks.

Mineral resources become concentrated in Earth's crust as a result of specific geologic processes associated with the formation of rocks. Many of these resources are associated with a group of minerals called nonsilicates that form another important class of earth materials. Nonsilicate minerals do not have the silica-oxygen tetrahedron in their chemical structure. One subcategory of nonsilicates, called oxides, includes minerals with oxygen in

---

> ### ✓ Checkpoint 7.21
>
> | ✓ basic | ✓ advanced |
> |---|---|
> | ✓ intermediate | ✓ superior |
>
> Cooking an egg could be seen as an analog for the formation of
>
>   a. Igneous rock.
>   b. Metamorphic rock.
>   c. Sedimentary rock.
>
> Concrete is formed by adding cement and water to a mixture of sand and gravel. This could be seen as an analog for the formation of what type of sedimentary rock?
>
>   a. Clastic
>   b. Chemical
>   c. Biochemical

Figure 7.36 The rock cycle.

their chemical makeup. Hematite (Fe_2O_3) is an important example that is mined to produce iron. Sulfides include materials bound with sulfur such as pyrite, or fool's gold. Carbonates include CO_3 in their chemical structure and form materials such as limestone used in the construction industry. Gypsum (see earlier mention in Chemical Sedimentary Rocks) is a sulfate that has SO_4 in its chemical structure. An important halide, or material with chlorine or fluorine, is halite, or common salt. Last, phosphate-based nonsilicate minerals are commonly used in agricultural products such as fertilizer. Exploration for minerals requires that geologists recognize the telltale evidence that signals the presence of useful mineral deposits. Geologic processes that result in the concentration of mineral resources can be divided into those associated with chemical reactions resulting from changing temperatures and movement of fluids through rocks and those formed by the physical rearrangement of earth materials during erosion, transport, and deposition. The raw materials may be formed during one part of the rock cycle but concentrated during a later stage. Consequently, it becomes important to know not only how rocks form but also what might happen to them in the millions of years that follow.

✓ Checkpoint 7.22

● basic   ● advanced
● intermediate   ● superior

Rock Cycle Diagram

The following diagram illustrates some of the interactions of the rock cycle. Match the lettered responses to the blank ovals on the diagram. (*Note:* Some letters are used more than once.) Example: If you believe that metamorphic rock is converted to magma by cementation and compaction, enter "a" in the top left oval.

a. Cementation and compaction (lithification)
b. Heat and pressure
c. Weathering, transportation, deposition
d. Cooling and solidification
e. Melting

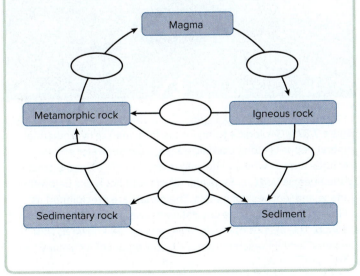

The Rock Cycle

Under the right conditions, any type of rock can become any other type of rock. Such changes involve a variety of interactions among the different components of the earth system, and they show that Earth is constantly changing. They also explain why geologists have found only a few rocks on Earth's surface that were formed close to the time the planet was created.

During their lifetime, minerals and elements in Earth's crust can be recycled through several different rocks. For example, follow the possible fate of a grain of quartz sand on a beach: As the sea level rises, that beach may become flooded, and the sand may be incorporated into a sedimentary rock such as sandstone. Over time, the forces of plate tectonics may cause the sandstone to be buried deep in the crust during a period of mountain building. The higher temperatures and pressures common in regional metamorphism may convert the sandstone to quartzite. Magma rising from the mantle may melt the adjacent quartzite, assimilating the quartz grain into the magma. The elements of the quartz grain (silicon, oxygen) may recombine with other elements to form new minerals as the magma solidifies to form granite within the crust. Erosion eventually exposes the rock, breaking it down to its constituent minerals and rock fragments (sediment), which are carried by streams to the coast, where the grains are deposited on a beach. And the cycle begins again.

Likewise, an igneous rock can remelt to form a new igneous rock, a sedimentary rock can re-erode and become another sedimentary rock, and a metamorphic rock can be heated and transformed over and over. Each cycle may last millions or even billions of years. Keep in mind that we are not adding new material to Earth, so any rocks forming today must be recycling elements from rocks formed at some time earlier in the planet's history.

Mineral Resources

The term *mineral resource* refers to nonfood, nonfuel resources such as metals (for example, aluminum, palladium) and industrial minerals (for example, gypsum, phosphate). The development of mineral resources depends on more than just the presence of a mineral deposit. The average concentration of minerals in the crust is insufficient to form an economically valuable mineral deposit. The concentration factor (CF) is the increase in the concentration of a mineral required to generate an economic deposit.

✓ Checkpoint 7.23

● basic   ● advanced
● intermediate   ● superior

Use information at the Minerals Education Coalition (www.mineralseducationcoalition.org/minerals) or the USGS Minerals Yearbook (http://minerals.usgs.gov/minerals/pubs/commodity/myb/) to make a list of 10 different minerals that are used to manufacture objects that you would use everyday. Try to find at least three minerals that you have not heard of before.

For example, copper makes up 55 parts per million (ppm, 0.0055 percent) of Earth's crust. Copper ores from the Bingham Canyon mine, Utah, are composed of 0.6 percent (6,000 ppm) copper, equivalent to a concentration factor of 109 (0.6/0.0055). Relatively rare minerals have large CF values (for example, gold, CF > 2,000), whereas more common elements have low CF values (for example, silicon, CF = 2).

High-Temperature Reactions. The high temperatures and abundant fluids associated with the formation of igneous and metamorphic processes can result in the concentration of a variety of metallic mineral resources. **A rock containing economic concentrations of metallic minerals is known as *ore*.** Many ores are

formed as a result of igneous or metamorphic processes associated with plate tectonics. Consequently, metallic deposits are often concentrated near current or former plate boundaries. Hydrothermal and magmatic processes are some of the more common ways to form economic deposits of minerals.

Magma and water are the key ingredients in the generation of *hydrothermal mineral deposits* (Figure 7.37). One location where water is readily available is the ocean floor. Magma rises along the oceanic ridges. Cold bottom waters from

Checkpoint 7.24

Element X and the Rock Cycle

☑ basic ☑ advanced
☑ intermediate ☑ superior

The following graph illustrates the partial life cycle of a sample of an element (X) over several million years. At some points during its life, the sample experienced temperatures similar to those found at Earth's surface (A, G), while at other times, it experienced high temperatures of approximately 1,000 and 800°C, respectively (C, E).

Using what you know about rocks and how they form, write the life history of element X with special reference to what happened at and between times A–G on the graph. Assume that the element passed through several rock types during its life. To get you started, we provide the first few steps:

A   Element X is weathered from a rock at Earth's surface and is carried in solution by a stream.
A–B   Element X combines with another element to form a mineral in a pile of sediment and becomes part of a chemical sedimentary rock.
B   Temperature is sufficient to convert the sedimentary rock to a metamorphic rock.

Figure 7.37 Geologic setting for the formation of hydrothermal mineral deposits along oceanic ridges. Cold seawater enters fractures formed where plates move apart in the rift at the crest of the oceanic ridge. **a.** Seawater reacts with hot rocks below the surface and leaches out metals. The metals are precipitated as sulfide deposits when the circulating seawater encounters the cold water. **b.** The plumes of precipitating metals are known as "hot smokers" as they resemble smoke billowing out of a chimney. 7.37b: USGS

the ocean floor can infiltrate the oceanic crust through networks of narrow fractures. The waters are heated to high temperatures (300 to 400°C; 570 to 750°F) by the rising magma, and chemical reactions occur that leach out metals from the surrounding rocks. Fluids transport these mixtures of dissolved metals (manganese, iron, copper, zinc) to the ocean floor where they meet much colder waters. The sudden decrease in temperature causes the minerals to precipitate from solution to form deposits known as massive sulfides that are incorporated into sediments deposited along the ocean ridge system.

Hydrothermal deposits also form on land when metal-rich fluids are expelled from magma chambers to form veins that fill fractures in the surrounding rock. Quartz-rich fluids discharged from magma chambers often form associations with gold deposits. Alternatively, economically valuable mineral deposits can become concentrated when heated groundwater reacts with rocks below volcanoes or when metamorphic processes result in reactions that cause minerals to recrystallize into other varieties.

The cooling processes inside magma chambers may cause minerals to be concentrated in narrow zones within igneous rocks. These *magmatic mineral deposits* (for example, platinum-group minerals, chromium) may form at high temperatures before the magma solidifies. Once those minerals crystallize, they are heavier than the surrounding magma, so they sink to the bottom of the magma chamber by a process called *crystal settling,* forming an enriched layer (Figure 7.38). During the final phase of cooling and solidification of the magma, the residual melt mixture may contain relatively rare elements, such as lithium and beryllium. These enriched magmatic fluids may cool very slowly as they form veins in narrow fracture zones around the magma chamber. The extra-slow cooling may yield a rock known as a pegmatite that contains giant crystals of individual minerals that can be up to several meters across.

Low-Temperature Reactions. The largest US iron deposits represent ancient (more than 2 billion years old) chemical sedimentary rocks that were precipitated from seawater when the chemistry of Earth's oceans was considerably different from today. The greatest concentration of these deposits is in the Upper Peninsula of Michigan and northern Minnesota. The iron must be concentrated by additional geologic processes before these deposits can be considered an ore. Two potential concentration methods are:

- The leaching of silica from the rocks (leaving iron behind in greater relative concentrations) by weathering processes that occur on Earth's surface.
- Metamorphism that causes an increase in grain size and the formation of new iron minerals that are easier to separate from the rock.

Water flowing through rocks on or near the land surface may remove soluble minerals to leave behind sufficient concentrations of economic minerals to form an ore. These ***residual mineral deposits* are formed most rapidly in areas with rapid weathering rates** such as the tropics. Iron- and aluminum-rich laterite forms as a result of leaching of minerals from thick soils in tropical regions. The world's principal source of aluminum ore is from a form of laterite known as bauxite. Most of the world's bauxite is mined in Australia, Guinea, and Jamaica.

a.

b.

Figure 7.38 Layering of different minerals in an igneous pluton. **a.** This image is based on the character of the Bushveld complex in South Africa. Heavy, early-formed minerals sank to the bottom of the magma chamber (crystal settling), resulting in a layer containing crystals rich in platinum and chromium. **b.** The metal-rich layer is less than 1 meter (3 feet) thick but holds the world's greatest reserves of platinum-group metals.
7.38b: © Stephen Reynolds

8.1

Chap

• Stu
   pro

• Stu
   en

• Stu
   ge

Wh
Andy Golc
posed of na
Goldsworth
into short-l
ings. Giant
or a long li
downstrear
into nature
are compos
oldest pres
Egyptian p
that were c

The
hours-days
centuries-m
understand
frame to wl
logic record
measured in
tific method
years old. I
commonly u
tinguishable
think in the
time is shap
Earth's rota
But what ab

Relative Time

The study of **relative time** involves placing events in the order in which they occurred. In the context of geology, it means describing the order or sequence of geologic events. We will examine interpretations of the ages of rocks in the Grand Canyon in Arizona to illustrate the methods and philosophies geologists have used to investigate relative time.

The Grand Canyon: Investigating Its Origin. The Grand Canyon is 15 to 30 kilometers (9 to 19 miles) across, and stair-steps downward over 1,500 meters (4,920 feet) from the canyon rim to the Colorado River below (Figure 8.2). Native Americans had been exploring the canyons of the West for many centuries, but the first person to make a detailed record of an expedition down the Colorado River and through the Grand Canyon was John Wesley Powell (Figure 8.3a). Powell was a truly exceptional man. He lost an arm in the battle of Shiloh during the Civil War, but thought nothing of leading a small team of men in three wooden boats down an unknown series of canyons in the most remote section of the West (Figure 8.3b). He actually did it twice, in 1869 and 1871, and made several stops along the way to document the landscape. Despite numerous hazards—the destruction of two of the boats, three of his men walking away from the expedition, and widespread reports of his death—Powell survived and was hailed as a national hero.

From observations made on his journeys, Powell produced publications that interpreted the geology of the Grand Canyon and the surrounding region. Powell's story, *Exploration of the Colorado River of the West,* was part adventure yarn, part scientific report, and a national best seller in its day. In his book, Powell describes one of the Native American beliefs that attributed the origin of the

canyon to a native god, Tarvwoats, who carved the canyon to serve as a trail between the land of the living and the land of the dead. Tarvwoats later filled the canyon with a river to cover his trail. In contrast, Powell interpreted the rocks in the canyon walls to be the result of long-lived geological processes that he believed had been operating "through ages too long for man to compute." This is the accepted scientific view of the history of the Grand Canyon and the one that we will present data to support in this chapter.

In a his book, *Grand Canyon: A Different View,* author Tom Vail offers a theology-based origin for the canyon. Vail's book suggested that the canyon's rock layers were deposited following Noah's flood, as recounted in the Bible, and are consequently only a few thousand years old. He went on to suggest that the canyon was formed in a few days by a rapidly eroding river that quickly carved through the relatively soft deposits left behind as floodwaters receded. Vail believed his interpretations were as valid as those of Powell and the many geologists who have subsequently studied the canyon. Vail has stated, "What they call science is theory just as what is in my book is theory. All the scientists here have as much a right to their opinion as anyone else."

A reader might consider these different views and then wonder: *Who is right?* It is important to recognize that these

Figure 8.1
lived structu
giant snowb
Goldsworthy
an installatio
were placed
midsummer,
creations qui
disappear. Lc
other hand, a
oldest structu
such as **b.** the
pyramids (26
c. the Great V
8.1a: © Jenny M
© Hisham F. Ibr
© Corbis

Figure 8.2   The Grand Canyon. The rock layers exposed in the Grand Canyon represent intervals of time stretching back thousands of millions of years. The canyon is up to 30 kilometers (19 miles) across in the well-layered sedimentary rocks but is much narrower in the metamorphic and igneous rocks of the inner gorge (center of image).
© Corbis

b.

Figure 8.3   John Wesley Powell and his Colorado River expedition. **a.** A 35-year-old Powell just before his first trip down the Colorado River. **b.** An illustration of the second expedition running a rapid in the Grand Canyon.
8.3a & 8.3b: National Park Service

contrasting ideas about the canyon's origins depend on the conceptual framework people bring to the debate. Perhaps someone you know would agree with Tom Vail's interpretation of the Grand Canyon. Perhaps there are Native Americans or others who believe in the Tarvwoats version. However, whatever one believes to be true does not make that belief *scientifically* valid. Vail's perspective differs from that of scientists. Following Vail's logic, all ideas about the formation of the canyon would be equally valid, including the Tarvwoats interpretation. In contrast, the **scientific method argues that some ideas are more-valid explanations for geologic phenomena than others and, more important, that those ideas can be tested over and over with similar results.** Science can also be used to predict future events or processes with some accuracy. The claim that the Grand Canyon originated from a flood cannot be considered valid within a scientific framework because it cannot withstand the scrutiny of a disciplined scientific process.

What can we learn from present-day natural processes that will help us unravel the Grand Canyon's history? How does the theology-based story for the age of the rocks compare to the scientific explanation? How would you begin to figure out the age of rocks in a feature such as the Grand Canyon? Before we try to answer these questions we must take a step back to learn more about the clues that help us determine the relative ages of rocks.

Steno's Geologic Principles. Nicholas Steno, a Danish physician, began the process of determining the age of Earth from a scientific standpoint in the late seventeenth century, just a few years after Archbishop Ussher published his chronology. Steno was studying anatomy in the Italian province of Tuscany when, during the dissection of the head of a shark, he noted that the shark's teeth resembled objects found in local rocks. The common view of that time was that such objects grew naturally in the rocks and that it was just a coincidence that they resembled shark teeth. But Steno believed that the objects actually *were* preserved shark teeth. Although he was not the first person to make this interpretation, he was the first person to take it further.

Steno began thinking about how shark teeth, or any other object, could end up in rocks on a Tuscan hillside. He hypothesized that such objects were preserved when they sank to the bottom of a body of water and were covered up by sediment. He reasoned that if rock layers were deposited in water, they must have originally been horizontal. This observation—that **sedimentary rocks are deposited in nearly horizontal layers**—became known as the **principle of original horizontality** (Figure 8.4a). A consequence of this principle is that sedimentary rocks that are no longer horizontal must have undergone an episode of deformation after their formation (Figure 8.4b). For example, sedimentary layers are often tilted during the uplift associated with mountain formation (see Chapter 6). The principle of original horizontality is the geologists' first rule for reconstructing Earth's timescale.

Sometimes papers pile up on your desk for several weeks before you have time to put them away. The oldest papers are at the bottom of the pile, while the most recent additions are near the top. The same principle holds true for sedimentary rocks. The lowermost objects (layers) placed down first are on the bottom of the pile. This simple idea became Steno's second contribution to determining the sequence of geologic events on Earth. His **principle of superposition** states that, in a series of sedimentary rock layers that are not deformed, **the rocks at the bottom of the stack are the oldest and the rocks at the top are the youngest.** The

same principle can be extended to many forms of volcanic igneous rocks. Volcanic eruptions can produce layers from lava flows or when ash and other debris fall back to Earth.

Steno applied his rules to some of the rocks in Tuscany. Prior to this time, most authorities saw Earth as a finished product that had once been covered in water by the biblical flood. Steno's work changed that perspective by using observations to create hypotheses that explained the landscape. His interpretations linked the origin of the layers to events that occurred before, during, and after the biblical flood. The fact that he based his rules on observations was critical to advancing science. In doing so, Steno's work over 300 years ago attempted to fit the components of our modern-day scientific method into a sequence of events constrained by the theological time frame of creation. Steno saw no contradiction between the formation of these rocks and the biblical timescale. But a century later, James Hutton introduced an alternative view.

Hutton's Geologic Principles. In the eighteenth century, the Scottish scientist James Hutton essentially invented our modern concept of geologic time. As was common for wealthy men in those days, Hutton was a gentleman farmer and dabbled in many branches of science. He noted that the landscape of his farmlands remained unchanged with the passage of time, even as rock was converted to sediment and sediment was washed from his fields into nearby streams. From this modest observation, he deduced two significant hypotheses. First, he suggested that the same slow-acting geologic processes that operate today must have operated in the past; therefore, **it must take a very long time for those processes (for example, weathering, erosion) to produce any significant change in the shape of Earth's surface.** This view of earth processes is now known as the *uniformitarian model,* and we will discuss it in greater depth in Section 8.5.

Second, Hutton noted that **all land should be worn flat unless some process acts to renew the landscape by forming new mountains,** which then renews the slow cycle of destruction. Hutton set out to find some examples of rocks that illustrated this idea of cyclical change. He found his evidence at an outcrop of rocks along the east coast of Scotland at a location known as Siccar Point. Here, he discovered gently tilted rocks overlying nearly vertical

Deposited last → C
B
Deposited first → A

Layers were originally deposited horizontally

Layers were tilted after their formation

a. b.

Figure 8.4 Steno's principles. Interpretations based on the principle of original horizontality are in blue; those based on the principle of superposition are in green. **a.** Layers A, B, and C must have been initially deposited as horizontal layers. A, as the lowest layer, must have been deposited first, and C was deposited last. **b.** Layers A, B, and C were tilted (deformed). Because they are parallel, it is likely that they were all deformed at the same time.

layers (Figure 8.5). He hypothesized that the lower layers of rock formed first (superposition) as flat layers (original horizontality) and were later pushed up to the surface and tilted. These layers were then worn down by erosion and buried beneath new horizontal deposits of sediment (Figure 8.6). Features in the geologic record that signal these sequences of deposition, erosion and renewed sedimentation are called *unconformities*. **Unconformities are evidence of breaks in the geologic record representing a gap in time.** Hutton thus had found an outcrop of rocks that illustrated his idea of cyclical processes.

Furthermore, in Hutton's example, the upper set of layers cuts across the lower layers and illustrates a third rule: the **principle of cross-cutting relationships.** This principle **states that older rocks may be cut by younger rocks or other geologic features** (Figure 8.6b). For example, when magma intrudes into a rock, it must be younger than the rock it cuts across (Figure 8.7), or when a fault cuts across rock layers, it must be younger than the layers themselves (Figure 8.7).

A variation on this theme is that younger rock units sometimes incorporate pieces of older rocks. For example, chunks of surrounding rock may collapse into a mass of magma as the molten rock forces its way upward through the crust. Often these chunks melt and become part of the magma, but sometimes the magma cools and solidifies to form an igneous rock before the other rock melts. These preserved chunks of rock are known as inclusions. **The principle of inclusions can be applied to identify older pieces of rock surrounded by younger igneous rocks** (Figure 8.8).

Hutton himself did not suggest an age for Earth, simply stating elegantly that the landforms he examined "had no vestige of a beginning—no prospect of an end." However, his message was clear: **Earth must be much older than the commonly accepted age of 6,000 years.** This was a hugely controversial idea at the time and was described in great detail in Hutton's thousand-page 1795 book *The Theory of the Earth*. Hutton's concept of an ancient Earth would not receive wide attention until after his death, when his ideas were summarized by a friend, John Playfair, and published in 1802 in a shorter form as *Illustrations of the Huttonian Theory of the Earth*. We will continue to trace the historical

development of the concept of geologic time in the section Fossils and Chronology, but let us first pause to apply the rules we have just learned to the example of the Grand Canyon.

James Hutton

The Grand Canyon: Applying the Rules. Identifying examples of the principles of original horizontality, superposition, cross-cutting relationships, and inclusions in nature allows geologists to reconstruct the geologic history of a rock sequence. Geologists do not stop at simply placing events in order; they also want to learn more about the conditions under which these rocks formed (see Chapter 7). Consequently, they must determine what types of rocks are present and what their characteristics tell us about their origins. For example, the rocks exposed in the Grand Canyon can be interpreted to reveal a complex geologic history.

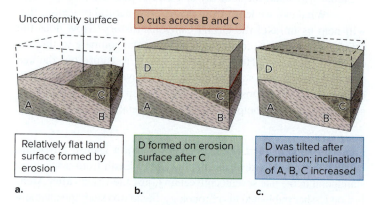

Figure 8.6 Hutton's principles. According to the principle of cross-cutting relationships, **a.** A, B, and C were subjected to weathering and erosion at Earth's surface. A significant time interval may have passed while erosion occurred. This eroded surface is known as an unconformity surface and is the physical expression of a gap in time. **b.** D lies above A, B, and C and cuts across these layers so it must have formed after C. **c.** D is no longer horizontal; therefore, all layers were tilted. Interpretations based on the principle of original horizontality are in blue; those based on the principle of superposition are in green; observations in red are made using the principle of cross-cutting relationships.

Figure 8.5 Outcrop of rocks at Siccar Point, Scotland. The gently tilted layers in the background lie on top of the vertical layers in the foreground.
© Dr. Clifford E. Ford

Figure 8.7 Cross-cutting relationships. The igneous pluton E formed after layers A–D and includes pieces of the layers (inclusions). F and G formed later because they lie atop all other rock units. The fault formed after F but before G. Finally, the river cut a valley through the upper part of G.

As we descend through the Grand Canyon, the rocks are defined by an obvious layering until we reach the inner gorge, close to the river. The walls of the inner gorge are composed of dark Vishnu Schist, which typically shows no layering—but is foliated—and lighter-colored Zoroaster Granite (Figures 8.9 and 8.10). The granite cuts across the foliations of the schist, so we can apply

Figure 8.8 An inclusion. A dark gray rock is enclosed in a larger outcrop of the lighter-colored igneous plutonic rock in a streambed in Vermont. Note quarter for scale.
© David McConnell

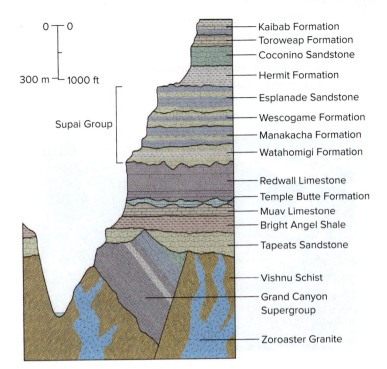

Figure 8.9 Sequence of rocks in the walls of the Grand Canyon.

Kaibab Formation
Toroweap Formation
Coconino Sandstone
Hermit Formation
Esplanade Sandstone
Wescogame Formation
Manakacha Formation
Watahomigi Formation
Redwall Limestone
Temple Butte Formation
Muav Limestone
Bright Angel Shale
Tapeats Sandstone
Vishnu Schist
Grand Canyon Supergroup
Zoroaster Granite

Supai Group

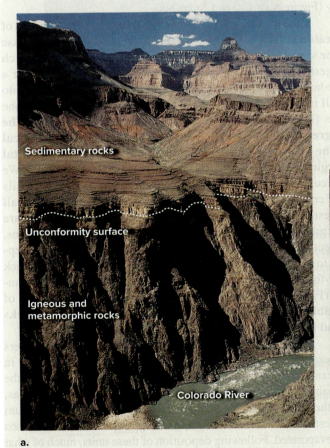

Sedimentary rocks

Unconformity surface

Igneous and metamorphic rocks

Colorado River

a.

b.

Figure 8.10 Igneous and metamorphic rocks in the inner gorge of the Grand Canyon. **a.** Metamorphic and igneous rocks of the steep-walled inner canyon lie below the horizontally layered sedimentary rocks. **b.** The pink Zoroaster Granite cuts across steeply inclined foliations in the dark Vishnu Schist in the inner gorge.
8.10a: © McGraw-Hill Education/Doug Sherman, photographer;
8.10b: © David Marx Photography

fossils and are similar to those produced by modern land-dwelling lizards, scorpions, and millipedes. Brachiopods and corals show up again in the Kaibab Limestone, along with occasional trilobites and cephalopods. Keep in mind that while some of the fossil types (for example, brachiopods) are found in multiple rock layers, the fossils in each layer are not exactly the same. That is, brachiopods make up a phylum, a large collection of organisms with similar traits. The phylum contains thousands of different species of brachiopods, and the ones found in the Kaibab Limestone are not the same species as those found in the Muav Limestone. (For comparison, humans are part of the phylum Chordata, which also includes fish, amphibians, snakes, and birds [all organisms with a backbone]—quite a diverse collection.)

8.3 Geologic Time

Learning Objectives

- Describe how and why Earth's history is divided into three long spans of time known as eons.
- Describe how the most recent eon differs significantly from previous eons.
- Explain the role of mass extinctions in defining transitions between eras in the most recent eon.
- Relate the fossil record to changes in the biosphere over much of Earth's history.

As discussed in Section 8.2, geologists use relative time to arrange geologic events in the correct sequence. But time can also be considered in purely numerical terms. Early in the twentieth century, scientists realized they could use the radioactive decay of isotopes to determine the actual age of some types of rocks and to put dates on the sequence of events. For convenience, we will incorporate the numerical times into this discussion of the geologic history of Earth, but we will hold the explanation of how these ages are determined until Section 8.4.

Evolution of Early Earth

Conditions on Earth more than 4 billion years ago would have been much different from today. Remember, Earth formed from accretion of materials orbiting the proto-solar system. This resulted in an Earth that was initially almost entirely molten. It was probably several hundred million years before Earth had a permanently solid crust. Scientists have determined that the oldest known rocks on Earth are nearly 4.3 billion years old and are found along the eastern shore of Hudson Bay, Canada. Even after the crust solidified, conditions on Earth would have been much different from today as heat left over from the planet's formation would have made conditions on Earth much warmer. There was little oxygen in the atmosphere, evidenced by the existence of minerals that would have broken down if oxygen had been present. Some scientists have interpreted the presence of extensive ancient volcanic rocks throughout the world to suggest that there was more volcanic activity than we observe today. To make matters worse, Earth was being regularly bombarded by asteroids and comets. In those early times, the environment in which life evolved was probably similar to the hot, caustic conditions we currently find in places such as the

hot springs of Yellowstone National Park. Those scorching, acidic waters harbor heat-loving bacteria. Because bacteria are some of the most primitive life-forms we observe on Earth today, scientists argue that early life was probably similar to them.

The Geologic Timescale

Earth's past is divided into three big chunks of time known as eons. Eons are the largest intervals of the timescale, each over 500 million years long, and they mark major changes in the earth system (Figure 8.14). The oldest known rocks were formed during the early Archean eon. The first part of the Archean used to be called the Hadean eon, and some people still use that term to represent the earliest part of Earth's history from its initial molten state 4.6 billion years ago until the first rocks formed. However, as we discussed earlier, science is provisional and subject to change. The expansion of the range of the Archean to include the former Hadean is an example of such a change that occurred as earth scientists learned more about Earth's earliest history and have been able to refine the timescale with radiometric dates. The origin of life is still enigmatic to scientists; remember, early Earth was not like it is today. Over long periods of time, hundreds of million of years, things began to change. Our first evidence of possible life comes in the form of enhanced carbon concentrations in crustal rocks from Greenland that are 3.85 billion years old. While these enhanced carbon signatures are not visible cells, they do indicate that organic compounds were present that long ago. Making the jump from organic molecules to replicating life-forms is where the edge of our current scientific understanding lies. A recent research breakthrough in the laboratory has been able to determine a process by which basic small molecules such as amino acids could have polymerized (combined in to larger chains of molecules) to form self-replicating RNA molecules. This process involved basic steps of evaporation, heating, and condensation that could readily have occurred in small warm ponds on an early Earth. Next scientists

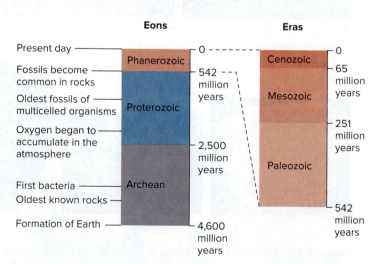

Figure 8.14 Eons and eras. Geologic time is divided into three long eons, the Archean, Proterozoic, and Phanerozoic. The Phanerozoic, the most recent eon, is characterized by abundant fossils, allowing it to be further subdivided into three eras. The eras are in turn broken down into shorter time intervals known as periods (see Table 8.1).

will attempt to apply this newfound knowledge to move toward a comprehensive working model of how life originated.

About 3.5 billion years ago the first continents would have been differentiated, and some of the rocks preserved from this time contain single-celled bacteria-like fossils that are composed of small rods and spheres. The first primitive life was dominated by blue-green algae, a group of bacteria that formed mound-like structures similar to modern organisms known as stromatolites (Figure 8.13a).

About 2 billion years ago, during the Proterozoic eon ("earlier life"), oxygen began to accumulate in the atmosphere in greater quantities (as interpreted by the presence of oxidized minerals in ancient rocks), and life evolved beyond primitive bacteria (see Chapter 14 for more on the evolution of the atmosphere). Multicellular life did not show up until about 1.7 billion years ago. The earliest things that we might call animals appeared in the fossil record about 580 million years ago in the Ediacaran time. This was a good 40 million years before the Cambrian period, which is famous for its so-called "explosion" of life. The Ediacara animals were weird-looking marine creatures that were mostly soft-bodied. They have been thought to be related to jellyfish, mollusks, and sponges. However, their forms are so strange-looking that it is difficult for paleontologists to reconstruct just what they were.

The beginning of the last eon, the Phanerozoic ("life revealed"), marks the break between older rocks in which fossils are rare and younger rocks that contain abundant fossils formed from organisms with shells and hard skeletons. Things got more recognizable, but not by much. Many of the major animal groups can be found in the Cambrian period, even though they did not look anything like their descendants do today. The Phanerozoic eon represents just the last 12 percent of geologic time. Geologists often use the informal term *Precambrian* to refer to all geologic time prior to the Phanerozoic.

Rock units of the Phanerozoic eon have been assigned to three shorter eras (Paleozoic, Mesozoic, and Cenozoic) of varying lengths based on the characteristics of their fossils (Figure 8.14; Table 8.1). Paleozoic is derived from the Greek for "ancient life," Mesozoic means "middle life," and Cenozoic represents "new life." Each of the eras is further subdivided into shorter time intervals known as periods (Table 8.1). The beginning of the oldest period, the Cambrian, signaled an explosion of diversity within the biosphere. On average, **the number of species represented by fossils increases from the Cambrian period onward.** Although all major phyla were derived by the Cambrian, the array of fossils preserved in rocks formed during the Phanerozoic eon suggest that the diversity of organisms has increased through time.

Table 8.1	The Geologic Timescale	
Era	**Period/Remarks**	**Time (million years ago)**
CENOZOIC	Quaternary - Large ice sheets centered on Earth's poles. Modern humans appeared. Some large mammals (for example, mastodons, mammoths) became extinct.	2.5–0
	Tertiary* — Neogene - Large mammals abundant (for example, mastodons, mammoths); earliest human ancestors appeared. North and South America were connected.	23–2.5
	Tertiary* — Paleogene - Age of mammals began. First large mammals on land and in the ocean (for example, whales, elephants, horses, bears); widespread expansion of echinoids.	65.5–23
MESOZOIC	Cretaceous - Dinosaurs dominated Earth. First grasses and flowering plants appeared; dinosaurs wiped out by an extinction event at the end of the period; ammonites abundant.	145–65.5
	Jurassic - Atlantic Ocean began to form. First birds appeared; dinosaurs and flying reptiles (pterosaurs) common.	200–145
	Triassic - Pangaea began to split apart. Reptiles dominated Earth; earliest mammals and dinosaurs appeared; first conifers; modern corals developed.	251–200
PALEOZOIC	Permian - Much marine life was wiped out in the most massive extinction known, at the end of the Permian.	299–251
	Pennsylvanian† (Late Carboniferous) - Appalachian Mountains formed when North America collided with Africa. Insects and early reptiles on land; first evergreen trees appeared, earliest forests.	318–299
	Mississippian† (Early Carboniferous) - Shallow tropical oceans covered much of the interior of North America. Marine fossils in limestone common.	360–318
	Devonian - Age of fishes. First land vertebrate animals (tetrapods: amphibians); first seed plants, trees, and forests. Insects on land as well.	416–360
	Silurian - First fish with jaws appeared; much of North America was under a shallow tropical sea with abundant reefs. First primitive plants occur on land (ferns, mosses).	444–416
	Ordovician - North America was near the equator. Abundant marine life.	488–444
	Cambrian - Explosion of organisms with hard skeletons (which can be easily preserved as fossils) occurred at beginning of Cambrian; trilobites flourished.	542–488
PRECAMBRIAN	Fossils rare in Precambrian rocks. Soft-bodied organisms present in the youngest Precambrian rocks; dominated by single-celled life.	4,600–542

* Authorities have recently recommended that the name Tertiary be replaced by two new periods, the Neogene and Paleogene. It will take several years before this change is widely accepted.
† These periods are used widely in North America but are combined to form the Carboniferous period internationally.

time of formation). As time passes, the amount of the parent isotope decreases, and the quantity of the daughter atoms increases (Figure 8.21). After one half-life, 50 percent of the parent remains, and the other 50 percent of the original atoms are converted to daughter atoms (Table 8.2). After two half-lives, the number of parent isotopes is again halved (25 percent), and the number of daughter atoms increases by an equivalent amount (to 75 percent).

The relative proportion (ratio) of parent isotopes and daughter atoms can be used to determine how many half-lives have passed since the formation of the mineral that contained the parent radioactive isotopes (Table 8.2; Figure 8.21). The half-lives for several of the most common radioactive isotopes are measured in millions or billions of years (Table 8.3). Isotopes with half-lives of billions or hundreds of millions of years can be used to determine the ages of the oldest rocks on Earth.

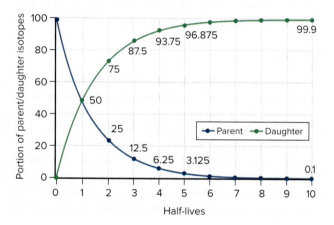

Figure 8.21   Change in percentage of parent and daughter isotopes undergoing radioactive decay. Notice that after about 10 half-lives, very little parent material remains.

Table 8.2	Proportion of Parent and Daughter Isotopes as a Result of Radioactive Decay	
Number of Half-lives	Parent Isotopes (% total isotopes)	Daughter Isotopes (% total isotopes)
0	100	0
1	50	50
2	25	75
3	12.5	87.5
4	6.25	93.75
5	3.125	96.875

Table 8.3	Half-Lives for Common Radioactive Isotopes	
Parent isotope	Daughter product	Length of half-life
Uranium-238 (^{238}U)	Lead 206 (^{206}Pb)	4.5 billion years
Uranium-235 (^{235}U)	Lead 207 (^{207}Pb)	704 million years
Rubidium-87 (^{87}Rb)	Strontium 87 (^{87}Sr)	48.8 billion years
Potassium-40 (^{40}K)	Argon 40 (^{40}Ar)	1.25 billion years
Carbon-14 (^{14}C)	Nitrogen 14 (^{14}N)	5,730 years

Let us look at an example of how dating can be done using half-lives: The calculated age of some rocks in northwestern Canada is approximately 4 billion years. The half-life for ^{238}U isotopes is 4.5 billion years; therefore, these isotopes would have experienced a little less than one half-life in a 4 billion-year-old rock. Such rocks should have a few more ^{238}U parent isotopes (55 percent) than ^{206}Pb daughter atoms (45 percent; see Tables 8.2 and 8.3). In comparison, the half-life for ^{235}U to ^{207}Pb decay is 0.7 billion years; ^{235}U isotopes in a 4 billion-year-old rock would have experienced nearly six half-lives and would have approximately 2 percent of the ^{235}U parent isotope remaining and 98 percent of ^{207}Pb. Not all radioactive isotopes decay over such long periods. The carbon isotope, ^{14}C, experiences a half-life every 5,730 years; ^{14}C is used to date archaeological artifacts such as cloth fragments from Egyptian pyramids and some very young sedimentary rocks (younger than 50,000 years of age) that have previously living material trapped within them.

Applying Both Relative and Numerical Time

Because sedimentary rocks are composed of recycled minerals from other rocks, radioactive isotopes found in rocks like sandstone cannot be used to date the formation of the sedimentary rock itself. The minerals in those rocks would simply yield the dates for their igneous or metamorphic source rocks. Sedimentary rocks are arranged in relative chronological order by using the relative dating methods discussed in Section 8.2. Therefore, **the numerical age of sedimentary rocks is most accurately estimated by using the cross-cutting relationships for igneous intrusions and the principle of superposition for layers of volcanic ash or lava flows** (Figure 8.22).

Figure 8.22   Relative time and numerical time elements can be used together to determine the ages of sedimentary rocks. From the information in this figure, we can estimate that beds 1 through 5 are greater than 70 million years old because they are cut by all the intrusions and must be older than the oldest intrusion. The bottom of bed 9 lies on an unconformity that eroded the oldest intrusion (70 million years old), but bed 9 is cut by a younger intrusion (55 million). Therefore, bed 9 must be 55 to 70 million years old. Bed 10 is younger than bed 9 (superposition), but the relationship between the 55 million years intrusion and bed 10 is unresolved because we can't tell from the information available if the bed truncates the intrusion or the intrusion cuts the bed. However, the bed is overlaid by a lava flow dated at 30 million years old, so we can estimate the age range for bed 10 as between 30 and 70 million years old.

Dates can be obtained for the igneous rocks that will provide minimum or maximum ages for the formation of the adjoining sedimentary layers. Although highly effective, such techniques are limited to regions where there are enough igneous rocks to provide a sufficient range of ages to accurately estimate the ages of the associated sedimentary rocks.

One of the greatest challenges for earth scientists who worked to establish accurate dates for geologic events was to determine when organisms with shells or skeletons first became abundant on Earth. These were the first animal remains to be readily preserved as fossils in rocks across the world. These rocks represent the beginning of the Cambrian period at the start of the Phanerozoic eon. Geologists have long sought to limit the age of these rocks by dating the youngest igneous rocks immediately below this boundary or the oldest igneous rocks cutting across units found above the contact between the oldest Cambrian and the youngest Proterozoic rocks. There are several locations around the world where fossil-bearing Cambrian rocks are interlayered with volcanic rocks. Individual minerals within these volcanic rocks were dated at 521 to 539 million years old. Ash deposits, lavas, and plutons in the youngest Proterozoic rocks are more than 543 million years old (Figure 8.23). Recent analyses of volcanic rocks immediately adjacent to the oldest Cambrian rocks have yielded an estimated date of 542 million years ago for the beginning of the Cambrian period.

Checkpoint 8.14

basic · advanced · intermediate · superior

Radioactive isotopes in clastic sedimentary rocks always predict an age that is

  a. older than the sedimentary rock.
  b. younger than the sedimentary rock.
  c. correct for the sedimentary rock.

The isotope of element X has 15 protons, 17 neutrons, and 15 electrons. The element has an atomic number of _____ and a mass number of _____.

  a. 15; 32
  b. 17; 15
  c. 17; 47
  d. 15; 30

If radioactive decay began with 400,000 parent isotopes, how many would be left after three half-lives?

  a. 200,000
  b. 100,000
  c. 50,000
  d. 25,000

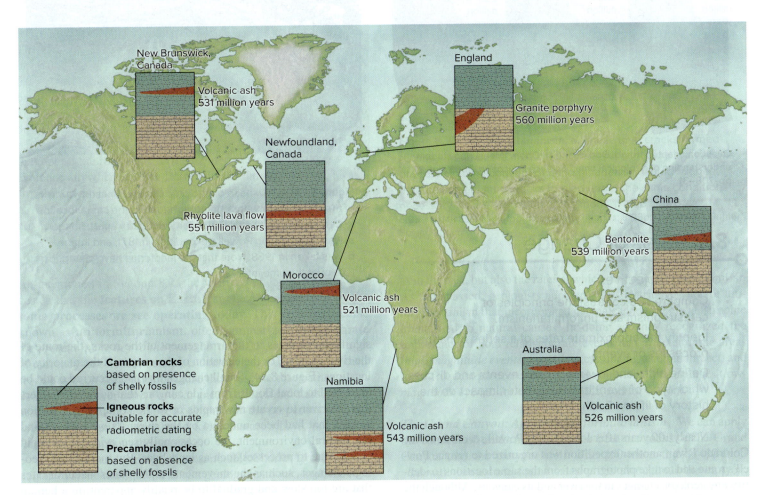

New Brunswick, Canada
Volcanic ash 531 million years

England
Granite porphyry 560 million years

Newfoundland, Canada
Rhyolite lava flow 551 million years

China
Bentonite 539 million years

Morocco
Volcanic ash 521 million years

Australia
Volcanic ash 526 million years

Namibia
Volcanic ash 543 million years

Cambrian rocks based on presence of shelly fossils

Igneous rocks suitable for accurate radiometric dating

Precambrian rocks based on absence of shelly fossils

Figure 8.23   Deciphering sequences. Relative time and numerical time elements can be used together to determine the ages of sedimentary rocks. From this information, we can establish the age of the start of the Cambrian period as 542 million years ago.

WEATHERING

FROST WEDGING
Rocks break apart when water freezes and expands in rocks.

LICHENS
Release acids that break down bare rock surfaces in mountains.

ACID MINE DRAINAGE
Oxidation of metallic minerals produces acidic waters.

DECAYING PLANT MATERIAL
Microscopic organisms break down organic material to release gases.

SINKHOLE

TREE ROOTS
Wedge apart fractures in the bedrock.

CAVES
Minerals are dissolved to form caves. Some of the dissolved minerals are precipitated to form stalagmites and stalactites.

PHYSICAL WEATHERING
Represents the disintegration of rocks and minerals into smaller pieces.

CHEMICAL WEATHERING
Represents the decomposition of rocks by the chemical breakdown of minerals.

BIOLOGICAL WEATHERING
Represents the actions of organisms that remove or break down rocks and minerals.

ACID RAIN
Rainwater in combination with gases released from burning fossil fuels.

UNLOADING
Pressure release cracks form parallel to the surface.

SALT WEDGING
Growing salt crystals push apart grains.

HYDROLYSIS
Feldspar converted to weaker clay minerals.

CHITON & SEA URCHINS
Excavate holes in bedrock under shallow marine conditions.

DISSOLVED MINERALS
Carried to sea by streams and groundwater.

"The term 'natural disaster' has become an increasingly anachronistic misnomer. In reality, human behavior transforms natural hazards into what would really be called unnatural disasters."

—Kofi Annan, former secretary general of the United Nations

© Juan Herrero/EPA/Newscom

Chapter

10

Landslides and Slope Failure

the big picture

Would you live here?

See The Big Picture box at the end of this chapter for the full story on this image.

267

LANDSLIDES

What type of landslide is this?

TOE

ROCKFALL
Rocks falling when dislodged from a cliff or steep slope as a result of physical weathering.

ROCKSLIDE
The rapid movement of a mass of rock downslope along a plane of weakness.

SCARP

Is this home in danger of a future landslide?

SLUMP
Downward rotational movement of blocks of unconsolidated material along curved surfaces.

CREEP
The very slow movement of unconsolidated material downslope under the influence of gravity.

11.1 Humans and Rivers

Chapter Learning Outcomes

- Students will recognize that water continuously cycles through the earth system.
- Students will connect the physical characteristics of a stream to the dynamic nature of stream flow and the underlying geology.
- Students will evaluate the risk to society from floods.

The year 2015 will go down in history as a bad year when it comes to flooding. In May of that year, Texas and Oklahoma were hit with historic flooding associated with a slow moving storm system. On October 18, 2015, Death Valley California received 2.7 inches of rain in one day—that is more than the area normally experiences in a year! Roads were washed out, historical monuments were damaged, and large sections of the park were closed for several weeks. Water from this event that pooled in normally dry lake beds is apparent in images taken after the torrential rainfall (Figure 11.1). Fortunately, no one was killed or injured during the Death Valley flash flood. Two months later, in late December, more than 18 million people along the Upper Mississippi experienced flooding that was in places as much as 4 feet higher than had ever been previously recorded. Twenty-four people died in this flood—most as they were washed away in their cars. Damage was in the billions of dollars. These cases illustrate the power of streams and the floods that come with them.

Geologists use the term *stream* to account for **any flow of water through a channel** (defined by its bed and banks), from the smallest creek to the largest river. As the stories above illustrate, rivers have the potential to cause death and destruction, but they also perform many positive functions. Streams bring water to irrigate crops, supply drinking water, provide coolant for power plants, transport barges, create ecosystems for wildlife, and serve as sites for recreation.

The Nile River: An Example of Stream Impact

The longest river on Earth, the Nile River, will serve as a model to illustrate the impact a stream can have on civilization. Stretching nearly 7,000 kilometers (4,350 miles) from the highlands of central Africa to the southern coast of the Mediterranean Sea, the Nile reaches across eight nations and 35 degrees of latitude (Figure 11.2a). For a single North American river to have similar length, it would begin near the Arctic Circle in Alaska and flow to the Gulf of Mexico. (It is a common misconception that the Nile is the only river that flows north. Rivers flow downhill, not from north to south or any other particular direction.)

Egypt is a desert state with just a few centimeters of rainfall each year. However, with ready access to waters from the Nile River (Figure 11.2b), Egypt became one of the earliest sites of civilization, characterized by organized agriculture, extensive cities, and a system of state governments. Even today, almost all of Egypt's agriculture and modern cities are located within the Nile valley. Most of the water that flows through Egypt originates as rainfall at elevations of around 1,800 meters (6,000 feet) much farther south, in other nations.

Civilization in Africa would have had a much rougher start without the Nile River. Humans began diverting water from the river to irrigate the desert more than 5,000 years ago. Farmers depended on seasonal variations that divided the year into times of flooding and times of low flow. Empires expanded and

Figure 11.1 A pair of images of Death Valley, California, illustrating contrasts between normal, dry conditions and flood conditions. The false color images emphasize aspects of the hydrogeology of the region. The areas in green and blue are the locations with greater moisture content. a) October 2014 when there was standard low precipitation throughout Death Valley; b) October 2015, note the abundant water covering in Badwater Basin in the eastern part of the image. This area is normally a dry lakebed.
11.1a & 11.1b: NASA

Figure 11.2a The most distant tributary of the Nile River is the Kagera River, just south of the equator in the nation of Burundi. The Kagera flows north through a series of lakes to become the White Nile and joins the Blue Nile and Arbara rivers in Sudan to form the Nile River. Lake Nasser formed behind a dam that was built at Aswan in southern Egypt to regulate the flow of the Nile for agriculture and domestic uses. The Nile is 6,825 kilometers (4,241 miles) long and drops 1,800 meters (5,905 feet) in elevation from its source to its mouth.

Figure 11.2b Egypt and the Nile River. The Nile River is the thin, sinuous ribbon that divides the eastern and western parts of Egypt. The delta along the southern coast of the Mediterranean Sea is the triangular dark-green region near the top of the image.
Jacques Descloitres, MODIS Land Science Team/NASA

the pyramids were built during times of ample water; during prolonged droughts, empires collapsed and the region experienced famine and sandstorms. (How would your life change if you did not have access to an adequate water supply?) The history of the Nile illustrates that societies are stressed when water becomes scarce. Climatologists note that increasing global temperatures will result in more evaporation, causing more precipitation at high latitudes and dryer conditions in some tropical areas. Some of the most significant future social changes will have more to do with declines in water supplies than with increasing temperatures.

Stream Management

While we are relatively powerless to prevent periods of drought or heavy rains, thoughtful management of our stream systems can help minimize the negative consequences of these events. The history of water in the twentieth century includes a record of extensive dam building where giant walls of concrete tamed big, wild rivers to provide steady water supplies, to control floods, and to generate electricity. Water power is by far the most widely used "green" energy source in the world. In the United States, it generates about

eight times as much power as solar and wind energy combined. Just as in Egypt, the control of western rivers has allowed agriculture and cities to flourish in states characterized by desert climates. This clean energy source and steady water supply is threatened by

Self-Reflection Survey: Section 11.1

Answer the following questions as a means of uncovering what you already know about streams and floods.

1. What river is nearest to where you live? Where does the river begin and where does it end?

2. How do people interact with streams? How do streams fit within the context of biology, economics, culture, politics, history, recreation, and aesthetics?

3. Use maps and other information sources to identify rivers that act as boundaries for US states or Canadian provinces. Find at least three examples of major rivers around the world that start in one country and end in another.

climate change that is expected to lower water levels in reservoirs throughout the southwestern United States.

In this chapter, we will discuss where the water in rivers comes from when it is not raining, why our eyes may deceive us when we interpret how fast streams flow, and what we need to know to avoid having to give birth in a tree. By the time we are finished, you will have a new way to classify rivers that transport one of our most basic resources, freshwater. As we will discover, large stream systems are influenced by plate tectonic processes. We will begin by examining how stream systems are connected to the rest of the earth system and gradually focus on processes within a single stream before putting it all together to review the impact of floods on society.

11.2 The Hydrologic Cycle

Learning Objectives

- Draw a labeled sketch of the water (hydrologic) cycle.
- Explain how water is distributed among different components of the earth system.
- Describe the origins of streams.

Where does the water in streams come from? It is tempting to think that stream channels just sit there waiting for rain to fall and fill them up. And yet, if you think about it, most of the time it doesn't rain. So how is it that there always seems to be water in large streams?

In reality, the distribution of water on Earth (and in streams) depends on the interactions among the four components of the Earth system: the geosphere, atmosphere, hydrosphere, and biosphere. **Water moves in and around the earth system, changing from one physical state (liquid, solid, water vapor) to another, in a sequence known as the hydrologic cycle** (Figure 11.3). Any form of water (rain, snow, ice) that falls from the sky is precipitation. Most precipitation falls into the ocean, but some falls on the land surface. Precipitation may run off the land (in streams) or move down through the surface (infiltrate) to become groundwater. A majority of precipitation reenters the atmosphere as water vapor through the process of evaporation. Evaporation occurs directly from any place where water is exposed to the air. Water vapor also enters the air indirectly when it is released from plants, a process called *transpiration*. Figure 11.3 summarizes these processes, and in Chapter 14, we will discuss how much energy must be absorbed or released for water to change from one state to another (such as when ice melts or water freezes).

Oceans are the ultimate source and final destination for all water on or below the land surface (Figure 11.4a). Most water on Earth (about 97 percent) is in the oceans, a smaller amount (about 3 percent) resides in and on land (Figure 11.4b), mostly as freshwater (water having less than 0.05 percent salts). A very small amount of water (water vapor or ice) is in the atmosphere (less than 0.01 percent). Since most water is exposed to the atmosphere in oceans, the bulk of evaporation (85 percent) occurs there (Figure 11.4c). The greatest evaporation rates occur over areas of warm temperatures at low latitudes. The circular path of the hydrologic cycle links evaporation, transpiration, precipitation, runoff, and infiltration.

Figure 11.3 The hydrologic cycle. Water evaporates from the ocean and rises through the atmosphere, where it condenses to form clouds. The clouds release precipitation that may flow overland in streams, sink underground as groundwater, or be absorbed by plants.
Adapted from Charles (Carlos) Plummer, *Physical Geology* 11e © 2007, reproduced with permission of McGraw-Hill Education

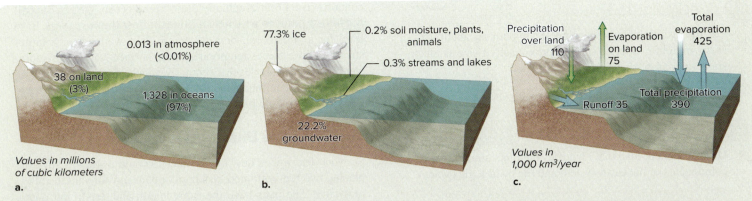

Figure 11.4 Water balance on Earth. **a.** The vast majority of Earth's water is in the oceans, with about 3 percent of Earth's water existing on land. **b.** Most of the water on land is stored as ice or groundwater. Streams and lakes are often supplied by water from melting ice in glaciers or from groundwater. **c.** Land areas receive more moisture by precipitation than they supply by evaporation. The difference is made up by evaporation from oceans. This excess water is returned to oceans by surface streams.

Figure 11.5 Where streams come from. **a.** The Colorado River begins as a small stream on the slopes of Rocky Mountain National Park, Colorado. **b.** The Allegheny (left) and Monongahela (right) rivers join to form the Ohio River in Pittsburgh.
11.5a: © Pat O'Hara/Corbis; 11.5b: © Andrew Harrer/Bloomberg via Getty Images

The Origin of Streams

Melting snow (known as meltwater) and rain run off the slopes of Long's Peak, Colorado, becoming a source of freshwater. That water flows into almost invisible depressions and carves out small channels that join to form a permanent stream, the source of the Colorado River (Figure 11.5a). More than 30 million people depend on water from the Colorado River, much of which originates as snowmelt. Earth's lakes store twice the freshwater of streams and serve as the source (or destination) for many rivers. For instance, the source of the Mississippi River is a relatively small lake (about 5 square kilometers; 2 square miles) in Minnesota.

Nearly one-third of the water falling as precipitation over land completes the circuit to the oceans by surface runoff in streams. (How long does it take for water that falls as precipitation to reach the ocean?) On average, **the residence time, the length of time that a given volume of water remains in streams,** is about 14 days. Because rates of water infiltration are slow compared to rates of runoff, only a small amount of the precipitation soaks into the surface to join water stored in rocks and sediment.

Checkpoint 11.1

basic · advanced · intermediate · superior

Sort the following 12 terms into six pairs of terms that most closely relate to one another. Explain your choices.

groundwater	plants	transpiration
stream	ice	infiltration
rainfall	precipitation	water vapor
gas	meltwater	runoff

Checkpoint 11.2

basic · advanced · intermediate · superior

Imagine that it rained continuously all over the world for a month. If we were to measure the depth of the oceans over a 5-day period near the end of the month, what would we observe? Explain your choice.

a. Ocean depths rise steadily.
b. Ocean depths fall steadily.
c. Ocean depths stay the same.

Checkpoint 11.3

Use the hydrologic cycle to suggest a hypothesis about why rivers in South America carry approximately twice as much freshwater as do rivers in North America.

 Checkpoint 11.4

Draw a concept map that identifies the links among the components of the earth system and the hydrologic cycle.

However, the volume of water stored in groundwater systems is about 70 times greater than the volume of water in streams and lakes (see Figure 11.4b). Likewise, ice is a huge freshwater storage system comprising more than three times the volume of water found in groundwater. Consequently, it is no surprise that many streams have their origins in glacial meltwaters (for example, the Brahmaputra and Ganges rivers, India; see Figure 7.24a) or in springs that bring groundwater to the surface (for example, the Nile River). Finally, we should note that many large rivers form when two smaller rivers join together, such as when the Allegheny and Monongahela rivers join in Pittsburgh to form the Ohio River (Figure 11.5b).

Regardless of where or how a stream begins, water flows downslope and joins, or is joined by, other streams to form a network. For instance, the Mississippi River's source (called the headwaters) is a small stream about 1 meter (3.3 feet) deep and 10 meters (33 feet) wide when it leaves Lake Itasca in Minnesota. Many more streams (some of them large rivers in their own right, such as the Missouri and Ohio rivers) join the Mississippi as it makes its sinuous trek southward. The Mississippi River ends as a large river 5 meters (16 feet) deep and over 500 meters (1,640 feet) wide as it empties into the Gulf of Mexico at New Orleans. Section 11.3 explores the interconnected nature of streams in drainage basins and how flow in one part of a stream system may affect streams in the rest of the basin.

11.3 Drainage Networks and Patterns

Learning Objectives

- Define what is meant by the terms *base level*, *drainage basin*, and *drainage divide*.
- Estimate the approximate locations of drainage basins and drainage divides on a regional map.
- Discuss the origin and persistence of drainage basins in the context of other geological processes.
- Explain how drainage patterns are governed by the underlying geology of a region.

Think about a stream or river near you that flows all year long. Where does the water flowing in that stream end its journey? Approximately two-thirds (68 percent) of Earth's land surface

contains rivers that eventually drain to the ocean. Geologists call the lowest point to which a stream can flow the *base level*. Base level for a stream can be a lake, reservoir, wetland, or a larger stream. Those places temporarily store surface water, thus representing a local base level. Eventually, much of the water in those locations will continue its journey to the ocean. The ocean is considered the ultimate base level.

The Drainage Basin or Watershed

So what determines how much water flows in a stream? Imagine placing two flat-bottomed, rectangular pans in your yard to collect rainwater. The pans have the same length and depth, but one is twice as wide as the other. How will the volume of water in each pan compare after the rain? Which one (if either) will have more water? The pan with the larger surface area will have a larger volume of water. The larger pan "drains" a larger area and so collects more precipitation. Similarly, the amount of water in a stream channel is related to the size of the area it drains and the average precipitation over that area.

The area drained by a stream and its smaller tributary streams is called a *drainage basin* or *watershed*. Drainage basins for the Mississippi and Nile rivers, two of the largest rivers in the world, occupy approximately 3 million square kilometers (1.16 million square miles), equivalent to one-half the land area of the United States. The two principal tributaries of the Mississippi are the Missouri and Ohio rivers. The drainage basins for each of these (and other) rivers are included within the larger Mississippi drainage basin (Figure 11.6). The Mississippi River is separated from streams that drain to the west by a drainage divide (a regional topographic high) located in the Rocky Mountains. Streams west of the Rockies flow to the Pacific Ocean or the Gulf of California. Streams in the eastern United States flow to the Gulf of Mexico or the Atlantic Ocean.

Drainage divides are found along the high ground separating drainage basins. For example, a drainage divide in Ohio separates streams that drain northward to the Great Lakes from streams that flow south to the Ohio River (Figure 11.7). The city of Akron in northeast Ohio straddles the divide between the Cuyahoga and Tuscarawas rivers. Most runoff directly north of Akron enters the Cuyahoga River, which drains to the north because it is part of the Great Lakes basin. Runoff south of Akron flows into streams that empty into the Tuscarawas River.

 Checkpoint 11.5

Why is the volume of water in the Mississippi River about 10 times greater than the volume of water in the Nile River?

a. The Mississippi River drainage basin is 10 times bigger than the Nile basin.
b. The Mississippi River drainage basin receives more precipitation.
c. The Mississippi River is a longer stream.
d. There is less vegetation to absorb precipitation in the Mississippi River drainage basin.

The Tuscarawas River drains south as part of the Muskingum drainage basin, and its water flows into the Ohio River, a major tributary of the Mississippi River. Clearly, precipitation falling on different sides of a drainage divide can take significantly different routes to its ultimate destination.

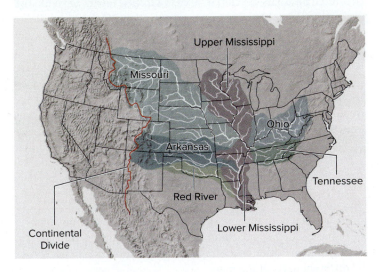

Figure 11.6 Mississippi River drainage basin. The different colors represent smaller drainage basins that make up the Mississippi River drainage basin.
Adapted from Charles (Carlos) Plummer, *Physical Geology* 11e © 2007, reproduced with permission of McGraw-Hill Education

Figure 11.7 A drainage divide and selected drainage basins, Ohio. Streams north of the drainage divide flow to the Great Lakes and ultimately through the St. Lawrence Seaway to the Atlantic Ocean. Streams south of the divide flow to the Ohio River, which joins the Mississippi River and flows to the Gulf of Mexico. Note that the drainage divide is also visible on the map of the Mississippi River basin in Figure 11.6.

Checkpoint 11.6

Rivers in Iowa flow to either the Mississippi River, which makes up the eastern state border, or the Missouri River, on the western state border. Draw the approximate position of the drainage divide for the Missouri and Mississippi basins in Iowa on the map provided here.

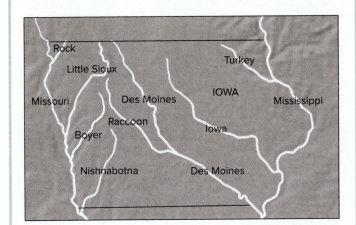

Checkpoint 11.7

Examine the following map of rivers in southern Georgia. Draw the approximate boundaries of the drainage basins for the named streams. Note that the Savannah and St. Marys rivers mark parts of the boundary of the state. Do not include the Savannah or St. Marys rivers' drainage basins.

Evolution of Stream Systems

Large drainage basins are long-lived features on Earth. We have previously discussed the great length of time needed for the Colorado River to carve the Grand Canyon (millions of years). The largest rivers often drain relatively stable highlands formed by uplift linked to slow, gradual plate tectonic processes. For example, the Amazon River originated in its present form approximately 10 million years ago as the Andes Mountains grew tall along the northern and western margins of the South American continent. The northern half of the continent was originally drained by a north-flowing stream system that carried water to the Caribbean Sea (Figure 11.8a). These streams deposited light-colored sediment eroded from granite rocks in the continental interior.

As the mountains rose, shutting off drainage to the north, the Amazon developed as one of a series of rivers eroding the Andes Mountains and feeding a vast expanse of interior lakes and wetlands (Figure 11.8b). These rivers transported material from metamorphic rocks exposed in the mountains. The tropical swamp conditions were ideal for the development of layers of immature coal deposits (lignite) that alternated with clastic sedimentary rocks. As the western side of the continent continued to rise, the wetlands were eventually drained by the east-flowing Amazon River and its tributaries. The Andes sediments began showing up in layers deposited along the Atlantic continental shelf about 10 million years ago (Figure 11.8c). Scientists have interpreted this to mean that it took several million years for the modern Amazon River drainage basin to become established.

Drainage Patterns

Streams typically follow the path of least resistance, forming valleys where rock is most readily eroded or following the steepest slope gradient. Consequently, geologists have identified four basic stream patterns that relate to the underlying geology: dendritic, rectangular, radial, and trellis (Figure 11.9). Dendritic (treelike) drainage patterns are characteristic of areas where the geology is relatively uniform, as when rock layers are horizontal or where the underlying rocks have similar properties. When viewed on a map, dendritic streams intersect with a characteristic V pattern with the tip of the V pointing downstream. Dendritic

patterns are typical of streams in the Mississippi River drainage basin and in Ohio streams (see Figure 11.7) and the Amazon River (Figure 11.8).

On a regional scale, streams sometimes intersect at right angles, forming trellis or rectangular drainage patterns. Trellis drainage patterns develop in areas of folded layers of weak and resistant rocks. Main streams occupy valleys, flowing parallel to the ridges, but occasionally cut across the geologic grain at water gaps. Smaller streams flow down the valley slopes perpendicular to the course of the main stream channel. Rectangular drainage patterns occur in regions where the streams are controlled by joints, fractures in the underlying bedrock. Streams exploit the fractures as planes of weakness in the rock. Finally, radial patterns are typically found on volcanoes, where streams flow directly downhill in all directions, away from the summit (Figure 11.9d).

✓ Checkpoint 11.8

What type of drainage patterns can you observe in this map of part of the Appalachian Mountains and adjacent areas? Explain any differences in patterns that you observe in different parts of the map.

Figure 11.8 The evolution of the Amazon River. **a.** Twenty-five million years ago, much of the northern half of South America was drained by a north-flowing stream system. **b.** About 15 million years ago, rivers fed an extensive interior basin filled with lakes and wetlands. **c.** The patterns of the modern river became established about 10 million years ago.
11.8a, 11.8b, 11.8c: Adapted from Carina Hoorn, "The Birth of the Mighty Amazon," *Scientifc American,* April 23, 2006, p. 57

a. Dendritic

b. Trellis

Ridge Valley

c. Rectangular

Fractures

d. Radial

Volcano

Figure 11.9 Stream patterns. The four principal types of drainage patterns are related to the underlying regional geology.
11.9a, 11.9b, 11.9c, 11.9d: Adapted from Charles (Carlos) Plummer, *Physical Geology* 11e © 2007, reproduced with permission of McGraw-Hill Education

11.4 Factors Affecting Stream Flow

Learning Objectives

- Discuss how the characteristics of a stream would change moving from its headwaters to its mouth.

- Explain how the stream gradient, channel roughness, and wetted perimeter influence stream velocity.

- Define the term *stream discharge* and explain how it is calculated.

Think for a moment about the last time you saw a stream or river. How would you describe that river to a friend? Chances are, you would mention the speed and color of the water, the size of the stream, and perhaps the landforms around the stream. Hydrologists, scientists who study water, do much the same thing when they study streams and rivers. They examine a variety of characteristics, including how fast the water flows and the size, slope, and roughness of the stream channel. All of these factors affect how the stream flows and consequently influence the evolution of surrounding landforms.

Stream Gradient

Why do streams flow, anyway? Water flows downhill, so streams simply carry water from higher elevations to lower elevations. **The slope of a stream (known as the stream gradient) is the change in elevation of the stream over a horizontal distance.** As you might expect, the highest gradients are found in steep-sided mountain stream valleys that may drop 40 to 60 meters for every kilometer (210 to 320 feet per mile) of stream length. The gradient would be 40 to 60 meters per 1,000 meters or, expressed as a percentage, 4 to 6 percent.

Where would you find the highest gradients for most streams—at the beginning (headwaters) or end (mouth) of the stream? The stream gradient gradually decreases along the length of a stream, from the headwaters (where it is steepest) to the mouth (Figure 11.10). Streams approaching the river mouth may decrease in elevation by as little as a few centimeters over 1 kilometer

Narrow V-shaped valleys

Steep gradient (~60 m/km)

Gentle gradient (<1 m/km)

Meandering stream channel

Delta

Figure 11.10 Variations in stream gradient. Maximum gradient occurs in mountainous areas located near the source; minimum gradient occurs at the stream's mouth.

(gradients of about 0.001 percent). The Nile River in Egypt has a gradient of about 8 centimeters per kilometer (5 inches per mile), and the last 1,000 kilometers (620 miles) of the Amazon River in Brazil drops just 1 centimeter per kilometer (less than half an inch per mile). So, how does the gradient relate to the stream velocity?

Stream Velocity

Anybody who has ever gone white-water rafting will tell you that much of the fun comes from being hurled from side to side as the raft bounces its way through the rapids of a mountain stream (Figure 11.11). That experience gives the impression that the raft, and hence all of the water in the stream, is moving fast. In contrast, if we walk or drive alongside rivers with low gradients, they appear to be moving relatively slowly—hence the phrase *lazy river*.

When we analyze the average velocity of water moving through a stream (the flow), it turns out that our senses are somewhat deceiving us. The average flow of water in the high-gradient part of a stream is typically slower than that through the low-gradient part of the stream. How can that be so? What makes a high-gradient stream different from a low-gradient stream? Look closely at Figure 11.11. How might the features you see there compare to what you would observe in a low-gradient river? The banks along the steep mountain streams typically have rocky channels. As one continues downstream, the streambed composition changes from gravel, to sand, and eventually to fine-grained silt and mud. This characteristic of the stream channel is called the *channel roughness*.

How do you think the presence of boulders would affect the flow of the stream? Think of the water as going through a maze. The boulders cause the water to become turbulent (chaotic). It goes forward, then sideways, and sometimes even backward before it moves forward again. Compare that to the flow through the

downstream portion of the river where the channel is smooth. Flow there is streamlined (laminar) because there are few obstacles in the channel to disrupt the flow. The channel roughness found in a high-gradient stream actually reduces the average stream velocity when compared a low-gradient stream.

But wait, there's more—a third characteristic, cross-sectional area, is involved in determining stream velocity. Cross-sectional area is determined by multiplying the stream's width by its depth, assuming a rectangular channel shape. (We are simplifying here. Most channels have more complex shapes that are carefully measured to determine cross-sectional area.) Consider cutting a slice through the cross section of a stream as represented in Figure 11.12. The bed and banks of the stream channel exert frictional drag on the passing water. The actual amount of drag is proportional to the length of the surface of the channel banks and bed in contact with the water (that distance is called the *wetted perimeter*). In streams

✓ **Checkpoint 11.9**

Explain why stream velocity would change along the same section of a stream at different times of the year.

Figure 11.11   White-water rafting. A raft passes through Dark Canyon Rapids in Cataract Canyon of the Colorado River. © Rob Howard/Corbis

Figure 11.12   Determining stream channel characterisctics. **a.** Cross-sectional area (xs) and wetted perimeter (wp) calculations for a rectangular stream channel. **b.** Three channels with the same cross-sectional area (360 square meters; 3,230 square feet) but different lengths of wetted perimeters. If all other factors were equal, velocity would be highest for stream C.

with the same volume of water and same channel roughness, the channel with the larger wetted perimeter exerts more frictional drag on the water and so has a lower stream velocity (Figure 11.12b). Generally, the larger the stream discharge, the less significant the effect of frictional drag from the wetted perimeter.

The combined effects of these factors (gradient, channel roughness, and wetted perimeter) generally determine the average velocity of a stream. In simple cases, the steeper the gradient, the smoother the channel, and the shorter the wetted perimeter, the faster the velocity. However, these parameters often offset one another. For example, high-gradient streams often have greater channel roughness, while the smooth banks and bed of low-gradient streams exert less friction on the flowing water.

At its source, the Mississippi River has a cross-sectional area of 10 square meters (99 square feet). In contrast, closer to its mouth, it has a cross-sectional area of 2,500 square meters (26,240 square feet). The velocity of the Mississippi River at its source is 0.5 meter per second (1 mile per hour), while it cranks along at a merry 1.3 meters per second (3 miles per hour) as it passes New Orleans on its way into the Gulf of Mexico. Typically, cross-sectional areas of streams increase downstream as more and more water is added to the stream through tributaries. At the same time, channel roughness tends to decrease because particles on the streambed are smaller. Thus, water in a stream moves faster as it moves downstream.

Stream Discharge

In addition to cross-sectional area, just described, there are other ways of assessing a stream's size. In terms of length, the Nile River (6,825 kilometers; 4,240 miles) beats out the Amazon (6,450 kilometers; 4,008 miles) and Mississippi rivers (6,260 kilometers; 3,890 miles). However, we can also measure the volume of water that flows out through the stream. *Stream discharge is the volume (cubic meters or cubic feet) of water that passes a given point in one second.* The discharge is calculated by multiplying the cross-sectional area by the velocity of the stream flow (Figure 11.13).

Discharge = width × depth × velocity

cubic meters per second (m³/s) =
[meters (m)] × [meters (m)] × [meters per second (m/s)]

Imagine placing an empty 1-gallon milk jug in your kitchen sink. Adjust the flow of the faucet so that it takes 1 minute to fill

the jug with water. The volume of water that flows out of the faucet in 1 minute is the discharge. If the gallon container fills in 60 seconds, the discharge from your faucet is 1 gallon per minute. Now repeat the experiment, but reduce the discharge from the faucet so that it takes 22 minutes to fill the gallon jug. You have just modeled the difference between stream discharge in the Amazon and Nile rivers. (For comparison, it would take about 10 minutes to fill the jug using the Mississippi River equivalent.)

Discharge from the Amazon River is about 200,000 cubic meters per second (7 million cubic feet per second) and accounts for 20 percent of all freshwater discharge from streams on Earth. To put it another way, 50 million gallons of water discharge into the Atlantic Ocean from the Amazon River every *second!* We will discuss how measurements of stream discharge become critical in tracking the potential for flooding in Section 11.6. But first, we will examine the impact of changes in stream velocity and discharge on a stream system.

Discharge = 5 × 2 × 10
= 100 m³/s

Figure 11.13 Stream discharge. Discharge is a measure of the volume of water that passes a point in a given time and can be calculated by multiplying the area of the stream channel (width × depth) by the distance traveled in a given time (velocity).

Checkpoint 11.11

Some scientists predict that global warming will result in a corresponding increase in evaporation from the oceans. How would this affect the discharge of the Amazon River?

a. Discharge would increase.
b. Discharge would decrease.
c. Discharge would stay the same.

Checkpoint 11.10

A stream channel narrows between support columns under a bridge. If discharge does not change, predict how stream velocity would be altered as water flowed under the bridge.

a. Velocity would increase.
b. Velocity would decrease.
c. Velocity would not change.

Checkpoint 11.12

Create a concept map that illustrates the connections among the factors that influence stream flow. Include the following eight terms and up to four more of your own choosing.

discharge	velocity	wetted perimeter
depth	gradient	channel roughness
cross-sectional area	width	

11.5 The Work of Streams

Learning Objectives

- Draw a labeled sketch of the materials that make up a stream's load and explain how they are transported.
- Explain why some streams transport more sediment than others.
- Describe where deposition is most likely to occur in association with streams.
- Interpret an image of a stream channel to identify locations of erosion and deposition, and predict the future path of the stream.

The action of streams is continually altering the land surface. Changes occur every day and may be almost imperceptible unless viewed on a timescale measured in decades or centuries.

Erosion

Streams wear away—or erode—the bed and banks of stream channels. Scientists observed an example of stream erosion following the eruption of Mount St. Helens in 1980 (see Chapter 6). That instance also served to answer the question: *Which comes first, the valley or the stream?* The eruption changed the landscape around the volcano, leaving behind thick deposits of ash and mud. Almost immediately, new streams began to form on the surface of these deposits. Initially, the streams cut relatively deep, narrow, channels due to stream erosion (Figure 11.14a). Subsequently, the stream erosion, in combination with mass wasting of the channel banks, gradually widened the stream channels (Figure 11.14b). While most stream channels are composed of much stronger material than the relatively soft unconsolidated volcanic debris, the nature of stream erosion and associated mass wasting is similar, serving to deepen and widen stream channels. Over geologic time, these actions sculpt valleys, gorges, and many other landforms we observe around us today. As we discussed in Chapter 8, the steady cutting of the Colorado River through the rocks of northern Arizona formed the Grand Canyon over millions of years.

All streams erode particles from channel beds and banks and carry them downstream. Larger sand and gravel particles that bounce or roll along the streambed are termed *bed load* (Figure 11.15). Finer particles that are light enough to be carried in suspension higher in the water column are termed *suspended load*.

A form of erosion known as *abrasion* results from the impacts of these large and small particles against the bed and banks of the stream channel. You can think of abrasion as waterborne sandblasting. The force of flowing water in the channel can also push through cracks in rocks to break off chunks of bedrock. At the other end of the spectrum, the water can simply dissolve soluble bedrock such as limestone or rock salt through chemical

a.

b.

Figure 11.14 Effects of erosion on a stream channel. **a.** Approximately 5 months after the eruption of Mount St. Helens, this channel of the upper Muddy River had formed on the soft, unconsolidated volcanic materials. **b.** Note the increase in width and depth of the channel 12 months later. 11.14a & 11.14b: USGS

Figure 11.15 Load carried by a stream. Bed load rolls gravel-sized particles along the bed of the stream channel and bounces sand-sized particles downstream by saltation. Suspended and dissolved loads are carried higher in the stream channel by the flowing water.
Adapted from Charles (Carlos) Plummer, *Physical Geology* 11e © 2007, reproduced with permission of McGraw-Hill Education

weathering (see Chapter 9.3). This material is usually invisible in the stream and is known as the *dissolved load*. **Erosion produces the combination of bed load, suspended load, and dissolved load that can be collectively referred to as the *stream load*.**

Transport

Erosion contributes to the stream load, and so does material delivered by discharge from tributary streams or from surface runoff from the land. How can you tell when a stream has a high sediment load? Think about what the water in a stream looks like after a heavy rain—it appears brown. That color is a result of fine material being transported as suspended load in the stream (Figure 11.16). The suspended load increases when discharge increases (Figure 11.17), because greater discharge is usually associated with higher average velocity that provides more energy to move particles. However, not all streams that carry a significant load are muddy. Streams in which the load is mainly composed of dissolved minerals may appear clear, just as a cup of hot water becomes transparent after a teaspoon of sugar dissolves.

Other factors that influence stream load are the geology and land use of the drainage basin. Some of the 300 million tons of sediment that the Mississippi River deposits in the Gulf of Mexico each year is stripped off fields as far away as Montana and Pennsylvania. You will recall that the Amazon River has 10 times the discharge of the Mississippi River, but it carries only three times the load. What factors might account for this difference? First, the soils of the Amazon basin are much different from those of the Mississippi River basin because of different parent materials and vastly different weathering histories (see Section 9.6). Also, much of the drainage basin of the Amazon is covered by natural vegetation, rain forest that protects the underlying soil from

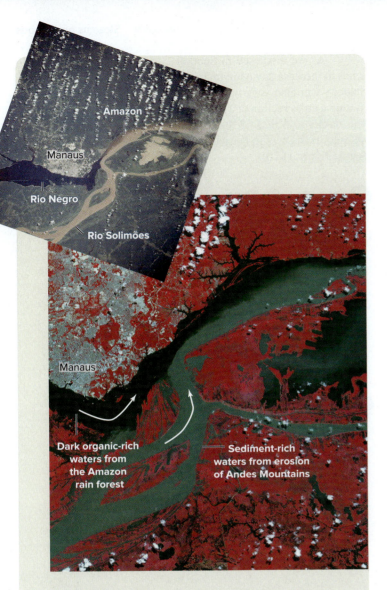

Figure 11.16 Evidence of stream load. View of the confluence of the Rio Solimões and Rio Negro to form the Amazon River near Manaus, Brazil (bottom photo). Note the contrast between the brown, sediment-laden Rio Solimões and its darker, sediment-light tributary. Flow is from the bottom of the image toward the top. Manaus is nearly 1,300 kilometers (800 miles) from the coast but is accessible by oceangoing ships because of the depth of the Amazon River. (Red colors in main image indicate vegetation imaged using the Advanced Spaceborne Thermal Emission and Reflection Radiometer.)
(inset) & (middle) NASA; (bottom) © Rob van Glabbeek

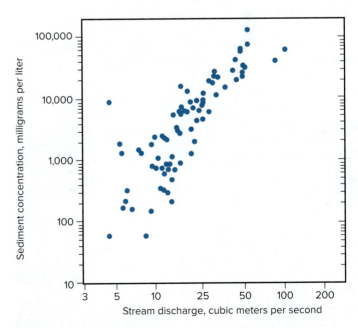

Figure 11.17 Sediment load versus stream discharge for the north fork of the Toutle River near Mount St. Helens, Washington. What is the relationship of sediment load to discharge?
Adapted from USGS Professional Paper 1573: http://vulcan.wr.usgs.gov/projects/sediment_trans/PP1573/HTMLReport/slide35.gif

erosion. In contrast, much of the Mississippi River drainage basin has been cultivated for more than a century, reducing the protection for the soil. Still, the Mississippi cannot match the sediment transport of China's Yellow River (Figure 11.18). That stream gets its name from its heavy load of fine, wind-deposited silt that is easily eroded and carried downstream. The Yellow and Mississippi rivers have similar lengths and discharges, but the Yellow River carries more than five times as much load as the Mississippi River.

Deposition

Now, imagine a stream that is flowing fast enough to transport all sizes of sediment, up to gravel-sized particles. What happens as the stream velocity decreases? Materials such as clay, silt, sand, and gravel that make up the stream load may be deposited in the streambed (just as on a windy fall day when blowing leaves settle from the sky as the wind dies down). Heavier particles drop out first, followed by lighter and lighter materials as the stream velocity decreases. When streams have heavy loads, sediment is often deposited in the stream channel itself to form sandbars or islands. These deposits cause the main channel to split into several smaller channels, forming a pattern known as a *braided stream* (Figure 11.19).

Other depositional features common near rivers are **the broad, flat lands alongside the stream channel, termed *floodplains*** (see Chapter Snapshot). Floodplains are areas along streams that are periodically covered with water and are actually part of the river system. Floodwaters carry sediments with them as they inundate the floodplain. When the water slows and finally recedes, this sediment is deposited in layers that, through time, leave broad, flat plains. Despite this threat of periodic flooding, floodplains have often been considered prime sites for urban and agricultural development because of their flatlands, fertile soils, and ready access to water. Though city zoning regulations often restrict development in floodplains, they are now home to more people and development than ever before. From a uniformitarian viewpoint, these are areas that will flood again, since past flooding formed them in the first place. As such, the areas are at high risk for damage or destruction when the stream inevitably exceeds its banks again.

As we discussed in Section 11.4, the velocity of a stream tends to increase downstream with increasing discharge. However, velocity can also vary within any section of the stream channel. Within a cross section of a stream channel, **velocity is lowest along the banks and bed of the stream, where friction is greatest; velocity is highest in the center of the stream, where friction is least** (Figure 11.20).

 ✔️ **Checkpoint 11.13**

✔ basic	✔ advanced
✔ intermediate	✔ superior

What statements are most likely true about a pebble found in a stream?

a. It formed from erosion of sedimentary rock in the streambed or bank.
b. It formed when sand and clay clumped together in the stream.
c. It is younger in age than the stream channel.
d. It may be composed of any type of rock.

 ✔️ **Checkpoint 11.14**

✔ basic	✔ advanced
✔ intermediate	✔ superior

Consider the consequences of constructing a dam on a river that has a large stream load, such as the Yellow River. Assume the dam and its reservoir are located about two-thirds of the way down the river. How would stream flow conditions be altered above and below the dam and its reservoir? What would be the implications for erosion, transport, and deposition?

Figure 11.19 Wrangell-St. Elias National Park, Alaska. A sediment-rich, braided stream channel. These streams begin as melted ice from glaciers.
© David Mott

Figure 11.18 The Yellow River, China. This river carries more sediment relative to its discharge than any other major river on Earth.
© Julia Waterlow; Eye Ubiquitous/Corbis

The lower elevations of most streams have numerous **broad curves, known as meanders** (Figure 11.21a and Chapter Snapshot). As water moves around these long bends, the velocity within the stream channel varies. Think about what happens when you try to drive or run around a corner too fast. Typically, your momentum carries you toward the outside of the corner. If you are going too fast, you will drive off the road or run into a wall. The same thing happens to water traveling in a stream. It is hurled against the outside of a meander, and the stream load bumps into the channel and erodes the outer side of the channel at a *cut bank* (Figure 11.21). In contrast, stream velocity is least on the inside of bends. This decrease in velocity may allow coarse sediment such as sand and gravel to be added to the inner bank of the meander, building a feature known as a *point bar* (Figure 11.21).

As a result of erosion and deposition on opposite sides of its channel, a stream can migrate across a valley floor in the direction of the erosion, a process called *lateral migration*. Or during flooding, the river may rapidly cut through the narrow neck of a fully developed meander, causing dramatic changes in the course of the stream. The meander may be preserved as an oxbow lake, isolated from the stream channel (see Chapter Snapshot).

Suppose you owned some land along a meander and the river served as your property boundary. What would happen if the migrating stream slowly cut away the banks of your property and carried it away? Would that eroded sediment still belong to you? Could you follow it downstream and claim ownership wherever it settled? Darrell Turner knows the answer to that question. In the last 40 years, he has lost half his farm, at the rate of an average of 0.4 hectare (1 acre) a year, to the migration of the Sauk River in Washington. Farther north, in Newtok, Alaska, the entire community of 340 residents will have to move to avoid being swallowed up by the Ninglick River, which is chewing up the river bank at a rate of 30 meters (100 feet) a year.

Despite the loss of property, real estate laws generally do not protect owners from the slow loss of land by erosion on the outer banks of meanders. What was previously considered real estate is

Figure 11.20 Variations in stream velocity. Stream velocity is greatest in the deeper parts of the stream and toward the outside of curves in the channel. Velocity is least on the inside of the curves. Adapted from Charles (Carlos) Plummer, *Physical Geology* 11e © 2007, reproduced with permission of McGraw-Hill Education

Figure 11.21 Meanders and channel migration. **a.** Erosion occurs on the outer bank of the channel, and deposition occurs on the inner bends. The channel gradually migrates across the floodplain as the meander becomes larger. **b.** Deposition and erosion around a curve in a stream in Ohio, showing relative positions of a point bar and cut bank and the direction of stream flow (arrow). The point bar is covered in sand and gravel that were deposited when the stream velocity decreased on the inside of the meander.
11.21a: Adapted from Charles (Carlos) Plummer, *Physical Geology* 11e © 2007, reproduced with permission of McGraw-Hill Education; 11.21b: © David McConnell

a.

b.

CHANNEL MIGRATION
IN THE MAMORÉ RIVER

2003 river

1990 channel

River straightened its channel by cutting across meander necks to form two new oxbow lakes.

No change in this segment. Channel may be controlled by underlying fractures.

5 km

Forest

Grasslands

Meander necks have narrowed.

N

View of a section of the Mamoré River, Bolivia, taken from the International Space station. An image of the 1990 channel has been superimposed over the 2003 river to illustrate channel migration.

② ④ ①

A section of the Mamoré River channel in 1990. Different parts of the stream channel represent different stages in channel migration (see numbered images below).

① Original river flow

E E

② Meander neck becomes narrower

Creation of an Oxbow Lake

Erosion on the inside of meander bends (1, E) causes the meander neck to narrow (2) before the stream cuts across the neck (3). Eventually, the meander bend is disconnected from the straight channel segment to form an oxbow lake (4).

South America

Mamoré River

Bolivía

④

③ Meander cut off

④ Oxbow lake

Mamoré River

Neck cutoff occurs

③

Oxbow lake

④

"We'll never know the worth of water till the well goes dry."
—Scottish proverb

"I derive more of my subsistence from the swamps which surround my native town than from the cultivated gardens in the village. . . . I enter a swamp as a sacred place, a sanctum sanctorum. There is the strength, the marrow, of Nature."
—Henry David Thoreau, US author/naturalist

(left) © David Aubrey/Corbis; (right) © Corbis

Chapter

12

Groundwater and Wetlands

the big picture

Which is better, bottled water
or tap water?

*See The Big Picture box at the end of this
chapter for the full story on these images.*

12.1 Meet Your Drinking Water

Chapter Learning Outcomes

- Students will describe geologic conditions necessary for ready access to groundwater.
- Students will discuss how water enters and leaves the groundwater system.
- Students will explain the connections among groundwater, wetlands, and human activity.

More than a dozen brands of water are on display in my local grocery store. Their labels proclaim artesian water, natural spring water, sparkling mineral water, or, more basically, indicate that the water is from a community water system. On the basis of the bottles on the shelves, there is apparently a great demand for the waters of Europe. Pricey bottles from such places as Norway, Germany, Italy, England, and France are available. An attractive glass bottle of Swedish water can be purchased for $4. That works out to about $20 for 1 gallon (3.8 liters). Alternatively, you could spend 99 cents to get a plastic gallon jug of American water from a nearby bottling plant.

Consumption of bottled water in the United States has grown faster than for any other beverage and has more than doubled in the last decade. Each day, we drink more bottled water than coffee or milk. This rapid growth is occurring despite the fact that most people have easy access to an alternative source of clean, fresh drinking water that costs about 1,000 times less. It is called *tap water*. You can just turn on the faucet in your kitchen and get 3 gallons (11 liters) of water for about 1 cent. If you like that tap water taste but prefer to carry a water bottle, you are in luck. About one-quarter of all water sold commercially in the United States is little more than bottled tap water.

Where Drinking Water Comes From

Our drinking water, whether we buy it in bottles or get it out of the tap, originally comes from streams and lakes on Earth's surface or from groundwater. **Groundwater is water that is present in rocks or unconsolidated materials below Earth's surface.** Groundwater forges a link between surface water systems and the materials in Earth's crust.

In some cases, water simply flows through gaps in the rock, and there is little or no interaction between the water and the minerals in the surrounding rock. In other cases, such as in areas underlain by limestone, the groundwater can dissolve away the rock. This groundwater is what we typically consider *mineral water* because it has a relatively high proportion of dissolved minerals. The total dissolved solids (TDS) represent a measure of how many minerals are present in the water. To be labeled mineral water, bottled water must contain at least 250 parts per million (ppm) TDS. In contrast, the most popular brands of US bottled water are labeled *purified water* and contain less than 10 ppm TDS, essentially indicating that the water is free of all minerals. If your bottled water says purified on the label, this is probably an indication that it is simply municipal tap water that has undergone a second round of water treatment prior to bottling. While some would argue that it is the presence of minerals in bottled water

that provides health benefits, there is little evidence to support this claim. Generally, it is not the natural minerals dissolved in the groundwater that we should be concerned about. Our water supplies are more likely to become contaminated as a result of corrosion from lead pipes or from manufactured chemicals leaking from storage tanks or washed off agricultural fields.

In the United States, there are more than 50,000 community water systems that annually provide clean water for domestic use from surface or groundwater sources. Despite safeguards, a small number of community water systems may become polluted, resulting in serious health risks. Cleanup of these sites can be especially challenging if the water comes from an underground source where the pollution is hidden from direct view. Unraveling the cause of these problems requires a thorough understanding of the geology of the groundwater system. Given the potential tragic consequences of groundwater contamination, it is perhaps no surprise that bottled water has become so popular.

A Case of Groundwater Contamination: Woburn, Massachusetts

Groundwater pollution was at the heart of a lawsuit filed on behalf of a group of families from Woburn, Massachusetts. Children of the families were diagnosed with leukemia after their mothers drank water from two polluted wells (wells H and G, Figure 12.1) while they were pregnant.

Figure 12.1 Potential sources of pollution at Woburn, Massachusetts. Leukemia cases were clustered in the Pine Street neighborhood south of wells H and G. Properties near the wells included W. R. Grace (chemicals), NEP (plastics), Olympia (trucking), UniFirst (dry cleaning), and Wildwood (a tannery).

In the Woburn case, the families' attorney tried to show that the wells were contaminated by industrial chemicals that had been dumped illegally by nearby companies. The case revolved around how quickly the different chemicals could travel through the groundwater system to reach the wells that supplied drinking water. Lawyers representing the plaintiffs (the families) and the defendants (the accused industries) called geologists as expert witnesses. What do you think these experts were asked to determine? The geologists tried to figure out how a variety of factors had influenced the flow of chemicals from different sources to the polluted wells. Scientists debated how factors such as the local geology, the distance of the industrial facilities from the wells, and the properties of the chemical products themselves could have influenced the rate at which the chemicals traveled through the groundwater to the wells. It was essential to understand the interactions between the rock and water systems to determine the source of contamination at Woburn and to make successful use of groundwater resources in general.

After more than 2 months of often complex scientific testimony, the Woburn jury found that chemicals from one company, W. R. Grace, could have contaminated the wells. Soon thereafter, Grace settled with the families for $8 million. But the other companies were not off the hook. Several of the surrounding landowners paid nearly $70 million to clean up the 133-hectare (330-acre) site (Figure 12.2). (For more on the Woburn case, consider checking out the movie *A Civil Action*.)

Though most municipalities test their water to ensure it is safe, the contamination in Woburn is not an isolated case; there have probably been similar sets of circumstances in your state (check www.epa.gov/superfund/sites/npl/). In 1980, Congress established the Superfund program to clean up the sites most contaminated by hazardous chemicals and where the original polluter was unknown or had gone out of business. Most of the companies responsible for pollution in Woburn could be readily identified, but some could not. More than 1,300 such sites are currently listed on Superfund's National Priorities List (NPL). In addition, there are thousands of other locations where the pollution is significant yet not extreme enough to make the NPL. Approximately one-quarter of the US population lives within 6 kilometers (4 miles) of an NPL site. Cleanup programs that seek to prevent dangerous levels of contamination from reaching drinking water wells must take into account the surrounding geology.

Groundwater is just one element in the hydrologic cycle (see Figure 11.3), which ties together the processes that cause water to change form as it moves between different components of the earth system. Abundant supplies of groundwater in regions that are today characterized by deserts are indicative of an ancient wetter climate. Population growth and global climate change will impact future groundwater supplies. In this chapter, we will describe creative ways to increase water for domestic supplies such as recycling toilet water. In addition, we will revisit some of the topics covered in previous chapters that discussed rocks and minerals, streams, and weathering processes. We will seek answers to such questions as: Is there groundwater below your feet? What does a naked aquifer look like? How did an effort to provide safe drinking water turn into the biggest mass poisoning in history? When is a swamp a good thing?

Self-Reflection Survey: Section 12.1

Answer the following questions as a means of uncovering what you already know about groundwater and wetlands.

1. Where are the nearest caves, springs, and wetlands to you?
2. How often do you drink bottled water? When at home, are you more likely to drink bottled water or water from the tap? Why do you make that choice?
3. List all the things you use water for in and around your home. In addition, what other ways is water used in your community?

a. b.

Figure 12.2 Evidence of contamination. **a.** Abandoned 55-gallon (208-liter) drum and **b.** excavation of contaminated soil, wells G and H site, Woburn, Massachusetts.
12.2a & 12.2b: EPA

12.2 Holes in Earth Materials

Learning Objectives

- Define the terms *porosity* and *permeability,* and describe how they might vary as sediment is converted to sedimentary rock.

- Explain how the properties of different rocks control the flow of groundwater.

Although it is impossible to get blood from a stone, we can get water from a rock. In fact, there is approximately 70 times more water below ground than in the lakes and streams on Earth's surface. In dramatic cases, water has flowed beneath the surface for so long that it has dissolved vast volumes of rock to form caves. Although caves can be created by other geologic processes, the dissolving of limestone by weakly acidic groundwater is the most common method of cave formation (Figure 9.10; see Section 9.3 for more on how limestone is dissolved). The world's largest known cavern system, Mammoth Cave in Kentucky, is an example of such a cave (Figure 12.3).

The flow of water through caves has also generated one of the most commonly held misconceptions about groundwater. Although parts of some cave systems contain streams (Figure 12.3b), most groundwater is *not* made up of a network of underground rivers and lakes. Rather, *most* **groundwater is present in billions of tiny spaces between mineral grains or in narrow cracks.** Caves are relatively rare, but groundwater resources are present nearly everywhere across the United States. In almost all cases, if you were to drill down below your present location, you would find groundwater, although sometimes not enough for your daily needs. The volume and composition of available groundwater at any given location are strongly linked to the physical characteristics of the materials below Earth's surface, particularly, their *porosity* and *permeability*. In most places, these earth materials will include bedrock and overlying unconsolidated materials that may be variously described as regolith (formed by weathering of underlying bedrock), sediment (weathered material that is transported and deposited), or till (glacial deposits).

Porosity

Where does the milk go when it is poured over a bowl of breakfast cereal? It fills in the gaps between the cornflakes or granola clusters or whatever is in your bowl that morning. The same thing happens when water enters rock or sediment. Nearly all earth materials contain fractures or small holes (pores) between grains. Those spaces can hold water. **Porosity is a measure of the proportion of a material that is made up of spaces** and depends on the size and arrangement of the grains and the density of fractures (Figure 12.4). Porosity tends to be highest where materials have been well sorted so that their grains are close to the same size and shape. Sorting occurs during transport and deposition of clastic sediment that is later lithified (see Chapter 7). In rare cases, as much as one-half of the total volume of a rock may consist of pores, making the porosity 50 percent.

Do you think porosity is higher in unconsolidated material or in its rock equivalent? Porosity is typically higher in unconsolidated earth materials, such as sand and gravel, than in the equivalent rocks (sandstone and conglomerate). As sediment goes through lithification to form sedimentary rock, its initial porosity

a.

b.

Figure 12.3 Features of Mammoth Cave, Kentucky. **a.** This wide passageway formed when the limestone bedrock dissolved away. **b.** A small stream, formed by the accumulation of groundwater in cave passages, exits a cave in Mammoth Cave National Park.
12.3a & 12.3b: Phil Stoffer, USGS

a.

b.

Figure 12.4 Porosity. Different grain sizes and packing arrangements result in different porosity values. **a.** Individual pore spaces decrease in size with decreasing grain size. **b.** Porosity varies with the packing arrangement of the constituent grains.

Porosity decreases

Figure 12.5 Porosity reduction during lithification. The process of converting sediment to rock reduces pore space through compaction, which squeezes the grains closer together, and cementation, which fills in some of the pore spaces.

is reduced (Figure 12.5). First, compaction squeezes grains closer together, collapsing air-filled spaces. Later, some of the remaining spaces are filled with cementing agents that hold the grains of the rock together (cementation). In some cases, porosity may later be enhanced by dissolution in limestone or by fracturing of the rock during faulting and mountain formation.

If you have ever picked a pebble out of a stream, you will have noticed a film of water on the surface of the pebble. Some water adheres (sticks) to the pebble surface. The same is true for grains and groundwater. Some of the groundwater sticks to the grain surfaces. The remainder of the water resides freely in the spaces between grains or in fractures. The volume of water **on the surface of the grains cannot flow through the material and is termed the** *specific retention* **of the material.** Even if water is allowed to drain from the rock, the water on the surface of the grains will remain. The volume of **groundwater that can drain from rock or sediment is known as the** *specific yield* (or drainable porosity) and is determined by subtracting the specific retention of a material from its porosity:

Specific yield = porosity − specific retention

Do-It-Yourself Porosity Demonstration. To visualize porosity, fill a 1,000-milliliter beaker with dry sand and pour 500 milliliters of water into another beaker (Figure 12.6a). Slowly pour the water into the sand-filled beaker. Allow the water to move through the sand. Watch the sand change color as it becomes wet. Continue to pour the water until all the sand is completely saturated and water rises to the surface of the sand. The water level in the second beaker will have declined by an amount equivalent to the

✓ **Checkpoint 12.1**

| ✔ basic | ✔ advanced |
| ✔ intermediate | ✔ superior |

Imagine that you have a box that is 15 centimeters (6 inches) square. You also have a never-ending supply of rectangular sticks measuring 15 × 3 × 3 centimeters (6 × 1.2 × 1.2 inches). How could you arrange the sticks in the box to model a system with very low porosity (~0 percent) and high porosity (~40 percent)?

a.

b.

Figure 12.6 Do-it-yourself porosity demonstration. **a.** Beakers are filled with 1,000 milliliters of sand (left) and 500 milliliters of water (right). Then the water is poured into the sand. **b.** The sand-filled beaker now contains an additional 350 milliliters of water. Water filled the original pore spaces, and 150 milliliters of water is left in the other beaker.

porosity of the sand. In our experiment, the water level declined from 500 milliliters to about 150 milliliters (Figure 12.6b). This indicates that 350 milliliters of water filled the pore spaces between sand grains. Therefore, 350 milliliters of the original 1,000-milliliter sand volume was made up of connected air-filled pore spaces (porosity = 35 percent). Can you suggest another similar type of bench-top experiment that would provide some additional data on porosity?

How could you determine the specific yield of the sand? If specific retention is very low, as is the case in a coarse-grained material such as gravel, specific yield is nearly equal to porosity because the surface film takes up a small proportion of the pore space. In contrast, yield is diminished when specific retention is high in fine-grained materials such as silt or clay because the surface films represent most of the water in the small pores between many grains. Consequently, **the specific yield of fine-grained materials is low even though their porosity may be high.** It turns out that the specific retention of even fine sand is similar to that of gravel so that the specific yield of sand is just a few points less than its total porosity. In contrast, clay can have a porosity approaching 50 percent, but almost all of that is accounted for by its specific retention, leaving just a few percent for specific yield.

Igneous and metamorphic rocks typically have low porosity because their grains grow together during rock formation, leaving no gaps. However, **the presence of fractures in some rocks produces properties suitable for the storage and flow of groundwater.** For example, thousands of meters of ancient lava flows are preserved as the Columbia plateau basalts in the northwestern United States (see the discussion of lava plateaus in Section 6.6). The tops of the lava flows were heavily fractured as the lava cooled during formation, and the bases of the adjacent flows often contain abundant gas bubbles. The fractures, and cavities left by the gas bubbles, create zones of high porosity that can yield more than 100 gallons (378.5 liters) of water per second. (For comparison, water flows from your kitchen faucet at 1 to 2 gallons [3.8 to 7.6 liters] per minute.)

Permeability

Think again about the porosity experiment we just described. Water was poured into the sand and passed easily through pore spaces between grains. Suppose you mixed cement with the sand and let the mixture dry. Would the water flow easily through the material? **Permeability is the capacity of water to flow through earth materials.** Water can flow readily through rocks that have connections (pathways) between the pore spaces or have many fractures. In contrast, sediments or rocks with fewer connected spaces have little or no permeability (impermeable). High permeability often goes hand in hand with high porosity and large grain size in unconsolidated materials and clastic sedimentary rocks. Permeability is also high in limestone where rock has been dissolved away and in all types of rocks that contain many fractures. High porosity on its own is no guarantee of high permeability. Connections between pore spaces are narrow in fine-grained materials (clay). Consequently, while clay and other fine-grained materials may have high porosity, the pore spaces are often not connected, and therefore water cannot flow through the material.

What can reduce permeability? Permeability is diminished by the same processes that reduce porosity: compaction and cementation, the processes that close the connections between pores. The surface water films (specific retention) in clay and other fine-grained materials may fill the narrow connections between pore spaces, blocking the passage of groundwater. For this reason clay is often used as a liner in ponds to keep water from seeping out into more permeable surrounding soils.

Groundwater Flow Rates. Groundwater flows much more slowly than the water in streams. While the water in streams typically travels 1 meter (3.3 feet) in a few seconds, **rapid groundwater flow in sand and gravel deposits is typically around 10 meters (33 feet) per day, and slow flow is in the range of 1 to 10 centimeters (up to 4 inches) per day.** Why does groundwater flow more slowly than water above ground? In Chapter 11, we explained that stream friction increases with increasing channel roughness and results in a decrease in average flow velocity. This situation is magnified for groundwater

✓ Checkpoint 12.3

◐ basic	◕ advanced
◑ intermediate	● superior

A large volume of liquid waste was dumped on the ground at Otis Air Force Base, Massachusetts. Both the base and the nearby city of Falmouth are located over the same deposit of sand and gravel. Examine the following diagram and predict the approximate length of time before the waste would begin to show up in the drinking water wells of the city of Falmouth, assuming a groundwater flow rate of 0.5 meter (1.6 feet) per day.

a. 10 months   c. 16 years
b. 8 years   d. 40 years

✓ Checkpoint 12.2

◐ basic	◕ advanced
◑ intermediate	● superior

Imagine that you have three identical containers (A, B, C) filled with flour, uncooked rice, and coffee beans, respectively. Predict what would happen if you were to pour water into each container. How would they rank in terms of permeability (from highest to lowest)? Explain your thinking.

✓ Checkpoint 12.4

✓ basic	✓ advanced
✓ intermediate	✓ superior

Your uncle learns that you are taking an earth science class and shows up at your house with three samples of sediment that he dug up from three different locations on his property. He tells you he is going to drill a well on the property and wants you to help him decide which location would make the best site for the well. How would you analyze the samples to help him pick the best site?

because the water must push through tiny spaces between grains, providing millions of opportunities for friction to slow down flow. This also explains why groundwater flows more rapidly through fractures and more slowly through pore spaces.

The slow velocity of groundwater flow can sometimes be a benefit. For example, if pollutants are discovered, there may be time to take corrective action, such as shutting down drinking water wells, before the contamination reaches underground water supplies.

12.3 Groundwater Systems

Learning Objectives

- Describe the principal characteristics of an aquifer, and identify the materials that make up most US aquifers.
- Explain how streams and groundwater are part of the same system in open aquifers.
- Draw and label a sketch to compare and contrast open and closed aquifers.
- Discuss how water flows into and out of aquifers.
- Describe some of the consequences of the overuse of groundwater.
- Evaluate the best and worst conditions for the development of a groundwater resource.

Water has been found in wells as deep as 9 kilometers (over 5 miles) into Earth's crust, but most usable fresh groundwater is relatively shallow. Unlike oil wells that are often thousands of meters deep, water wells tend to be drilled down less than 100 meters (330 feet) in unconsolidated materials or may be up to several hundred meters deep in bedrock. Deeper waters are more expensive to retrieve and are more likely to contain high concentrations of dissolved minerals.

What types of earth materials make the best sources of groundwater? To identify the best place to drill a well, we need to know the important physical characteristics of groundwater systems (see Chapter Snapshot). Groundwater is stored in bodies of rock or sediment called aquifers. **An aquifer is composed of sufficient saturated permeable material to yield significant quantities of water.** There is nothing particularly mysterious about the rock or sediment that is found in aquifers. The same materials

that make up an aquifer can be found in "naked aquifers," sand and gravel deposits and rock outcrops at the surface (Figure 12.7).

The best aquifers hold large amounts of water (high porosity) and allow water to flow freely from one location to another (high permeability). In such aquifers, we can pump out the water we need, and nearby groundwater will flow in to quickly recharge (refill) the aquifer. The most productive aquifers are typically found in unconsolidated earth materials. Aquifers composed of sand

a. b.

c.

Figure 12.7 Naked aquifers. Such aboveground formations are representative of underground aquifers, which can be composed of a variety of materials, including **a.** sand and gravel, **b.** sandstone with good porosity and permeability, and **c.** fractured rocks.

and gravel (Figure 12.8a) account for approximately 80 percent of all groundwater withdrawals in the United States. Dissolved openings in limestone, fractures in the volcanic rock basalt, and pore spaces in sandstone account for the high permeability of other common aquifer systems (Figure 12.8b, c, d, e). Thick rock units that have lower permeability can be equally productive as a thinner aquifer with high permeability. Many locations have more than one available aquifer. A stacked sequence of multiple aquifers is often interlayered with materials with low permeability.

A low-permeability material such as clay, shale, or an unfractured igneous or metamorphic rock can act as a barrier to groundwater flow and is known as an *aquitard*. The properties of the materials that make up an aquifer or an aquitard may change from place to place so that flow within aquifers may be

a.

c.

b.

d.

Figure 12.8 Common surface aquifer materials in the United States. The maps show where the major aquifer materials make up the shallowest principal aquifer. Only small areas of some aquifers are illustrated, as they may be covered by other shallower aquifer systems. Other, more productive aquifers may lie below those shown, while some local aquifers are too small to appear on these maps. Most carbonate aquifers are composed of limestone. What types of aquifers are present in your state?

12.8a: Adapted from http://capp.water.usgs.gov/aquiferBasics/uncon. html#anchorA; 12.8b: Adapted from http://capp.water.usgs.gov/ aquiferBasics/sandstone.html#anchorA; 12.8c: Adapted from http:// capp.water.usgs.gov/aquiferBasics/carbrock.html; 12.8d: Adapted from http://capp.water.usgs.gov/aquiferBasics/volcan.html#anchorA; 12.8e: Adapted from http://capp.water.usgs.gov/aquiferBasics/sandcarb. html#anchorA

e.

compartmentalized. As a result, some parts of a producing aquifer may become depleted, while other nearby parts of an aquifer can continue to produce water.

The first half of this section describes the natural flow of water in groundwater systems, and the second half considers how natural systems are altered by use.

Aquifers

Aquifers may be open to direct recharge from above (open aquifers) or closed to direct recharge (confined or artesian aquifers).

***Open (Unconfined) Aquifers.* In an open aquifer, water infiltrates through overlying rock or earth materials directly into the aquifer below.** The water passes downward until it reaches the saturated zone of the aquifer, where all the pore spaces are filled with water. The top of the saturated zone is called the water *table* (Figure 12.9), because in areas of low relief, it is relatively flat over distances of hundreds of meters.

In general, the water **table follows the shape of the land surface**—for example, it is higher under hills and lower in valleys. Groundwater flows down the slope of the water table (called the hydraulic gradient) from areas where the water table is high to areas where it is low (Figure 12.10). The actual flow paths may vary from straight lines to long curves, depending on the geology (for example, rock type, fractures). When the water table intersects

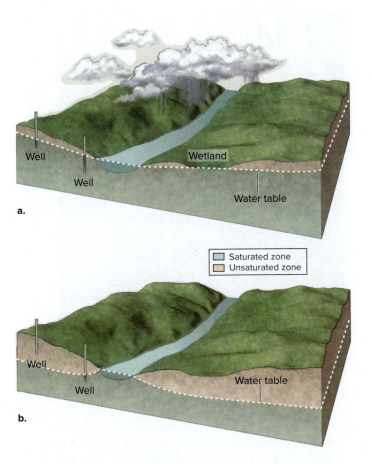

a.

b.

Figure 12.9 An open aquifer. Water levels in an open aquifer vary during periods of wet and dry climate. **a.** Wetlands and springs are present where the water table lies at the ground surface. **b.** Some wells may become dry if they do not penetrate far enough into the saturated zone of the aquifer.

Checkpoint 12.5

✓ basic   ✓ advanced   ✓ intermediate   ✓ superior

The following cross section simplifies the groundwater sources in a county in a midwestern state. Which location would have the potential for the best groundwater production?

a. Location A
b. Location B
c. Location C
d. Location D

Sand and gravel   Shale   Sandstone

Liquid hazardous waste is disposed of by pumping it down injection wells. On the following diagram, which well location would be most suitable for use as an injection well?

a. Well A   b. Well B   c. Well C

Permeable rocks
Impermeable rocks

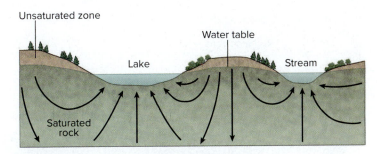

Figure 12.10 Groundwater flow paths. Groundwater flows through rock down the hydraulic gradient (slope) from locations where the water table is at high elevations to places where it has low elevation. The shape of the flow path varies, depending on the local geologic characteristics.

Adapted from Charles (Carlos) Plummer, *Physical Geology* 11e © 2007, reproduced with permission of McGraw-Hill Education

the land surface, a stream, lake, or spring will occur. As the depth of the water table fluctuates over time, water levels in these associated surface features will also change.

Why does the depth of the water table change? The water table drops during prolonged dry periods and rises again when precipitation is plentiful. Therefore, wells must be drilled far enough into the saturated zone to ensure a year-round supply of water. If a well is not drilled deeply enough, it may stop producing water in periods of drought. On the other hand, money is wasted if the well is drilled too deeply (beyond where the well could conceivably ever go dry).

Confined (Artesian) Aquifers. **A confined aquifer is enclosed above and below by impermeable materials** that prevent water from infiltrating directly down into the top of the aquifer or from flowing out the base of the aquifer. ("Artesian" bottled water comes from a confined aquifer.) Because the surrounding rocks are aquitards, water can enter the aquifer only if the rock layer is exposed at higher elevations, called the *recharge zone* (Figure 12.11). The water flows down into the aquifer layer, partially filling it up.

Have you ever let air out of a tire (or had a flat)? Why does the air escape? It escapes through a hole because the air inside

✓ Checkpoint 12.6

A ruling in a court case between Kansas and Nebraska required farmers in southern Nebraska to pump less water from wells in order to maintain water levels in the Republican River as it flowed south to Kansas (see map). The two states have long had an agreement that prevents Nebraska from drawing too much water from the river because Kansas farmers need to use the river water for irrigation. Nebraska authorities claimed that they were not taking water from the river but from underground aquifers and that this practice is not prohibited by the agreement. Explain why the court ruled that groundwater should be considered part of the surface water system. Draw a sketch to show how the consumption of groundwater would have changed water levels in the river.

⊘ basic ⊘ advanced
⊘ intermediate ⊘ superior

NEBRASKA

COLORADO

Republican River basin

KANSAS

a.

b.

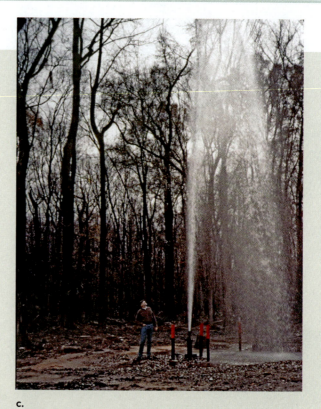

c.

Figure 12.11 A confined aquifer. Changing water levels in a confined aquifer can affect wells drilled into the aquifer. Pressure from the water higher in the aquifer can cause the water table to rise **a.** above the level of the aquifer in a specific well or **b.** all the way to the surface without pumping. **c.** An artesian well shoots water far above the land surface. 12.11c: USGS the tire is at a higher pressure than the outside air. If geologic conditions are right, water is actually forced out of wells drilled in confined aquifers because that water is under pressure. How is this possible beneath the ground? The weight of the overlying water column in the confined aquifer increases pressure deeper in the aquifer (just as happens to your eardrum when you dive into a pool). That pressure on the aquifer is sufficient to cause the level of water in a well to rise above the depth of the aquifer. Water in these *artesian* wells can rise high enough above the aquifer to reach the surface under pressure and flow without pumping (Figure 12.11c).

In the early 1900s, settlers on the plains of South Dakota observed water gushing to 30 meters (100 feet) above the ground from artesian wells drilled into a confined aquifer composed of the Dakota Sandstone. The aquifer was recharged by rainfall in mountains to the west. More than 10,000 wells had been drilled into the aquifer by 1915. Inevitably, as water was withdrawn from the aquifer, the pressure dropped, requiring many of the wells to add pumps to bring water to the surface.

Recharge zones for confined aquifers do not have to be located in mountainous regions. The Floridian aquifer represents the largest limestone aquifer in the United States and yields 4 billion gallons (15 billion liters) of water each day to supply agriculture and cities throughout Florida. The aquifer is made up of a sequence of limestone and other rocks that are over 900 meters (2,950 feet) thick in places. The limestone layers in the aquifer tilt gently southward and can be traced from outcrops in northern Florida to depths of more than 300 meters (980 feet) below the southern tip of the state. Recharge occurs where the aquifer rocks come to the surface in Florida and neighboring Georgia and in places where fractures or sinkholes allow water to pass downward through the overlying confining layer.

Natural Groundwater Budget: Inflow Versus Outflow

What would happen to a city or town whose main aquifer stopped producing water? Could that even happen? The answer is yes, and such questions are currently being asked about some of the largest cities in the southwestern United States. To understand these issues, you must understand the water budget of groundwater systems—the balance between inflow (recharge) and outflow (discharge).

Aquifers are usually recharged when precipitation infiltrates into the ground or when water from streams moves into the groundwater system. But there is one other source for groundwater;

it may already be present in aquifers, left over from a wetter time in the past. Aquifers in presently dry regions may have stored large volumes of water during previous periods of especially wet climates.

Not only do humans extract water from aquifers, but water eventually leaves the groundwater system at discharge points such as streams, springs or wetlands, and the ocean.

Recharge Through Infiltration. Rainfall infiltrates through permeable earth materials to replenish open aquifers or through recharge zones for closed aquifers. The volume of water that enters an aquifer depends on how and where precipitation occurs and the properties of the aquifer. Steady rainfall replenishes an aquifer more than torrential storms because it takes time for water to infiltrate. If the ground becomes saturated during a storm, water runs off to streams instead of flowing below ground. Likewise, rain falling on farm fields or natural lands is more likely to infiltrate below ground than rain falling in urban areas (why?). Seasonally, relatively small changes take place in the elevation of the water table. The water table can drop significantly during prolonged droughts or rise during periods of long-term (months) precipitation.

Recharge Through Streams. In areas with dry climates, the water table is often far below the ground surface. Streams in dry regions may lose water to groundwater through high-permeability materials that make up the stream banks and streambed. Such streams are termed *losing streams* (Figure 12.12a and Chapter Snapshot). For example, water is lost from unlined canals and surface reservoirs built to supply irrigation waters from the Colorado River in the desert Southwest. Water may flow directly from the stream into the aquifer's saturated zone or may infiltrate down through the unsaturated zone in a disconnected losing stream that holds water temporarily due to infrequent precipitation (Figure 12.12b).

Stored Groundwater. The climate of North America is warmer and drier today than it was prior to the end of an interval known as the *last glacial maximum* (ended 12,000 years ago), when much of the northern half of the continent was covered in a thick sheet of ice.

✓ **Checkpoint 12.7**

✓ basic   ✓ advanced
✓ intermediate   ✓ superior

Draw a labeled diagram that summarizes how water flows into and out of an open aquifer.

Water table

a. Losing stream

Water table

b. Losing stream (disconnected)

Water table

c. Gaining stream

Figure 12.12 Losing and gaining streams. **a.** Losing streams lose water to the saturated zone of an underlying aquifer or **b.** to the unsaturated zone in dry regions. **c.** Water enters gaining streams from the saturated zone of an adjoining aquifer.
12.12a, 12.12b, 12.12c: Adapted from Charles (Carlos) Plummer, *Physical Geology* 11e © 2007

A great deal of the water in North American aquifers today began as precipitation during the wetter climate of the glacial maximum. In much of the western United States, there is relatively little present-day inflow of water from precipitation or streams. Since those ice sheets are now gone, the water in many western aquifers represents a finite resource that cannot be replaced once used.

Discharge Through Streams. Groundwater may flow into streams in areas with relatively high water tables. These streams are termed *gaining streams* (Figure 12.12c). Depending on the region, groundwater can account for much of the normal daily flow of a stream. For example, the Sturgeon River, Michigan, flows over permeable

sands and gravels, and receives approximately 90 percent of its minimum discharge from groundwater (Figure 12.13). One benefit of this arrangement is that the river rarely runs dry, even during prolonged periods of low rainfall. In contrast, streams with no input from groundwater may experience periods of low flow conditions (Figure 12.13b).

Discharge Through Wetlands and Springs. Water may flow out at the ground surface from **a spring or wetland located where the water table intersects the ground surface.** Springs may form in a single location where fracture systems or cave systems (Figure 12.14a and b) bring water to the ground surface to form

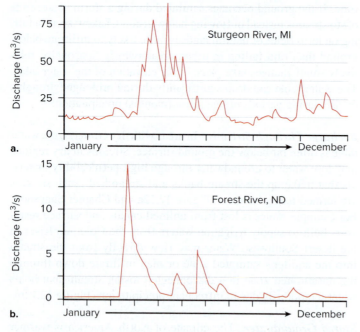

a.

b.

Figure 12.13   Hydrographs illustrating the influence of groundwater on stream flow. **a.** Base flow in Sturgeon River does not approach zero, as the river receives much of its discharge from groundwater, making it a gaining stream. **b.** The Forest River flows over less permeable silt and clays, and it receives little of its flow from groundwater sources. The stream discharge is reduced during periods of low precipitation. In what season do both streams have their maximum discharge?
12.13a, 12.13b: Adapted from USGS

✓ Checkpoint 12.8

| ● basic | ● advanced |
| ● intermediate | ● superior |

Student Answer Evaluation Exercise

In response to the question "Is there groundwater below lakes?" a student wrote the following response. The instructor gave the answer a score of 7 out of 10 and wrote the following comment on the assignment: "You have clearly understood some of the concepts related to this topic, but you have not yet pulled all the key ideas together. Think a little more about the difference between surface water and groundwater systems."

Review the student's answer, and write a more complete response.

Is there groundwater below lakes?
No, there should not be any groundwater below lakes. Groundwater is present in rocks with good porosity and permeability. If rocks under a lake had good porosity and permeability, then the water from the lake would drain out into the groundwater and the lake would be gone. But if the rock under the lake had bad permeability and porosity, then water could not move into it and the lake would not lose water.

a.   **b.**   **c.**

Figure 12.14   Groundwater reaches the land surface at springs and wetlands. Springs form where **a.** fractures or **b.** cave systems intersect the land surface. **c.** Wetlands may form where several small springs distribute water over a region underlain by a low-permeability material such as clay or shale.
12.14a, 12.14b, 12.14c: Adapted from Charles (Carlos) Plummer, Physical Geology 11e © 2007

a stream. Alternatively, many small springs may form in multiple locations where a permeable layer overlies an aquitard, so that surface flow is dispersed over a wide area to form a wetland (Figure 12.14c).

Discharge to the Ocean. Saltwater along the coast will infiltrate the subsurface just as freshwater does (Figure 12.15). Consequently, a layer of fresh groundwater (called a *lens* because of its shape) is typically found floating above denser saltwater in coastal regions. Groundwater flows into the ocean where the lens of freshwater meets the coast. Coastal cities can extract water from the upper part of the freshwater lens. However, if the freshwater is pumped faster than it is naturally replenished, saltwater may enter water wells and pollute the water supply, creating a situation known as *saltwater intrusion* (Figure 12.15b). Saltwater intrusion has been a problem for communities in Long Island, New York, and along the southeast coast of Florida. Consequently, many coastal urban communities rely on surface sources of freshwater (reservoirs, streams, and lakes).

Consequences of Human Actions

Groundwater systems were operating for millions of years before humans got involved. Water levels rise during wetter periods and fall during droughts. The rapid growth in human populations and the development of technology to extract large volumes of groundwater place greater stress on underground water sources. In some locations, the groundwater cannot be replenished fast enough to replace water consumed for human activities. This produces a situation known as **groundwater overdraft, where groundwater extraction occurs more rapidly than recharge.** In such circumstances, water is essentially a nonrenewable resource; in fact, some suggest that water has the potential to become the "new oil"! However, some ingenious solutions are also on offer to replenish groundwater supplies. One of these involves the reuse of water that is flushed down toilets in Orange County, California, before being reprocessed and coming back as drinking water. More than 2 million residents will be supplied by this water from the world's largest water purification project. Despite the apparent "yuck factor," the resulting drinking water is of better quality than water flowing from taps in many cities. The domestic wastewater goes through a series of cleaning treatment steps in which it is purified and filtered before being

pumped into lakes. The water then infiltrates into groundwater through the permeable sediment and rock at the bottom of the lakes. Finally, years later the groundwater is pumped back to the surface and into homes throughout Orange County.

Two negative consequences of the overuse of groundwater are local changes in the water table and a drop in the land surface as grains in the aquifer collapse when water is removed.

Changes in the Water Table. Although aquifers can store vast volumes of water, the physical flow of water to a well can limit the volume of water available. Extracting groundwater from an aquifer is a little like trying to suck up spilled soda from a table using a straw. Ever tried it? You get the soda that was close to the straw, but most of the drink remains on the table surface (Figure 12.16). As previously stated, groundwater moves very

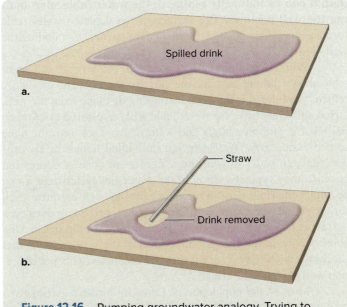

Figure 12.16 Pumping groundwater analogy. Trying to pump groundwater is like sucking up a spilled drink from a table. No matter how big a straw you use, most of the drink stays on the tabletop.

a.

b.

Figure 12.15 Saltwater intrusion. **a.** A layer of fresh groundwater above the layer of saltwater can be pumped and used for domestic water supply. **b.** However, if the freshwater is pumped out faster than it can be replaced, saltwater may intrude and pollute the water supply.

Figure 12.17   Cone of depression. When the water table **a.** is drawn down by overpumping of a well, a cone of depression forms **b.** around the heavily pumped well. Smaller wells located near the cone of depression may lose their water supply. Note that a considerable volume of groundwater remains beyond the reach of the most productive well.

slowly, and it can only come from the earth materials immediately surrounding the well. **If groundwater is pumped out faster than it can be naturally replaced, the water table surrounding the well declines.** The surface of the depleted water table forms a cone shape, called a *cone of depression,* surrounding the well (Figure 12.17).

When overpumping has occurred, the level of the water table shows the greatest decline close to the well and gradually returns to regional levels with increasing distance from the well. Given sufficient time, the water table will be restored to its original level if pumping stops and water flows back into the cone of depression area. Wells serving individual homes rarely produce enough water to generate a significant cone of depression because they typically do not take water out faster than it can be replaced. However, large-volume wells, such as those used for irrigation, are entirely different. These wells may develop a permanent cone of depression as long as they remain in use. One consequence of the change in the level of the water table is that shallower wells may go dry if they are too close to deeper irrigation wells (Figure 12.17b).

Most of the population of Libya in northern Africa lives in cities along the southern coast of the Mediterranean Sea. Unfortunately, excessive groundwater use has resulted in saltwater intrusion into drinking water wells over large areas of the coast. Without the discovery of a giant aquifer up to 4,000 meters (13,000 feet) thick under the desert, Libya's citizens would have been forced to migrate. The aquifer, called the Nubian Sandstone aquifer, extends into the neighboring nations of Egypt, Chad, and Sudan (Figure 12.18). The aquifer stores water that fell as rain during the last glacial maximum. The water supply in the Libyan sector of the aquifer is equivalent to several times the volume of the North American Great Lakes. However, much of this water cannot be used because two widespread cones of depression have developed around the source wells. These cones of depression formed when Libya drilled two great well fields into the aquifer to extract groundwater and pump it 1,600 kilometers (990 miles) to the north to supply its major population centers. The two cones of depression are gradually increasing in size as great volumes of water are extracted and almost none is replaced. Given the huge volume of water available in the aquifer, it will be several decades before water supplies begin to diminish, but much of the water will remain untapped unless Libya establishes more wells in other locations in the desert.

✓ Checkpoint 12.9

Two wells (A and B) are drilled in rocks that have the same porosity, but the rocks around well A have a higher permeability than those around well B. Suppose both wells are pumped at the same rate. Which statement is true?

   a. Well A will have a larger cone of depression.
   b. Well B will have a larger cone of depression.
   c. The cone of depression will be the same for both wells.

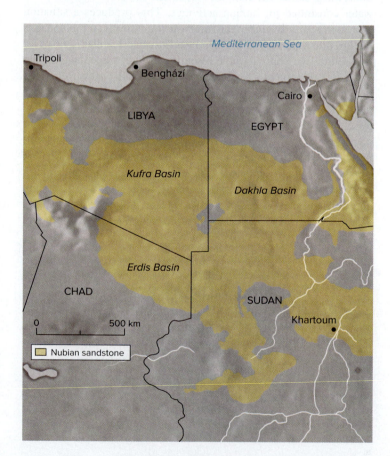

Figure 12.18   The Nubian Sandstone aquifer. The aquifer covers a region that includes parts of Egypt, Libya, Chad, and Sudan. Adapted from www.grida.no/aeo/fg2e5.htm

Figure 12.19 Subsidence due to groundwater withdrawal. **a.** Some of the world's largest cities have experienced subsidence. For example, over 12,000 square kilometers (4,630 square miles) were affected in Houston, making some subdivisions vulnerable to subsequent flooding. **b.** Groundwater withdrawal caused more than 10 meters (over 30 feet) of subsidence in some parts of the San Joaquin Valley since the 1920s. 12.19b: USGS

Ground Subsidence. What happens to a sponge left to dry on your kitchen sink? It shrinks, making the pores smaller. A similar thing happens underground when groundwater is rapidly removed from an aquifer made of loose earth materials such as sand or gravel. **When groundwater is removed from unconsolidated material, the grains that make up the sediment may compact as the pore spaces that once held water close.** But, unlike the sponge, once sediment pore spaces collapse, they do not increase in size again when more water is present. Normally, the weight of the overlying material (including structures such as roads and buildings) is supported by both the mineral grains and the water in the pore spaces in the aquifer. Pressure from water on the surrounding grains keeps the pore spaces open. When the water is extracted, mineral grains may collapse inward on the empty pore spaces. This causes a decrease in the volume of the underlying sediment and can result in *subsidence* (a drop) of the ground surface.

Ground subsidence due to groundwater depletion is an international problem (Figure 12.19a). Subsidence induced by groundwater withdrawal from thick sand and gravel deposits has occurred at rates of about 10 centimeters per year (4 inches per year) over an area of approximately 13,000 square kilometers (5,000 square miles) in the San Joaquin Valley, California (Figure 12.19b). Much of that water was used for irrigation and subsequently lost to evaporation or runoff.

Cities built over weak, unconsolidated, fine-grained sediments associated with environments such as floodplains, deltas, or lake beds often show evidence of subsidence. Roads, pipelines, and large buildings may fracture or collapse as a growing city's population extracts more and more groundwater.

✔ Checkpoint 12.10

| ● basic | ✓ advanced |
| ✓ intermediate | ✓ superior |

Who Owns the Groundwater?

Bob, a rancher in Arizona, drilled a 50-meter-deep (164-foot) well that readily supplied all the water his family needed for 10 years. Then his new neighbor, Glen, drilled a network of new and deeper (150-meter-deep; 492-foot-deep) wells to irrigate a cotton crop in response to a jump in cotton prices. Within 6 months, Bob's well had gone dry.

Bob took Glen to court, arguing that rocks stay but water moves—essentially, that a landowner cannot own the groundwater any more than he or she can own the air above the land. Glen's lawyer disagreed, claiming that Glen owned everything under the ground surface of his ranch and could pump as much groundwater as his crops needed as long as the well was on his property.

Draw a diagram to illustrate why Bob's well went dry. How could Bob ensure a continuous supply of groundwater?

✔ Checkpoint 12.11

| ● basic | ✓ advanced |
| ✓ intermediate | ✓ superior |

In the case described in Checkpoint 12.10, the court ruled in Glen's favor. What are the implications of this ruling for dry western states that are currently experiencing rapid population growth that puts a strain on water used for both agricultural and domestic supplies?

GROUNDWATER

RECHARGE ZONE
Location where water enters the exposed rock layer of the confined aquifer.

WETLANDS

WATER TABLE
Top of the saturated zone.

INFILTRATION
Water seeps slowly into the groundwater system through the soil and underlying bedrock.

IRRIGATED FIELDS
More than two-thirds of US groundwater is used for irrigation of crops.

OPEN AQUIFER
Typically composed of sand and gravel with high porosity and permeability.

ARTESIAN WELL
Pressure from overlying water in the confined aquifer forces water to rise toward the surface without pumping.

CONFINING LAYERS
Aquitards that prevent the flow of water into or out of the confined aquifer.

CONFINED AQUIFER
Composed of saturated, permeable materials and surrounded by low-permeability confining layers.

PRECIPITATION
Steady, low-intensity rainfall provides
more water for infiltration than
short-duration, high-intensity storms.

RUNOFF
Excess rainfall is more
likely to enter streams
than groundwater.

EVAPORATION

WATER TABLE

GAINING STREAM
Water from the groundwater
system enters the stream.

WELLS
Groundwater is artificially
brought to the surface by
pumping water from wells.

LOSING STREAM
Water from the stream enters
the groundwater system.

SUBSIDENCE
Excessive groundwater use can cause
the elevation of the land surface to drop.

CONES OF DEPRESSION

WATER TABLE

✓ Checkpoint 12.12

● basic	● advanced
● intermediate	● superior

Groundwater Evaluation Rubric

You are asked to help locate a new aquifer that will supply your town with water. In examining the potential sites, you recognize that several different factors will influence groundwater availability and at no single site are all of the factors optimal. You decide to create a scoring scheme to evaluate the most important factors that will influence the availability of groundwater. The location that scores the highest according to the rubric will be selected for the well field. One factor is included as an example in the table below; identify five more.

Factors	Poor (1 point)	Moderate (2 points)	Good (3 points)
Depth to water table	Deep	Intermediate	Shallow

12.4 A Case Study: The High Plains Aquifer

Learning Objectives

- Describe the location and properties of the High Plains aquifer.
- Sketch a graph showing changes in groundwater levels in the aquifer from 1900 to 2000.
- Explain why the High Plains aquifer water table is dropping in some places but is not changing in other locations.

Approximately two-thirds of all the fresh groundwater pumped from aquifers in the United States is used for irrigation (Figure 12.20). Much of this irrigation occurs in agricultural lands west of the Mississippi River in the region occupied by the Great Plains states of Texas, Oklahoma, Kansas, and Nebraska and their neighbors. Early explorers dubbed this region the "Great American Desert." Leading an expedition through the area in 1819, Major

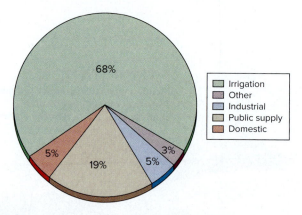

Figure 12.20 Uses of fresh groundwater in the United States, 2005. More than two-thirds of the 80 billion gallons (303 billion liters) of groundwater used per day was used for irrigation; most of this was used in western states with relatively dry climates. Public supplies may be used for community drinking water or other uses. The "other" category represents use of groundwater by mining, utilities, livestock, and aquaculture.
Adapted from USGS

Stephen Long considered it as "almost wholly unfit for cultivation and . . . uninhabited by people depending upon agriculture for their subsistence." Nevertheless, with additional water supplied by irrigation, the Great Plains states have successfully developed an agricultural economy. For example, Nebraska produces about one-half of the nation's corn, but few locations in the state can rely on precipitation alone to produce maximum crop yields. Corn needs approximately 63 centimeters (25 inches) of water to thrive during its 5-month (May–September) growing season in Nebraska. However, the eastern half of the state receives about 45 centimeters (18 inches) of precipitation over this time, while the dryer western part of the state averages just one-half of the necessary rainfall.

Irrigation waters for many of the Great Plains states are taken from a single large aquifer, the High Plains (also known as the Ogallala) aquifer (Figure 12.21). The High Plains aquifer is mainly composed of sand and gravel with some underlying sandstone. Because it is an open aquifer, it is partially recharged by water infiltrating from above, typically from rainfall and snowmelt. The aquifer's saturated zone is up to 425 meters (1,400 feet) thick, reaching its greatest thickness below Nebraska, but aquifer dimensions vary along its length (Figure 12.21a). The water table is relatively shallow, typically less than 100 meters (330 feet) below the surface.

Groundwater from the aquifer was first widely used in the 1930s in Texas. Today, the aquifer produces more water than any other groundwater source in the nation. It supplies almost one-quarter of the water taken from all US aquifers and 30 percent of groundwater used for crop irrigation. More than 170,000 wells draw water from the aquifer throughout an area of 450,000 square kilometers (174,000 square miles), making it the largest area of irrigation-sustained cropland in the world.

Just as with the Nubian Sandstone aquifer in Libya, there is no contemporary source for water to recharge the whole aquifer. The aquifer is partially replenished through the limited recharge that occurs from precipitation and losing stream inflow. But most of the water in the aquifer entered the groundwater system thousands of years ago during a wetter climate interval associated with the end of the last glacial maximum. This "fossil" water is being

a.

b.

c.

Figure 12.21 Characteristics of the High Plains aquifer. **a.** Thickness of the saturated zone in the aquifer. **b.** Changes in the depth of the water table prior to 1980. **c.** Variations in annual precipitation within the regions of the High Plains aquifer. (Compare with Figure 12.8a.) Note: Values presented in feet and inches on the basis of the original US Geological Survey maps.
12.21a: Adapted from http://water.usgs.gov/GIS/brose/ofr00-300_sattk9697.gif; 12.21b: Adapted from http://ne.water.usgs.gov/html/highplains/hp98_web_report/fgures/fg2.htm; 12.21c: Adapted from http://ne.water.usgs.gov/html/highplains/highfg1.htm

used up faster than it is replenished, and the water table is dropping over large regions of the aquifer (Figure 12.21b). More than 10% percent of the total groundwater supply has already been extracted.

Over the life of the aquifer, the groundwater overdraft has caused the water table to drop by up to 70 meters (more than 200 feet) in parts of the Texas panhandle (Figure 12.21b); areas in western Kansas have consumed nearly 40 percent of their groundwater. Both of these locations are in areas with some of the lowest annual rainfall totals for the region (Figure 12.21c).

In contrast, Nebraska has more than 60 percent of the aquifer underlying the state, recharge from the Platte River, less intensive groundwater consumption, and some of the region's highest precipitation (Figures 12.21 and 12.22). Researchers are currently studying the High Plains aquifer to determine how long it can continue to produce groundwater at current consumption rates.

Figure 12.22 Areas of recharge in the High Plains aquifer below Nebraska. The water table in the High Plains aquifer has risen along stretches of the Platte River in central Nebraska, in contrast to a regional decrease in water table elevation.
Adapted from USGS

✓ **Checkpoint 12.13**

Which is the *best* explanation for the changes in water level in the High Plains aquifer? Explain why you chose your answer.

a. The distribution of present-day rainfall on land above the aquifer
b. Differences in the thickness of the rocks that make up the aquifer
c. The number of water wells that have been drilled into the aquifer
d. The type of crops supplied by irrigation from the aquifer

✓ **Checkpoint 12.14**

basic | advanced
intermediate | superior

Sketch and label a graph illustrating changes in water levels in part of the High Plains aquifer from 1900 to 2000. Assume that wells were not drilled into this part of the aquifer until 1940 and that the area is currently experiencing groundwater overdraft.

feat resulted in greater flood control, expanded agriculture, and improved water supply (Figure 12.28). Although these changes benefited landowners, they often had negative consequences for the local ecosystems. The loss of 50 percent of the original wetlands destroyed fish and wildlife habitats, including feeding and nesting areas for wading birds along the transition from dry land to the wetlands.

Scientists belatedly realized that the Everglades wetland system was very sensitive to the duration that surface water stayed in the wetlands, its depth, and its flow patterns. Changes to the natural system increased the amount of time it took for water to flow through the Everglades, decreased water depth, and changed flow patterns. Lower water levels in the wetlands and underlying open aquifer allowed saltwater to intrude (migrate landward) along the east coast. The lower water levels also lowered groundwater recharge rates, which reduced the flow of water to Florida Bay along the southern coast. The water entering the Everglades became more polluted with fertilizers carried by runoff from agricultural operations such as sugarcane production that had replaced much of the original wetlands (Figure 12.28).

The loss of water to the wetlands caused soil formation to slow, and lower water levels exposed peat (compressed vegetation matter that forms thick layers in wetlands) that then dried. Ironically, those dry materials later created fire hazards during drought conditions in what were formerly saturated wetlands.

To start reversing these negative trends, the South Florida Water Management District has purchased thousands of hectares of land to be used to enhance groundwater recharge in an effort to halt the decline of groundwater levels. The federal government has also purchased large tracts of land in and around the Everglades in an effort to restore it to more natural flow conditions.

Figure 12.28   Land use patterns in southern Florida from 1900 to the 1970s. Original wetlands represented by marsh and sawgrass were replaced by agriculture dominated by sugarcane. Additional environmental stress was placed on the ecosystem by the increase in population represented by the expansion of coastal cities.
Adapted from http://www.cotf.edu/ete/modules/everglades/egremote2.html

✓ Checkpoint 12.23

basic · advanced · intermediate · superior

Venn Diagram: Wetlands and Groundwater Compared

Use the Venn diagram provided here to compare and contrast wetlands and groundwater systems. Write features unique to either group in the larger areas of the left and right circles; note features that they share in the overlap area in the center of the image. Identify at least 12 features.

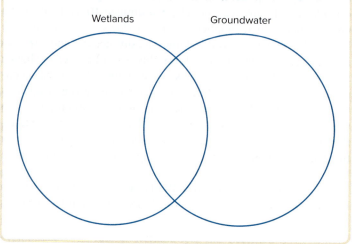

✓ Checkpoint 12.24

basic · advanced · intermediate · superior

Wetland Management Exercise

After graduation, you begin work for your local county government. A few years later, you are put in charge of the local park system. One park includes a wetland area in the meander bend of a river. A boardwalk has been built out into the wetland so that residents can admire the water lilies, watch for fish, and sit at dusk to look for beaver and ducks. Part of your job is to monitor the health of the wetland. There is some concern that it is starting to be overrun by a type of water grass that will push out other vegetation and change the character of the area. What kinds of factors could you survey annually to measure the general health of the wetland? How could you monitor the status of the water grass?

the big picture

One central theme for this chapter has been the social value of science. The cases of contamination we described for aquifers in Massachusetts and Bangladesh illustrate the effect of the earth system on people. Depletion of the water supply from the Nubian Sandstone and High Plains aquifers, as well as the environmental changes resulting from development of the Everglades, show how people can affect the earth system. These cases remind us that we have to be careful in how we interact with Earth's most basic resources. The water in Chapter 11 was surface water—pretty familiar to everyone and easy to observe. Groundwater, as discussed in this chapter, is more obscure and, unless they rely on a well for drinking water, is easily overlooked by most people. This out-of-sight, out-of-mind situation makes it more difficult to identify when groundwater systems are contaminated or are undergoing significant changes.

What priorities should we consider when determining the appropriate use of water resources? Do crops get higher priority than cities? Should development be put on hold if needed wells would reduce water levels in natural springs that are home to endangered species? Who will answer the questions and make the decisions? The issues are complex and involve many components in the earth system. Scientists' findings add to the public debate, but many of the considerations for these decisions lie outside the realm of science and depend on informed citizens balancing economic, cultural, environmental, and scientific factors.

Americans drink more than 9 billion gallons (34 billion liters) of bottled water each year. Whether it is from a bottle or the tap, our drinking water comes from the same place and typically undergoes similar processing and monitoring to make sure it is safe. At the beginning of the chapter we asked: Which is better, bottled water or tap water? In almost all cases, there is no real difference in their quality. The Environmental Protection Agency regulates community water systems, while the Food and Drug Administration regulates bottled water. Municipal water systems are typically required to test for more potential contaminants and to do so more frequently than water bottling plants.

One area of clear difference is cost. For the cost of 1 gallon (3.8 liters) of that fancy Swedish mineral water mentioned at the start of the chapter, you could buy thousands of gallons of tap water from most municipal water systems. But beyond the purchasing cost, there are additional costs for Americans to import water from Sweden or any other foreign nation. Consider the fossil fuels consumed to transport crates of water on trucks and ships from the Swedish bottling plant to a grocery store in your city. Or think about the 1.5 million tons (1,361,000 metric tons) of plastic used annually to manufacture water bottles—most of which are not recycled. Of course, tap water is not as portable as bottled water . . . unless you pour it into an empty water bottle.

Groundwater and Wetlands: Concept Map

Complete the following concept map to evaluate your understanding of the interactions between the earth system and groundwater and wetlands. Label as many interactions as you can using the information from this chapter.

"Look when the clouds are blowing
And all the winds are free:
In fury of their going
They fall upon the sea;
But though the tempest raves
The deep immense Atlantic
Is still beneath the waves."
—*Frederick W. H. Myers,*
English poet

Chapter

13

Oceans and Coastlines

the big picture

Would you wait out a hurricane in one of these hotels?

See The Big Picture box at the end of this chapter for the full story on this image.

13.1 Our Changing Oceans

Chapter Learning Outcomes

- Students will recognize that Earth is the water planet.
- Students will describe physical characteristics of the seafloor and how they affect oceanic circulation.
- Students will describe the influence of oceanic currents, circulation patterns, and tides on the earth system.
- Students will predict the consequences of human development along shorelines.

Over two-thirds (about 71 percent) of Earth is covered in seawater (Figure 13.1), but we know relatively little about the oceans and seas in comparison to our knowledge of the continents. More than 4,000 people have stood on the summit of Mount Everest, the highest point on Earth (8,850 meters; 29,035 feet), but only three people have dived to a depth of approximately 11,000 meters (33,000 feet) in the deepest region of the ocean floor along the Mariana Trench in the western Pacific Ocean. The oceans represent the final frontier for research on Earth.

Few fields of earth science exhibit as much diversity in research methods as the study of oceans. For example, right now, somewhere in the northern Pacific Ocean, there is an elephant seal with a few electronic data sensors glued to its back. Although elephant seals are big, lumbering creatures that often weigh more than a ton (Figure 13.2), they can do a better job of collecting certain kinds of detailed information about the oceans than sophisticated satellites and research vessels. Satellites yield information about the surface conditions of the oceans, and widely spaced stationary buoys collect data about water conditions at different depths. However, neither of these technologies can provide detailed information about changes in the properties of shallow ocean waters. In contrast, as elephant seals migrate from California to Alaska and back, they dive as deep as 600 meters (1,980 feet) up to 50 times each

Figure 13.2   Researchers at rest. A female elephant seal with an electronic data sensor (background) relaxes with her seal pup on a California beach. There is no evidence that sensors harm the animals or interfere with their behavior.
© Chris Howarth/South Georgia/Alamy Stock Photo

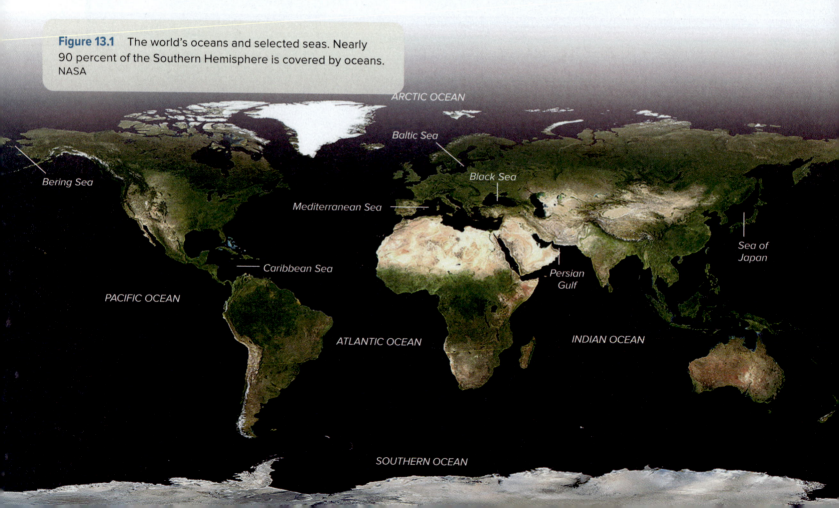

Figure 13.1   The world's oceans and selected seas. Nearly 90 percent of the Southern Hemisphere is covered by oceans.
NASA

ARCTIC OCEAN

Baltic Sea

Black Sea

Bering Sea

Mediterranean Sea

Sea of Japan

Caribbean Sea

Persian Gulf

PACIFIC OCEAN

ATLANTIC OCEAN

INDIAN OCEAN

SOUTHERN OCEAN

day and can provide temperature and salinity (saltiness) profiles of the ocean at relatively closely spaced locations. This adds up to thousands of data sets during a single season. The animals then return to the same California beach, where they molt, shedding their old layer of skin and leaving the data sensors to be collected and downloaded. Alternatively, data may be transmitted directly to satellites when the animals come to the surface.

==Thousands of pelagic marine animals from many different species collect data for the Tagging of Pacific Predators (TOPP) pro-gram.== (The term *pelagic* refers to the open ocean.) The TOPP data can be used to better understand the physical character of the oceans and animal behavior. The TOPP program and similar research provide scientists with data to help them better understand oceanic circulation systems that help regulate global climate patterns.

The Dynamic Nature of Oceans and Coastlines

The oceans are not just big bathtubs of water that separate the continents. They are dynamic environments in which water is constantly in motion. ==Water in the oceans moves both horizontally and vertically because of the rotation of Earth, the influence of the moon's orbit, the push of the winds on the surface, and variations in the properties of seawater.== Small changes in the positions of the continents can change these circulation patterns to bring warm temperatures to one location or allow massive ice sheets to form in another. **==Oceanic and atmospheric circulation patterns redistribute Earth's heat and play a crucial role in controlling our planet's climate.==**

Along the margins of the oceans, coastlines advance or retreat, depending on the balance between the supply of sediment from continental interiors and the removal of material by wave erosion. Daily tides and seasonal variations in stream flow and storm activity cause short-term changes in the position of the coastline. Climate cycles measured in decades, centuries, or millennia may result in increasing or decreasing sea levels, shrinking or growing the size of the landmasses. Finally, tectonic cycles measured in thousands or millions of years may revitalize coastlines by periodic episodes of uplift, as exemplified by the contrast between the sandy beaches of the Atlantic shore (passive margin) and the rocky headlands of the Pacific coast (active margin).

Besides these natural processes, human activities have an impact on oceans and coastlines—and vice versa. Today, many people enjoy living along the coastlines of major oceans; more than one-quarter of the US population lives in counties along the Atlantic and Gulf coasts. Many citizens and organizations take a keen interest in preserving the natural coastlines. For example, to document coastline changes caused by either natural processes or human development, Ken and Gabrielle Adelman set out to create a photographic record of the entire California coast from Oregon to Mexico. Gabrielle flew a helicopter a few hundred meters offshore as Ken snapped thousands of photographs, which he then loaded onto their website (www.californiacoastline.org) for anyone to see (Figure 13.3).

The Adelmans had hoped their site would be used by environmental organizations, state agencies, and anyone with an interest in the coastline. But how could they make sure people even knew it existed? That problem was soon solved. Without knowing it, one of the images the Adelmans had snapped showed the back side of a palatial home in Malibu owned by celebrity

Barbra Streisand. This displeased the famously reserved singer, who promptly filed a $50 million lawsuit for invasion of privacy. Well, this was big news; suddenly, the website with all those views of the California coast was itself famous and receiving thousands of visitors each day. The suit was eventually thrown out of court, Streisand was required to pay the Adelmans' legal expenses, and the photograph is still on the website.

Legal conflicts are perhaps inevitable when private property borders public coastlines. It is a similar story in international disputes over who owns the seafloor where access to potential mineral and energy resources is up for grabs. In this chapter, we will take a walk across the floor of the Atlantic Ocean to consider various features found there. Later, we will learn how the chemistry of the oceans is slowly changing and how bath toys can be used as a research tool in the Pacific Ocean. Finally, we will focus on how natural processes and human activities affect the ocean margins.

Self-Reflection Survey: Section 13.1

Answer the following questions as a means of uncovering what you already know about oceans and coastlines.

1. How have you interacted with the world's oceans, either directly or indirectly?
2. What are the advantages and disadvantages of living along the coast?
3. Does the ocean have an effect on you where you live today? If so, how? If not, why?

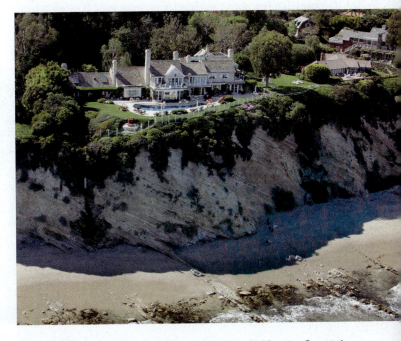

Figure 13.3 Coastal development, Malibu, California. Coastal California communities are home to some of the most expensive real estate in the United States. The rock layers in the cliff provide greater resistance to coastal erosion than sand dunes present along the Atlantic Coast.
© WENN Ltd/Alamy Stock Photo

13.2 Ocean Basins

Learning Objectives

- Identify the locations of seafloor features associated with active and passive continental margins.
- Interpret the topography of the ocean floor within a major ocean basin.
- Describe the characteristics and processes associated with the four major depth zones of the ocean floor.

The physical characteristics of the ocean basins result from plate tectonic processes. As you learned in Chapter 4, the configurations of most of the wide oceans and smaller seas (for example, Atlantic Ocean and Red Sea) evolved as slabs of continental lithosphere moved apart, while others were created by the closure of previously larger basins (for example, Mediterranean Sea). Even now, oceans and seas continue to slowly increase or decrease in size as plates diverge or converge.

From what you learned about plate tectonics in Chapter 4, would you expect the depths to be the same throughout the world's oceans? **The depth from the ocean surface to the ocean floor varies from 0 meters along shorelines to a maximum of nearly 11 kilometers (7 miles) along the Mariana trench in the western Pacific Ocean.** The highest landform on the continents, Mount Everest, would sit comfortably in the trench with more than 2,000 meters (6,600 feet) to spare. The average elevation of the land surface is less than 1 kilometer (0.6 mile), but the average depth of the oceans is approximately 3.8 kilometers (2.3 miles). The volume of water in the oceans is nearly 10 times the volume of the dry land that lies above sea level. Consequently, if erosion were to lower the level of the continents to sea level, the eroded material would fit easily in the ocean basins with room to spare.

Sea Level

The **elevation at the surface of the ocean is known as sea level** and is assumed to be 0 meters. Sea level has changed by hundreds of meters in the geologic past. Why do you think sea level fluctuates that much? Sea level can rise or fall because of changes in the shapes of the ocean basins or as a result of long-term climate changes that trap water in ice caps or cause ice to melt.

Recent studies have revealed that the sea surface is not actually "level." Rather, it has bumps and low spots, fluctuating by as much as 100 meters (330 feet). National Oceanographic and Atmospheric Administration (NOAA) scientists determined this by precisely measuring the distance from an orbiting satellite to the ocean surface (Figure 13.4a). This technology has also been used to determine the depth and topography of the ocean floor. We will discuss the link between rising global temperatures and sea level rise in Chapter 17.

Bathymetry of the Ocean Floor

The elevation of the surface of the ocean varies because the elevation of the ocean floor varies (Figure 13.4). The **measurement of the depth to the ocean floor, and the mapping of its features, is termed** *bathymetry*. It took more than a century for ship-based research tools such as sonar (sound wave imaging) to accurately reveal the detailed bathymetry of some parts of the ocean floor. However, satellite technology has made it possible for researchers to determine the character of large sections of the ocean floor much more rapidly. Satellites are needed to unmask the features on the ocean floor because it is impossible to simply guess how deep a segment of ocean floor will be, and it would take many decades to map the ocean floor using shipborne instruments alone. Data collected by ships and submarines are combined with satellite data in locations such as the remote Arctic Ocean, which is shrouded in ice. These data reveal that the ocean floor has changes in elevation similar to the mountains, valleys, and plains found on land.

a.

b.

Figure 13.4   Measuring the elevation of the ocean surface. **a.** The gravitational attraction of a volcano on the ocean floor causes water to pile up above the volcano. A satellite measures the difference between the elevation of the ocean surface and the neighboring ocean. **b.** On the basis of the altimeter data, the gravitational field of the ocean basins can be calculated. Those gravity data correlate closely with ocean floor topography.
13.4b: NOAA

As we will learn in Section 13.5, the gravitational attraction of the moon causes the sea level to rise and fall, producing tides. A similar gravitational attraction for water can be attributed to features on the seafloor. For example, masses of rock that comprise volcanoes or ridges on the seafloor exert an additional gravitational pull on the water in the ocean, causing it to pile up and form a mound on the ocean surface (Figure 13.4a). Similarly, trenches on the ocean floor are much deeper than surrounding areas. Since the trench is filled with water rather than rock, the gravitational attraction there is lower. This slightly lower gravitational field causes a depression in the overlying ocean surface. These changes in the elevation of the ocean surface can be accurately measured by radars on satellites and used to calculate variations in Earth's gravity field. The gravity data reveal differences in the bathymetry of the ocean floor (Figure 13.4b).

A Walk Across the Ocean Floor: The Four Major Depth Zones

If you set out to walk across much of North America, you would be walking over relatively flat ground in the center of the continent, while regions in the east and west would force you to climb up and down mountains. What would you encounter if you set out to walk from west to east along the floor of the Atlantic Ocean from New Jersey to Portugal along latitude 40°N? (A lot of people yelling and trying to drag you out of the water.) OK, it's a hypothetical walk. What does the ocean floor look like as we cross from one continent to another? Beginning at the edge of the continents, we can recognize four major depth zones in the oceans: continental shelf, abyssal plain, oceanic ridges, and oceanic trenches (Figure 13.5 and Chapter Snapshot).

Continental Shelf. Entering the water along the shore of Normandy Beach, New Jersey, the first depth level you would encounter would be the continental shelf, the shallow ocean floor immediately adjacent to continental landmasses. **The continental shelf is actually submerged continental crust that slopes away from the coast with maximum water depths of a few hundred meters.** The width of the shelf varies, depending on whether it is adjacent to a passive continental margin, where no plate boundary is present (for example, Atlantic coast), or to an active margin adjacent to a trench (for example, south coast of Alaska). The continental shelf is

✓ Checkpoint 13.1

✓ basic	✓ advanced
✓ intermediate	✓ superior

On the following map, label three active continental margins with an A and three passive continental margins with a P. (The X–Y line is used in Checkpoint 13.3.)

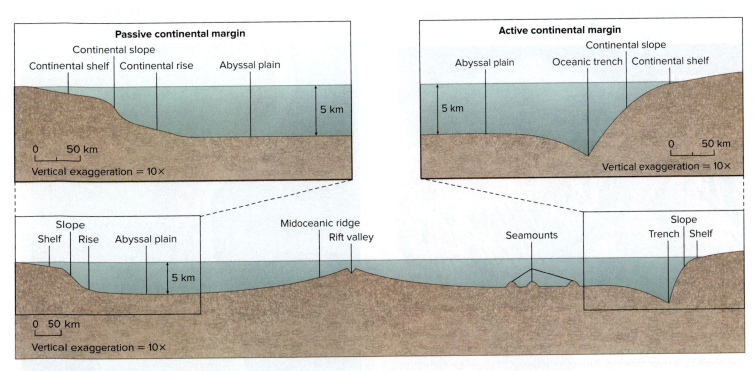

Figure 13.5 Ocean depth zones. The principal topographic elements of the ocean floor include an elevated oceanic ridge; deep, narrow trenches; and a gradual rise of the ocean floor to the continents along passive margins. Note the narrow shelf along the active margin and the broad shelf along the passive margin.

Adapted from Charles (Carlos) Plummer, *Physical Geology* 11e © 2007

Sketch a profile of ocean floor bathymetry along line X–Y on the following map. (Hint: See Figures 4.5 and 4.6.)

13.3 Ocean Waters

Learning Objectives

- Identify and explain processes that produce global surface salinity patterns.

- Describe the pattern of surface temperatures in the world's oceans.

- Explain why the salinity and temperature of oceans are variable at the surface but relatively uniform at depth.

- Discuss how and why ocean water salinity, temperature, and density change from the ocean surface to the ocean floor.

Where did all that water come from? You will recall from Chapter 2 that Earth's surface was a hostile, hot mass of nearly molten rock for millions of years after its formation. Widespread, violent volcanic activity released gases, including lots of water vapor, into the early atmosphere. This "outgassing" is still observed today, though at a much reduced rate, and is the likely source for most of the water on Earth.

The early atmosphere was much warmer than today and capable of retaining much more water vapor. As the planet slowly cooled, this moisture condensed to form thick clouds, which eventually released their water as precipitation. The rain immediately evaporated from the warm surface, causing a continuous loop of evaporation, condensation, and precipitation. These dismal, hot, overcast, and rainy conditions continued until the planet's surface was sufficiently cool for water to collect in hollows—the beginnings of oceans. The original oceans grew steadily as the planet cooled so that scientists now think that they were mostly in place by 4 billion years ago. As we mentioned in Chapter 3, considerable

modern research is aimed at finding out whether comets could also have contributed water to the young Earth. Data from space missions targeting comets were inconclusive because at least some of the small population of studied comets contained water with different properties from that found in Earth's oceans. However, recent examination of a comet originating from the Kuiper belt (see Chapter 3) has revealed water with a similar chemical signature to that of seawater, strenthening the hypothesis that a significant proportion of our oceans were delivered from an extraterrestrial source.

Water Chemistry

Have you ever gone to the ocean and gotten a mouth full of seawater? Do you remember what it tasted like? Seawater contains dissolved salts and minerals (Table 13.1). Where did all those dissolved materials come from and which are most common in seawater? Rainwater falling on the continents reacted with rocks, dissolving minerals that were carried to the oceans in rivers and groundwater. Rivers and streams in the United States alone are estimated to deliver several hundred million tons of sediment and dissolved solids to the ocean annually. It is likely some salts and minerals were also added to the oceans from volcanic activity and atmosphere-ocean reactions. Over time, the dissolved solids increased the salinity of the oceans. For the purposes of this section, think of seawater as a combination of freshwater plus salt (note that common salt, NaCl, accounts for about 85 percent of the dissolved solids in seawater). **Salinity is the measure of the concentration of salt in seawater.** All water contains some salt. Even water in streams on land is likely to have a salinity of up to 0.5 percent. The average salinity of much of the world's major ocean basins is 3.5 percent (35 parts per thousand).

Recently, scientists have noticed a troubling trend in ocean chemistry, *ocean acidification.* The average pH value of the world's oceans has declined from 8.16 to 8.05 since the Industrial Revolution. The pH is slowly decreasing and is expected to decline to around 7.8 by the end of this century (see Chapter 9 for a fuller discussion of pH). The oceans are absorbing some of the excess carbon dioxide from the atmosphere. This forms a weak carbonic acid in the oceans. The more carbon dioxide absorbed, the greater the acidification. Note that the oceans are not technically acidic. A pH value of 7 is neutral; lower values signal acidic solutions, and higher values indicate alkaline solutions. However, pH values are now sufficiently low that the shells of some tiny

Table 13.1	Major Elements and Compounds Found in Seawater
Element or Mineral	Percent of Dissolved Material by Weight in Seawater
Chloride (Cl)	55.0
Sodium (Na)	30.6
Sulfate (SO$_4$)	7.7
Calcium (Ca)	1.2
Magnesium (Mg)	3.7
Potassium (K)	1.1
Mn, Pb, Au, Fe, I	Trace

marine organisms are dissolving and other organisms are growing more slowly; the lower pH levels also can be toxic to eggs and larvae of some species.

Salinity Patterns. The salinity of the warm, well-mixed surface waters over much of the world's major ocean basins ranges between 3.3 and 3.7 percent (33 and 37 parts per thousand; Figure 13.8). **Increasing the amount of salt in seawater increases the density of the seawater.** What patterns of salinity concentration do you observe in Figure 13.8, and what variables might influence those patterns? Salinity in the world's oceans varies because of temperature and the mixing action of ocean currents, plus freshwater input from rain, streams, and melting ice. The big picture is that salinity is generally highest around 25°N and 25°S latitude (Figure 13.8), where temperatures are relatively high but skies are clear as a result of global air circulation patterns (see Chapter 14). The higher temperatures and lack of rainfall result in more evaporation, which removes water but leaves the salt it contains behind.

Why are the salinity values not highest right at the equator, where annual temperatures are highest? Salinity values are a little lower at the equator because more precipitation occurs over equatorial regions, delivering a regular supply of freshwater. Salinity values are a few points lower at higher latitudes (for example, 60°N and 60°S; Figure 13.8), where cooler temperatures mean less evaporation and where melting ice adds freshwater. Note that the difference in salinity between the coast of Antarctica and the ocean around the Hawaiian Islands in the north-central Pacific Ocean is less than 0.2 percent (2 parts per thousand; ppt). As we will learn in Section 13.4, the ocean currents provide an effective mixing mechanism to even out some of the differences in salinity in the open ocean.

If we want to find more-extreme salinity values, we need to look in smaller, restricted ocean basins and seas. For example, look at the Mediterranean Sea in Figure 13.8. Salinity values

Checkpoint 13.5

✓ basic	✓ advanced
✓ intermediate	✓ superior

Examine the following map of mean salinity for the Indian Ocean. Provide an explanation of why salinity values are lower for the tropical Bay of Bengal (east of India) than for the cold waters of the Southern Ocean just north of Antarctica.

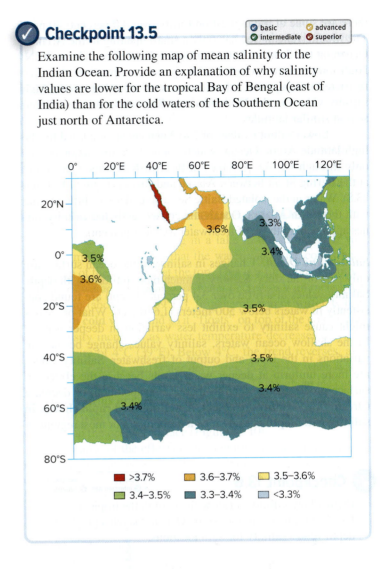

Legend:
- >3.7%
- 3.6–3.7%
- 3.5–3.6%
- 3.4–3.5%
- 3.3–3.4%
- <3.3%

Annual salinity at the surface (quarter-degree grid)

Figure 13.8 *Salinity and latitude. Numbers represent annual mean salinity values in parts per thousand (10 parts per thousand equals 1 percent) using a scale known as the Practical Salinity Scale (PSS). Salinity in the open ocean is greatest in tropical regions and decreases in polar regions, being lowest in the isolated Arctic Ocean. Note that salinity values are consistently higher in the Atlantic Ocean than in the Pacific Ocean. NOAA*

GLOBAL CIRCULATION AND TOPOGRAPHY

DESCENDING CURRENT
Cold, salty water sinks to form the North Atlantic Deep Water (NADW) current that drives the thermohaline circulation.

UPWELLING CONDITIONS
The continent serves as a barrier for deep currents and the prevailing winds direct surface currents away from land.

North America

GULF STREAM
Warm, salty water moves north from the tropical Atlantic Ocean.

Atlantic Ocean

Pacific Ocean

South America

DEEP CURRENTS
The NADW travels south 2-4 km below the ocean surface.

Oceanic trench ②

Abyssal plain ④

① Oceanic ridge

③ Continental shelf

OCEANIC RIDGE
The youngest part of the oceanic floor; high heat flow due to rising magma contributes to higher elevations along the oceanic ridge system.

OCEANIC TRENCH
The deepest section of the ocean floor marks a plate boundary where the oceanic lithosphere of the Nazca plate descends a subduction zone adjacent to South America.

CONTINENTAL SHELF
Shallow, submerged section of continental crust present along *both* passive and active margins but wider at passive margins.

ABYSSAL PLAIN
Some of the oldest and flattest sections of the ocean floor, located landward of the oceanic ridge.

Not to scale.

Europe

Asia

Africa

UPWELLING CURRENT
Hundreds of years after sinking in the North Atlantic, the deep waters return to the surface in the Indian and Pacific oceans.

Indian Ocean

Australia

THERMOHALINE CIRCULATION
A global oceanic conveyer belt that transports water through the world's oceans as a result of density contrasts related to differences in temperature and salinity.

Warm surface current

Cold deep current

Checkpoint 13.11

basic ✓   advanced ✓
intermediate ✓   superior ✓

A fish tank is filled with water at room temperature. Cold water is added on one side of the tank, and warm water is added to the other side. The water at each temperature is dyed a different color to show its movement through the tank.

1. Predict what will happen when warm and cold water are added to the tank simultaneously. Briefly describe your prediction and sketch it in the drawing of the tank.
2. Label the diagram with features that serve as analogs for the low latitudes, the high latitudes, and the thermocline.

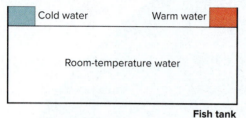

Cold water   Warm water

Room-temperature water

Fish tank

Checkpoint 13.12

basic ✓   advanced ✓
intermediate ✓   superior ✓

Predict what would happen to thermohaline circulation if there were catastrophic melting of the Greenland ice sheet. What would be the consequences for northern Europe?

upwelling currents. Drought occurs in western Pacific nations, and warm air moves farther north over parts of North America. El Niño results in altered climate patterns around the world. Some of the results are negative (droughts, floods), whereas others may be positive (milder winters).

Eventually, conditions may either return to normal or swing in the opposite direction, forming La Niña conditions (Figure 13.18b). In the latter case, cold conditions are even more pronounced than usual, causing droughts in South America and the western United States and more severe weather than normal in the western Pacific. These changes are also reflected in atmospheric air pressures with rising pressures in the east accompanied by falling pressures in the west, and vice versa. The flip-flop in pressure is measured between sites (Darwin, Australia, and Tahiti) in the Southern Hemisphere, thus representing the "SO," or southern oscillation, aspect of ENSO. The cause of the El Niño/La Niña cycle is still under investigation, but it clearly results in a series of interactions among all components of the earth system. Solar radiation from the sun (exosphere) heats the ocean (hydrosphere), which interacts with the atmosphere (winds).

The ocean and weather conditions that result affect the biosphere and, through weather-induced erosion and landslides, can also impact the geosphere.

13.5 Tides

Learning Objectives

- Discuss the cause of tides, including the difference between spring and neap tides.
- Describe the type of tidal patterns associated with different coastal environments.
- Predict how tidal conditions on Earth would change if we had two moons.

We now turn our attention from the open ocean to the interactions between the oceans and the continents. Interactions among all components of the earth system (hydrosphere, geosphere, atmosphere, biosphere, and exosphere) play important roles where oceans encounter continents. Next we will examine how wave action can modify the coastline (see Section 13.6). But first, we will discuss another phenomenon that contributes to the impact of wave action—the tides.

Have you ever noticed that the beach appears wider—that is, more land is exposed—at some times of day than at others? Such daily changes are the result of tides. ==Tides are changes in the sea surface height caused by the gravitational attraction of the moon and the sun.== These changes are visible along the shoreline, where the beach is alternately wider when the tide is out and narrower as part of it is flooded when the tide is in.

Why Tides Occur

Recall from Chapter 2 that the moon orbits Earth and Earth orbits the sun. Figure 13.19 illustrates the relative positions of the moon and Earth during the moon's phases. Tides primarily occur because of the force of gravity exerted by the moon and the sun on the

Figure 13.19 Phases of the moon. It takes 29.5 days to complete a cycle of the phases of the moon. The new moon occurs when the moon is between Earth and the sun, and the full moon represents times when Earth is between the sun and the moon.
Adapted from Konrad Krauskopf, *The Physical Universe* 11e © 2006

Views of the moon from Earth

1 New moon   2 New crescent   3 First quarter   4 First gibbous   5 Full moon   6 Last gibbous   7 Last quarter   8 Old crescent

a.

b.

Figure 13.20 Spring and neap tides. **a.** The gravitational pull on the oceans is greatest during spring tides, when the moon and sun are aligned (during the new or full moon). **b.** The gravitational pull is least during neap tides, when the moon and sun pull the oceans in different directions. Not to scale.
13.20a, 13.20b: Adapted from Carla Montgomery, *Environmental Geology* 7e © 2006

oceans. Since the oceans are a fluid, they are free to respond to gravity. In the simplest case, the side of Earth closest to the moon at any given time during a day will experience the greatest gravitational attraction. Gravity "pulls" water toward the moon, forming a tidal bulge on the side of Earth closest to the moon. (Although the sun has 27 million times more mass than the moon, it is much farther away from Earth, so its gravitational pull is about one-half the pull of the moon.) The tidal bulge on the opposite side of Earth (Figure 13.20) cannot be explained using just the force of gravity from the moon. That equal, but opposite, tidal bulge can be compared to what would happen if you were to take a water balloon tied with a string and swing it around in a circle (trouble just waiting to happen). The water would deform the balloon on the side opposite to where the string is attached. The balloon keeps the water from flying off in one direction (which it would do if the balloon broke). In the case of tides, the tidal bulge occurs through a complex balancing of forces. The gravitational attraction from the moon, coupled with rotation of the Earth-moon system about a common center of mass called the *barycenter,* and rotation of the Earth on its axis combine to cause the bulge. See http://oceanservice.noaa.gov/education/kits/tides/tides02_cause.html for a more detailed explanation.

Tides are greatest when the moon and the sun align, which occurs at the new moon and the full moon. At these points, gravity forms the largest tidal bulges, which are known as *spring tides,* regardless of the time of the year (Figure 13.20a). In contrast, during the first and last quarters of its orbit, the moon is oriented at right angles to the sun, generating the smallest differences between tides, which are called *neap tides* (Figure 13.20b).

Because Earth rotates much faster than the moon orbits, the location of the tidal bulge changes; the moon is not always over the same spot on Earth. We can visualize the effect of Earth's and the moon's rotation by considering the moon to be stationary while Earth rotates on its axis. Imagine the tidal bulges as stationary, with Earth rotating below the bulges. A coastal site would rotate below two tidal bulges on opposite sides of Earth each day. It would also pass through two minima (where the lowest tides would be recorded). In the simplest case, every coastal location on Earth would experience two clearly defined maximum and minimum tides of similar heights occurring 12 hours apart each day. The moon, however, is not stationary—it moves slowly around Earth. A complete cycle of the moon's orbit takes 29.5 days from one new moon to the next. This means that successive high or low tides are separated not by 12 hours but by 12 hours and 25 minutes.

✓ Checkpoint 13.14

☑ basic ☑ advanced ☑ intermediate ☑ superior

Which tidal pattern is represented by the tide data for San Diego, California, shown here?

a. Semidiurnal   b. Diurnal   c. Mixed

✓ Checkpoint 13.13

☑ basic ☑ advanced ☑ intermediate ☑ superior

What would happen to spring tides if the moon were farther away from Earth?

a. Tides would be higher.
b. Tides would be lower.
c. There would be no change to spring tides.

a single storm. In contrast, the redeposition of the beach typically occurs much more slowly over several following months of calmer weather.

Rip Currents. Irregularities such as sandbars and channels along the coastline cause differences in the height of the waves moving parallel to the beach in the surf zone. These variations in the surf zone can drive the backwash of water to generate dangerous currents known as rip currents. **Rip currents are narrow currents of water flowing through gaps in sandbars lying just offshore** (Figure 13.27a). Look at Figure 13.27b. Do you notice a location that might prove dangerous if you were swimming there? Do you think you could see the danger if you were on the beach? Similar currents can form adjacent to human-made structures that act as obstacles to funnel water back to the sea. Rip currents can flow

at Olympic swimmer speeds of more than 2 meters (6.6 feet) per second and can carry people far offshore. These currents are typically faster where wave size is greater. It is estimated that rip currents annually cause the deaths of about 100 people in the United States, double the average number killed by tornadoes. If you get caught in a rip current, the best escape strategy is to allow the current to take

 Checkpoint 13.18

basic ● advanced ● intermediate ● superior

Standing on a beach, you observe a red ball floating about 15 meters (50 feet) offshore. You notice that some well-defined waves are approaching the shore and are just about to reach the ball. The waves are not breaking until they are 5 meters (16 feet) from the beach. Where will the ball be when the last wave reaches the beach?

  a. Closer to shore
  b. At about the same position
  c. Farther offshore

 Checkpoint 13.19

basic ● advanced ● intermediate ● superior

What are the potential advantages and disadvantages of the adoption of wave energy technology along the North American coastline?

 Checkpoint 13.20

basic ● advanced ● intermediate ● superior

One summer, you get a job as a lifeguard on a beach in a southern state. The previous year, four people died because of rip currents, so the state has mandated that a lifeguard from each section of beach attend a half-day training session on rip currents. During the training session, the participants are divided into teams and asked to create a scoring scheme to estimate the daily risk from rip currents along a stretch of beach. The teams are given one factor (included as an example in the following table) and asked to identify four more. Complete the scoring rubric by adding other factors and identifying the characteristics that make them high-, moderate-, or low-risk.

Factors	Low risk (1 point)	Moderate risk (2 points)	High risk (3 points)
Wave height	Low (<1 meter)	Medium (1–2 meters)	High (> 2 meters)

Figure 13.27 Rip currents. **a.** Rip currents form because of irregularities in waves in the surf zone that are often related to changes in the morphology of the seafloor. **b.** Rip current on a Florida beach after Hurricane Jeanne, 2004. The channel in the center of the beach represents the location of a rip current. Water funnels into the head of the channel along the upper part of the beach and flows back to the ocean. The plume of water on the oceanward side of the channel is often an indicator of the presence of a rip current.
13.27a: Adapted from Woods Hole Oceanographic Institution; 13.27b: NOAA

Waves are refracted toward headlands

Bay   Headland

Refracted waves accelerate headland erosion

Deposition   Erosion

Eroded material deposited in sheltered bays

Erosion
Deposition

Continued erosion straightens the coastline

Erosion

Figure 13.28 Erosion and deposition. Wave erosion is concentrated on headlands, and bays become areas of deposition. Eventually, the coastline becomes straightened as the headlands are eroded back and the bays are filled with sediment.
Adapted from Charles (Carlos) Plummer, *Physical Geology* 11e © 2007

you past the sandbar or structure responsible for its formation. Once past the sandbar, swim parallel to the beach away from the rip current and then back toward shore. People often drown when they get worn out from trying to swim directly against the current.

Headlands and Bays. If there are irregularities in the shoreline or changes in the seafloor, waves approaching a coastline may change direction. The seafloor shallows more rapidly toward headlands that protrude from the coast than toward beaches in adjoining sheltered bays. The wavelength and velocity of the leading edge of the wave decrease because of friction. That part of the wave in deeper water continues to move forward at its original speed. **This causes a "bending" of the wave toward the shore that is called** *refraction* (Figure 13.28). Waves approaching a rugged coastline are refracted toward the resistant headlands, causing erosion to be focused on the headlands and ultimately resulting in the straightening of the coastline.

Wave Energy

New energy companies have proposed harnessing the perpetual motion of waves, tides, and ocean currents to provide a clean, renewable energy source for coastal locations around the world. The principle behind these installations is much the same as that employed by wind energy sites on land. But instead of using moving air to rotate blades, these devices convert the movement of water into electricity (Figure 13.29). A variety of technologies have been proposed that would be anchored to the seafloor or tethered to floating buoys at or near the water's surface in near-shore or offshore locations. Given the characteristics of ocean circulation patterns, it is expected that commercial projects along the East Coast would rely on the powerful flow of the Gulf Stream located just 25 kilometers (15 miles) from the Florida coast. In contrast, West Coast installations would be more likely to utilize the more

Figure 13.29 The rotating blades of current turbines look like underwater wind turbines and produce power in much the same way. A generator converts the rotational energy of the blades into electricity that is incorporated into an electrical grid.
© alex-mit/iStock/Getty Images

intense waves of the Pacific Ocean rather than rely on the weaker California current. Projections suggest that a wave farm could generate 50 megawatts of electricity (an average coal-fired power plant generates 600 megawatts).

13.7 Shoreline Features

Learning Objectives

- Identify the principal types of coastlines.
- Discuss how the sediment budget is related to erosion or deposition along coastlines.
- Assess the impact of human activity and climate on sediment budgets.

Depending on the geologic characteristics of the continental margin, coastlines may be characterized by rocky cliffs or sandy beaches. Rocky shorelines form where resistant rocks meet the shore, and cliffs are often present where the land is rising relative to sea level. This may be the result of several factors, including an active plate boundary (for example, much of the west coast of North America) or gradual uplift following the removal of a large ice sheet (for example, the eastern coast of Canada). A cliff may be like a wall, or alternating headlands and bays may form from more- and less-resistant rocks (Figures 13.28 and 13.30a, b). Beach shorelines are typically backed by sand dunes, which can range from a few meters to as many as 25 meters (82 feet) in elevation (Figure 13.30c). Much of the Atlantic and Gulf coasts are fronted by long, narrow sandbars known as barrier islands (Figure 13.30d).

The Changing of Coastal Landforms

Wave action causes the erosion, transport, and deposition of sand and other materials along coastlines. These processes not only give rise to certain characteristic landforms, but also lead to physical changes in coastlines over time.

a.

b.

Figure 13.31   Moving the Cape Hatteras lighthouse, North Carolina. **a.** The lighthouse was just 40 meters (130 feet) from the shore on June 17, 1999 before **b.** it was moved 800 meters (half a mile) to a final location that was 500 meters (1,640 feet) from the shore (move completed on July 9, 1999). Do you think the lighthouse will need to be moved again?
13.31a & 13.31b: North Carolina Department of Transportation

Figure 13.30   Types of coastlines. Depending on the geological history of the coast, coastlines may be rocky (a, b) or sandy (c, d). **a.** Giant's Causeway, Northern Ireland, is a rocky coastline dominated by cliffs and low-lying wave-cut platforms. **b.** White Park Bay, Northern Ireland, features rocky headlands separated by beaches. **c.** Strathy, Scotland, is a low-lying coastline dominated by long beaches backed by dunes. **d.** Assateague Island, Maryland, developed a narrow, elongated barrier island that separates the ocean from the mainland.
13.30a & 13.30b: © David McConnell; 13.30c: © Atlantide Phototravel/Corbis; 13.30d: © Kevin Fleming/Corbis

a.

b.

Erosion. Waves approaching a rugged coastline are refracted toward the resistant headlands. Sediment eroded from those headlands is typically deposited in the relatively quiet waters of the nearby bays to form beaches. ==Over time, the coastline is straightened as erosion wears away the headland and the bays are filled with sediment== (see Figure 13.28).

In 1999, the National Park Service had to pick up the nation's tallest lighthouse at Cape Hatteras, North Carolina, and move it to a site farther inland (Figure 13.31). Why move such a large structure? The lighthouse was originally built 500 meters

(1,640 feet) from shore, but erosion had slowly worn away the distance between the waves and the building. One big storm could have destroyed it.

Structures in other coastal regions of the United States are frequently in danger because of eroding shorelines. Twenty-six of the 30 states bordering an ocean or one of the Great Lakes are experiencing a net loss of usable land along their shorelines (Figure 13.32). The Federal Emergency Management Agency estimates that 10,000 structures are located on coastal lands that will erode within a decade. The most rapid erosion rates are along the

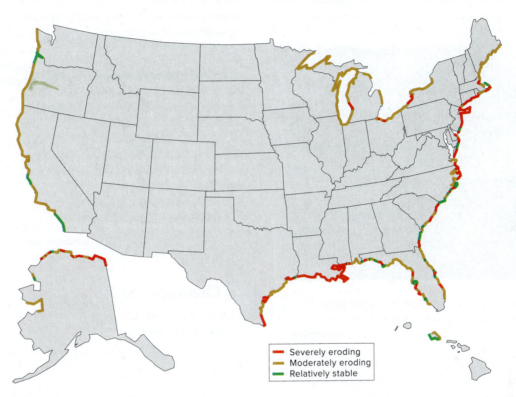

Figure 13.32 US coastal erosion rates. Erosion rates are highest (red) for sections of the Gulf and East coasts.
Adapted from Coasts In Crisis, USGS

c.

d.

Gulf coast (Louisiana) and the Atlantic shore (South Carolina, Maryland, New Jersey), where erosion eats away an average of at least 1 meter (3.3 feet) of coastline each year.

Erosion is most effective on loose, unconsolidated sediment such as sandy beaches or on adjacent sand dunes; rocky shorelines are less susceptible. Erosion is accelerated by the actions of surges and violent waves associated with large storms. Hurricanes and other storms generate waves that can destroy beaches and dunes in a matter of hours. Erosion is particularly severe in association with winter storms (Figure 13.33), especially those occurring during times when the sea level is higher than normal. Typical erosion

rates for sections of shoreline in parts of California and Massachusetts are 0.2 meter (8 inches) per year, but winter storms can cause these same shorelines to recede 5 to 10 meters (16 to 33 feet) in a single season, up to 50 times faster than normal.

Transport and Deposition. While much of a coastline is undergoing erosion, other sections, sometimes along the same stretch of beach, may be growing wider because of deposition of eroded sand.

The relative orientation of the waves to the shoreline controls the distribution of erosion and deposition. Waves that strike the beach head-on move sand up and down the beach. During severe storms waves may transport sand offshore and deposit it in sandbars. If waves strike the beach at an angle, they transport the sediment along the beach in a zigzag pattern. These oblique-angle waves generate **a longshore current, which transports sediment parallel to the beach in the surf zone** (Figure 13.34). As a result, sediment on the beach is transported parallel to the shoreline in the direction of wave motion. Both mechanisms—the physical transport of sand particles along the beach and the movement of sand in suspension in the longshore current—ensure that sediment is transported along the length of the beach and may be added to widen and extend some sections of beaches (Figure 13.35).

The beach is not the final resting place for coastal sediment; it is an intermediate stop on a longer journey. Sediment is transported to the coast by streams and redistributed along the

Figure 13.33   Retreating shorelines. Twelve homes were condemned at Pacifica, California, when the cliff retreated 10 meters (33 feet) during a series of winter storms (1997–1998). © Corbis

✓ Checkpoint 13.21

○ basic   ○ advanced
○ intermediate   ○ superior

Make a labeled sketch or diagram illustrating the four principal types of coastlines.

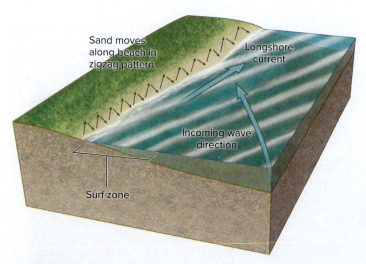

Figure 13.34   Longshore currents. Waves strike a beach obliquely, generating a longshore current that transfers sediment along the shore (left to right, in this case). Sand particles are moved along the beach in a zigzag pattern. Adapted from USGS

Figure 13.35   Sand deposition. **a.** The bay at Puget Sound, Washington, has been almost cut off from the ocean by a narrow spit that may become a baymouth bar. The spit formed from sediment transported along the coastline by longshore currents from the lower part of the image. **b.** Tomales Bay, north of Point Reyes, California, prior to storms. **c.** After storms, sand was transported along the shoreline from the left side of the image. 13.35a: © D.A. Rahm/ Easterbrook Photo/ Image Center; 13.35b & 13.35c: USGS   **a.**

Checkpoint 13.22

basic | advanced
intermediate | superior

Examine the section of coastline in the following image taken at Santa Barbara, California. Storms erode sand from the cliff and carry it along the coast. In what direction does the sand travel along the beach?

    a. Left (north)   b. Right (south)

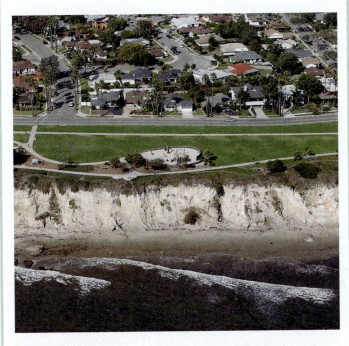

© 2002–2009 Kenneth & Gabrielle Adelman, California Coastal Protection Network

shoreline by longshore currents and wave action. Some of that sand may form offshore sandbars, while some of the remainder may be lost when it flows through submarine canyons off the edge of the continental shelf. Sediment transported along the beach can give rise to certain characteristic landforms when it is eventually deposited in the calm waters of adjoining bays. Sandbars that partially block the mouth of a stream channel form a *spit* (Figure 13.35a), or they may completely block the channel to form a *baymouth bar.*

The Sediment Budget

Shoreline processes are influenced by the **sediment budget, the balance between the material that is deposited on the shore and the material that is eroded from it.** Sediment added to the shore comes from headland erosion or is delivered to the coast by stream flow. Much like a financial balance sheet, the sediment budget remains in a state of equilibrium as long as the volume of sediment coming in is equal to the material being lost.

The natural cycle of shoreline erosion, transport, and deposition along the coast—and hence, the sediment budget—is often disrupted by human activity, as exemplified by changes to the Louisiana coastline.

Natural and Human-Induced Coastline Changes: The Mississippi River Delta. The Louisiana coastline serves as an example of the combined effects of natural processes and human activity on a coastline. Erosion rates for some sections of the Louisiana coast are as high as 20 meters (66 feet) per year. Much of the eroded sediment is deposited in a delta at the mouth of the Mississippi River (see Figure 11.22). The delta is submerged below sea level as the sediment becomes compacted. The subsidence rate, the rate at which the land surface is sinking, is approximately 1 meter (3.3 feet) per century. In the past, this subsidence was compensated

b.

c.

by additional sediment supplied during floods. However, the construction of flood control levees along the river's channel has robbed the delta of its primary source of sediment, which is now deposited in the Gulf of Mexico. The delta subsides as sediment is no longer being added, and the coastline moves landward, flooding what was once usable land.

Loss of sediment from the continental interior to the coast accelerates coastal erosion and removes a buffer that previously slowed down hurricanes, reducing their power. The construction of dams on major rivers reduces the volume of sediment reaching the coast, resulting in sediment starvation because sediment that would once have been deposited along the shoreline is trapped in upstream reservoirs. For example, dams within the Mississippi River drainage basin (see Chapter 11) have reduced sediment supply to the Mississippi delta by approximately one-half. Natural drought conditions may also reduce stream flow and thus diminish the amount of sediment transported to the shore by streams. Coastal development may further modify the system by resulting in the construction of structures designed to reduce erosion or to prevent the deposition of sediment in navigation channels.

✓ Checkpoint 13.23

| ⊘ basic | ⊘ advanced |
| ⊘ intermediate | ⊘ superior |

Draw a diagram to illustrate how sand is transferred among the components of a shoreline.

✓ Checkpoint 13.24

| ⊘ basic | ⊘ advanced |
| ⊘ intermediate | ⊘ superior |

Venn Diagram: Streams and Coastal Systems

Use the Venn diagram provided here to compare and contrast erosion, transport, and deposition in stream systems and in coastal systems. Write features unique to either group in the larger areas of the left and right circles; note features they share in the overlap area in the center of the figure. Identify at least eight features.

Stream systems   Coastal systems

13.8 Shoreline Protection

Learning Objectives

- Compare and contrast the impact of seawalls, breakwaters, and other shoreline protection measures on beach development.
- Evaluate the best prevention and adjustment strategies that could be used to diminish erosion along coastlines.
- Discuss the case for and against funding of beach nourishment programs in states with severe coastal erosion.

Suppose you lived on a shoreline that was eroding at such a rate that it threatened to destroy your house within the next few years. Short of selling your home, what could you do? Three basic strategies are available to the desperate homeowner: (1) stop the cause of the erosion by keeping the waves from reaching the shoreline; (2) change the characteristics of the location to encourage deposition rather than erosion; or (3) add more sand to replace that being lost. While these measures may temporarily provide some protection, you should keep in mind that recent research suggests that many of these measures do more harm than good. In some cases, local governments are taking steps to prevent landowners from making any changes to the shoreline.

Erosion Prevention Strategies

A survey of conditions along a section of Florida shoreline revealed certain natural shoreline features that are most likely to protect coastal residents from erosion:

- Tall, continuous dunes behind beaches protect against large storms.
- Wide, stable beaches (30 meters [100 feet] or more across) absorb and deplete wave energy.
- Exposed offshore sandbars absorb the force of breaking waves before they arrive on the beach.

These features were not universal. Although more than 80 percent of the beaches in the study were wide enough, only about half had dunes, and just 7 percent had an exposed sandbar.

Artificial Barriers. When natural shoreline features do not exist to deter erosion, engineers have sometimes designed artificial barriers to mimic the desired natural effects. For example, a little more than one-tenth of the California coastline is armored with structures intended to prevent erosion. Three types of structures may be built on or near coasts—seawalls, groins, and breakwaters—and a process called *artificial beach nourishment* has also been employed to offset the consequences of erosion.

Seawalls built of rock or concrete are used to protect property from receding shorelines (Figure 13.36). Such a wall serves as a barrier between waves and the shoreline, and it prevents wave erosion from undercutting cliffs below buildings. Waves that strike the seawall are reflected from the wall onto the adjoining beach. Unfortunately, although seawalls may protect the shoreline behind the wall, they typically hasten beach erosion adjacent to the wall because of wave refraction. Erosion is often exaggerated where the seawall ends, causing the shoreline to recede more rapidly on either side of the structure.

Groins are wall-like structures built perpendicular to the shoreline along beaches to act as barriers to longshore currents (Figure 13.37). These structures are intended as obstacles to the transport of sediment along the beach. As a longshore current meets the groin, it loses speed, causing the current to deposit part of its sediment load. That material is deposited on the upcurrent side of the groin, thus building up the adjacent beach. However, as the current passes the groin, it picks up additional sediment on the downcurrent side of the structure, causing local erosion there.

✓ **Checkpoint 13.25**

● basic   ● advanced
● intermediate   ● superior

Compare and contrast seawalls and breakwaters.

✓ **Checkpoint 13.26**

● basic   ● advanced
● intermediate   ● superior

Analyze Figure 13.31 and explain why the shoreline erosion/deposition processes at the site of Cape Hatteras required that the lighthouse be moved.

Breakwaters are barriers built offshore to protect part of the shoreline (Figure 13.38). They prevent erosion by slowing the waves and allowing the beach to grow behind the structure. However, the beach behind the breakwater often grows at the expense of the adjacent unprotected shoreline, which is more quickly eroded.

Figure 13.37 A groin. This structure was built adjacent to Los Angeles International Airport, California. What is the direction of sediment transport along the shoreline?
© 2002–2009 Kenneth & Gabrielle Adelman, California Coastal Protection Network

Figure 13.36 A seawall. This wall constructed of rock was placed at the base of an eroding cliff north of Monterey, California. Note how erosion is exaggerated where the seawall ends.
USGS

Figure 13.38 Breakwaters. These structures are intended to protect the south shore of Lake Erie at Sims Park, Ohio.
US Army Corps of Engineers

Ultimately, the best buffer for protecting valuable real estate along the shore is a big, wide beach. Waves lose energy and erosive power as they cross a wide beach. **Artificial beach nourishment involves dredging sand offshore and then pumping it onto the beach** (Figure 13.39). The beach will grow as long as material is added faster than natural processes remove it. Artificial beach nourishment is often a temporary fix because the sand continues to erode and must be replaced again a few years later. For example, material added to many East Coast beaches remained for fewer than 2 years before the beaches returned to their pre-nourishment state. However, a successful beach nourishment effort occurred at Miami Beach, Florida, in the 1970s. The city spent $64 million to stabilize and expand nearly 16 kilometers (10 miles) of beaches to meet the needs of the booming tourism industry. Much of that beach remains today.

Erosion Adjustment Strategies

As we have just pointed out, nearly all of the artificial methods for preventing erosion have shortcomings. Although they may work in the short term, their effects are local and often lead to other problems elsewhere. Therefore, during the past decade, many coastal states have focused on adjustment strategies instead. For example, state governments have passed legislation aimed at controlling construction adjacent to eroding coasts rather than trying to prevent the erosion. Florida has introduced strict regulations that require buildings constructed near the shoreline to meet rigorous standards so that they are strong enough to resist storm surges and high winds. In 1995, when Hurricane Opal struck the coast, none of the buildings that had been constructed to meet these standards failed, while 56 percent of all other habitable buildings in the storm's path were heavily damaged.

Figure 13.39 Beach nourishment. A pipeline pumps sand collected offshore by a dredge onto the Ocean City Beach in Maryland. Note the wider beach near the bottom of the image. US Army Corps of Engineers

✓ **Checkpoint 13.27**

basic   advanced
intermediate   superior

In Chapter 11, we discussed the difference between prevention and adjustment for flooding. Describe the difference between prevention and adjustment strategies for shoreline protection.

✓ **Checkpoint 13.28**

basic   advanced
intermediate   superior

The US Army Corps of Engineers is responsible for spending money to protect the nation's beaches that are most susceptible to erosion. The Corps is funded by the federal government. Should your tax money be used to maintain wide beaches in places such as South Carolina, Florida, or Texas? What are the consequences of not funding these programs? Identify arguments for and against continued federal funding of artificial beach nourishment programs in states with severe coastal erosion.

the big picture

To ancient people, the ocean was a frightening place—a massive, endless body of water possibly inhabited by large, scary monsters and too daunting to even try to investigate. Scientists do not think this way. Science assumes that Earth *is* knowable. People can discover patterns in all of nature by using their intellect, their senses, and instruments that extend their senses. At the same time, scientists realize that the quest for knowledge is eternal, especially in parts of the planet that are hostile or almost unreachable to humans. Exploration and discovery will continue to expand our knowledge of the deep oceans as long as people ask questions and answers remain to be found.

In the United States, over 123 million people live in coastal counties, and coastal property values continue to rise to levels far above those in most of the rest of the nation. Some folks are glad to spend millions to live in elegant homes with a view of the ocean. However, the weather patterns that bring balmy breezes on warm summer days can also generate devastating storms and even hurricanes.

In 2004, four hurricanes (Frances, Ivan, Jeanne, and Charley) barreled across Florida, causing extensive damage and substantial coastal erosion. Damage was greatest where the timing of the hurricane coincided with the local high-tide conditions. Hurricane Ivan entered the Gulf of Mexico as a strong hurricane that made landfall in Alabama. That storm moved inland across the southeast United States, went back out to sea, and then moved south as a tropical storm that also affected southern Florida before entering the Gulf of Mexico again. The associated storm surge eroded beaches and dunes, causing the hotel pictured in the chapter opening photo to collapse when the underlying sand was eroded (Figure 13.40).

However, if you want to take your chances in the beautiful but dynamic coastal environment, home lots are still available. You might have to plunk down a few million dollars just for a lot to build on, but if you leave the lot undeveloped, it may save you the inconvenience of rebuilding after hurricane season.

Oceans and Coastlines: Concept Map

Complete the following concept map to evaluate your understanding of the interactions between the earth system and oceans and coastlines. Label as many interactions as you can with information from this chapter.

Figure 13.40 Damage from Hurricane Ivan. The five-story building collapsed when the dune below it was eroded by waves generated by the hurricane.
USGS

"For the first time in my life, I saw the horizon as a curved line. It was accentuated by a thin seam of dark blue light—our atmosphere. Obviously, this was not the 'ocean' of air I had been told it was so many times in my life. I was terrified by its fragile appearance."

—Ulf Merbold, astronaut

Top: Wrigley Field, Chicago Bottom: Coors Field, Denver
(top): © Sandy Felsenthal/Corbis; (bottom): © Dave G. Houser/Post-Houserstock/Corbis

the big picture

Why is it easier to hit home runs at Coors Field than Wrigley Field?

See The Big Picture box at the end of this chapter for the full story on these images.

14.1 Science and Skydiving

Chapter Learning Outcomes

- Students will apply physical principles to predict atmospheric temperatures.

- Students will apply principles governing the exchange of water in the atmosphere.

- Students will describe and apply the processes and consequences of atmospheric circulation.

Early in the morning of August 16, 1960, while many people were settling in for breakfast, Captain Joe Kittinger was plummeting to Earth at nearly 1,000 kilometers per hour (620 miles per hour). Captain Kittinger was completing the world's highest skydive as part of a special US Air Force (USAF) program designed to test potential escape methods for pilots of high-altitude jets. The USAF was investigating whether pilots could bail out of a troubled aircraft flying at extreme altitudes in the farthest reaches of the atmosphere. Captain Kittinger had already made jumps from ultrahigh altitudes, but on August 16, his balloon capsule ascended above the New Mexico desert to an unprecedented height of 32 kilometers (20 miles). (For reference, Mount Everest is less than 9 kilometers [6 miles] high, and commercial jets fly at about 11 kilometers [7 miles] of altitude.)

As he drifted slowly upward in the capsule, Captain Kittinger watched the balloon expand and the surrounding sky turn black. He wore a pressurized suit to protect him from the subfreezing temperatures that dominated most of his ascent. The thin air also required that his equipment supply him with oxygen. From the target altitude, he could observe the curvature of Earth on the horizon. After a few minutes, he shuffled to the door of the capsule, said a short prayer, and stepped out (Figure 14.1), beginning his long fall. He dropped rapidly, reaching speeds of 980 kilometers per hour (609 miles per hour) with only a small stabilizing parachute to prevent him from going into an uncontrollable spin that would have resulted in certain death. For several minutes, he rushed toward the ground, gradually slowing as he fell. A second, larger parachute opened at about 5 kilometers (16,400 feet) and carried him safely to his landing site, less than 14 minutes after he stepped out of the capsule far above Earth's surface.

Kittinger's skydive record stood for 52 years until Felix Baumgartner jumped from 39 kilometers (24 miles) on October 14, 2012. Baumgartner was supported by a crack team of experts (including Kittinger), the best of modern technology, and commercial sponsors. He fell so quickly during the early part of his descent that he broke the sound barrier, reaching speeds of more than 1,350 kilometers per hour (840 miles per hour). Unlike the original Kittinger jump, which is memorialized by only a few photographs, Baumgartner's jump was filmed by an array of cameras and watched live on YouTube by millions of viewers. Watching the video clips online will give you a feel for what Earth looks like when viewed from the stratosphere. Why did Captain Kittinger's balloon get bigger as he rose higher, even though the atmospheric temperature had decreased? Why is the sky blue at low altitudes and black at high altitudes? Why did Baumgartner fall at the highest speed early in his dive and then *slow down* as he came closer to the surface of Earth? How close was he to being in space and where exactly *is* space? Joe Kittinger and Felix Baumgartner would have known the answers to these questions before they stepped into the balloon capsule. You will resolve these and other questions by the end of the chapter.

The atmosphere consists of the air around us and serves as a protective layer against the many potential hazards from space. For example, Earth's atmosphere blocks harmful radiation from the sun and protects us from incoming space projectiles that burn up before reaching the planet's surface (see Chapter 3). **The lower boundary of the atmosphere touches Earth's surface; its upper boundary is a gradational transition into space.** Ironically, to get a clear picture of the thin veil of atmosphere that surrounds Earth, we need to leave the planet. When viewed from beyond Earth, the atmosphere appears as just a thin cover that stretches upward a few hundred kilometers before giving way to the airless vacuum of space (Figure 14.2).

a.

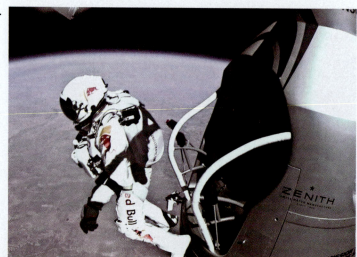

Figure 14.1   **a.** Captain Joe Kittinger steps out of the balloon capsule 32 kilometers (20 miles) above Earth's surface. **b.** Felix Baumgartner drops from his capsule at an altitude of 39 kilometers (24 miles). The experiences of these men help us to think about Earth's atmosphere at different altitudes.
14.1a: US Centennial of Flight Commission; 14.1b: © Red Bull Stratos/ZUMA Press/Newscom

Mesosphere. The mesosphere, the th
atmosphere, is **characterized by decr**
that reach a minimum at the boundary
mosphere (Figure 14.6). Temperatures
this layer because there are fewer and
be warmed by incoming UV radiation.
gen molecules are rarer at these altitud
present to heat up incoming space deb
near-Earth objects burn up in this layer (
mesosphere boundary, the mesopause, i
minimum in the atmosphere.

Thermosphere. The outermost layer
thermosphere, blocks much of the ha
including X-rays, gamma rays, and sc
reaching Earth's surface. All of this i
raises temperatures in the upper ther
1,000°C (1,830°F), but the number of g
at this altitude that heat energy is very

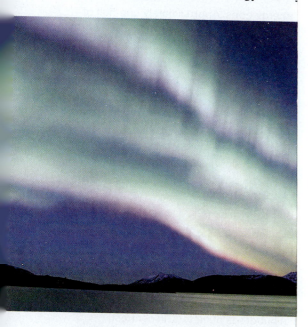

Figure 14.7   Aurora borealis (the Northerr
Anchorage, Alaska. A similar aurora know
is visible in the Southern Hemisphere.
© Hinrich Bäsemann/dpa/Corbis

The interaction of the thin layer of air with solar radiation determines the distribution of global temperatures. **Solar
radiation supplies the energy necessary for cloud formation,
precipitation, and wind generation.** There are extensive plans
to harness wind power to provide a larger share of US energy in
the future. Most of the nation's current energy needs are supplied
by fossil fuels (oil, gas, coal) that carry with them concerns about
energy security and global warming. Renewable energy in various
forms produces 11 percent of US energy generation. Renewable
energy sources such as wind power have the potential for minimal
environmental harm and freedom from reliance on foreign suppliers. Wind power currently supplies about 2 percent of our energy
needs and over 4 percent of US electricity.

Figure 14.2   Earth's
atmosphere. **a.** View of
Earth with heavy cloud
cover in the Southern
Hemisphere, as seen by
the *Apollo 17* crew.
b. A slice through
Earth's atmosphere as
viewed from the space
shuttle is termed the
limb of Earth. Large
clouds rise over the
Pacific Ocean.
14.2a & 14.2b: NASA

a.

Atmosphere

Stratosphere

Troposphere   Clouds

b.

14.2 Air Evolves

Learning Objectives

- Describe the composition and relative proportions of
  gases in the atmosphere.
- Explain how the composition of the atmosphere
  changed during Earth's history.
- Discuss how the proportion of oxygen in the
  atmosphere is tied to the evolution of Earth's
  biosphere.

The specific mix of gases in Earth's atmosphere is termed
air. **Two common gases, nitrogen and oxygen, make up about
99 percent of dry air** (Figure 14.3). Air also contains trace amounts
of other gases (argon, carbon dioxide, methane, nitrous oxide, and
ozone), as well as microscopic particles and chemicals (collectively
called aerosols). The amount of water vapor in the air can range
from next to nothing over dry deserts to up to 7 percent (by weight)
in humid climates. We observe changes in the atmosphere above
our heads in the form of daily weather patterns and seasonal trans-
formations. These relatively short-term changes assume a uniform
atmospheric composition—that's what makes them predictable. In
contrast, as we will discuss in Chapters 16 and 17, most scientists
agree that the composition of the atmosphere is gradually changing
because of human activity and that such alterations will inevitably
lead to modifications to other components of the earth system.

An Atmosphere Evolves

Atmospheres are common around other planets and even some
moons in our solar system. The presence of an atmosphere
is an important consideration in the search for life beyond
Earth. As described in Chapter 2, our guardian atmosphere has been largely responsible for the development
and protection of life on our planet. The early atmospheres of both Venus and Earth began with an initial hot, hazy mixture of gases that included abundant

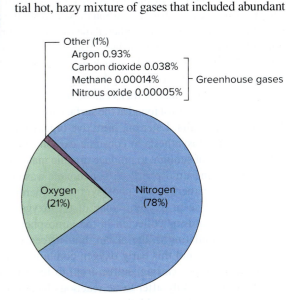

Other (1%)
Argon 0.93%
Carbon dioxide 0.038%   } Greenhouse gases
Methane 0.00014%
Nitrous oxide 0.00005%

Oxygen
(21%)

Nitrogen
(78%)

Figure 14.3   Gases in Earth's atmosphere. Concentrations of
gases in dry air. Greenhouse gases trap heat near Earth's surface.

✓ Checkpoint 14.5

What is the difference between heat and

a. Heat deals with total kinetic energ
    average kinetic energy.
b. Heat deals with average kinetic en
    with total kinetic energy.
c. There is no difference, since they l
    kinetic energy.

Next in this section, we will use
temperature to identify vertical changes in
atmosphere. In Section 14.5, we will discuss
ported within the atmosphere and how that a

The Four Layers of the Atmospl

How could you measure the temperature of
many decades, such measurements were taken
on balloons. Today, as a result of advances in
also collected using rockets and satellites.
alternating layers of increasing and decreasing
the atmosphere. These layers are the tropo

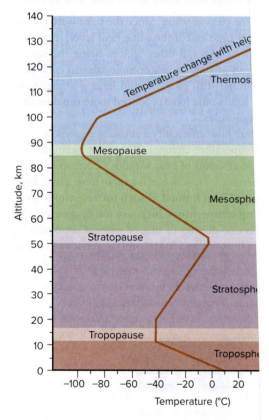

Figure 14.6 The four layers of the atmosphere
three areas of higher temperatures (the ground
stratosphere, and thermosphere) and two of low
(upper troposphere and upper mesosphere). Th
continues upward to approximately 500 kilomet
from Earth's surface.

---

represents the number of wave crests that pass a point in 1 second. The sun's EMR spectrum spans wavelengths from short-wavelength (0.000001 micrometer; 0.00000000004 inch) gamma rays to long-wavelength (hundreds of meters) radio waves (Figure 14.8a).

The supply of energy used daily by natural earth systems is derived almost completely from solar radiation. Visible light (wavelength 0.4 to 0.7 micrometer) makes up a little less than one-half (47 percent) of the solar radiation reaching Earth's atmosphere (Figure 14.8b). Infrared radiation is associated with heat and makes up another 45 percent of the sun's radiation emissions that make it to Earth. About 8 percent of solar radiation is short-wavelength UV radiation. All short-wavelength radiation (including X-rays, gamma rays, and UV rays) is invisible to the naked eye. Humans cannot affect the amount or type of radiation emitted from the sun. However, any human action or natural event that indirectly affects the amount of solar radiation reaching the surface (or that is retained by the atmosphere) may have an effect on life on Earth.

Earth's Energy Budget

What do you think can happen to EMR that reaches Earth? There are three possible fates for incoming solar radiation as it passes through Earth's atmosphere on its way to the surface of the planet (Figure 14.9):

1. ***EMR can be scattered.*** When EMR strikes small particles and gas molecules in the atmosphere, it may change direction, a process known as *scattering*. Have you ever wondered, "Why is the sky blue?" It is blue because the wavelength of blue light is about the same

✓ Checkpoint 14.9

✓ basic	● advanced
✓ intermediate	● superior

The ozone hole over Antarctica actually represents a thinning of the ozone layer. What are the consequences of the loss of ozone?

---

as the size of the oxygen atoms in the atmosphere. Consequently, blue light is scattered more easily than other colors with longer wavelengths. The blue light that is scattered toward Earth's surface reaches our eyes, making the sky appear blue. In higher sections of the atmosphere with fewer gas molecules to scatter incoming solar radiation, the sky would appear black, much as it appeared to Felix Baumgartner when he reached an altitude of 39 kilometers (24 miles).

2. ***EMR can be reflected.*** Think of your reflection in the mirror. What about your reflection on a shiny table or a brick wall? Different surfaces reflect different amounts of light, making it possible to see our reflection in some surfaces but not in others. In contrast to scattering, **reflection in the atmosphere causes incoming radiation to be returned to space.** Solar radiation is reflected from gas molecules and other particles in the atmosphere as well as from features on Earth's surface. Some natural surfaces on Earth are more reflective than others. For

Figure 14.9 Where solar energy goes. Approximately one-half of the incoming solar radiation heats Earth's surface.

Figure 14.8 The electromagnetic spectrum. **a.** Wavelengths of radio waves can be measured in hundreds of meters. In contrast, wavelengths for visible light are less than 1/1,000 millimeter across but 1 million times larger than the wavelengths of gamma rays (1 million micrometers = 1 meter). **b.** The relative proportions of solar radiation reaching Earth. Infrared and visible light make up more than 90 percent of solar radiation at Earth's surface.

instance, more solar radiation is reflected from light-colored surfaces than from dark-colored surfaces. **The measure of the reflectivity of a surface is termed** *albedo* (see Chapter Snapshot). The average albedo of Earth is about 31 percent, though it varies with the region. Light-colored, reflective surfaces and thick cloud cover have a high albedo (80 to 90 percent) because these features reflect more sunlight. In contrast, dark surfaces (for example, forests, water) in the absence of cloud cover have low albedo values (less than 25 percent). Pollution from light-colored aerosols, which are carried downwind from populous regions, produces a haze that blocks and reflects incoming EMR, producing a regional cooling effect. Some have suggested taking advantage of this phenomenon to inject aerosols high into the atmosphere in an effort to reduce incoming radiation and slow rates of global warming.

3. *Some forms of EMR are absorbed* in the atmosphere or at Earth's surface. Absorption is the process whereby radiation interacts with a material and is converted to some other form of energy (for example, heat). For absorption, the wavelength of the incoming solar radiation is important because different atmospheric gases absorb specific wavelengths. Forms of radiation with the shortest wavelengths (X-rays, gamma rays) are absorbed by gases in the thermosphere, while ozone in the stratosphere absorbs UV radiation (Figure 14.10). This characteristic of ozone effectively blocks the majority of incoming UV rays that can cause skin cancer. Forest fires and industrial pollution may add black soot particles that will absorb EMR and heat the surrounding

atmosphere. These dark particles are a significant contributor to global warming (see Chapter 17). Dark soot particles that are washed out of the atmosphere by rain can reduce the albedo and increase heat absorbed on the ground. These particles are blamed for some of the increase in melting rates of Himalaya glaciers. Visible light and longer-wavelength infrared radiation reach the troposphere. Water vapor and carbon dioxide in the troposphere absorb infrared radiation in the range of approximately 1 to 2 micrometers. As we will learn in a moment, it is this property that causes the atmosphere to warm up.

Visible light, some ultraviolet rays, and infrared radiation (what we feel as heat from the sun) may reach Earth's surface (see Figure 14.9). **Sunlight is absorbed by land and oceans, warming the planet's surface.** Some of this energy is used to evaporate water or to heat objects on the surface. The energy that is absorbed (regardless of its wavelength) is converted to thermal energy (heat). To illustrate this process, suppose you go the beach in the morning of a sunny summer day. You park your car on a black asphalt parking lot and walk in your bare feet down to the beach. No problem. Later that day, you decide to return to your car—in your bare feet. What will you experience as you walk (or maybe run) across the parking lot? Big problem! The parking lot surface is hot, much

✓ Checkpoint 14.10

✓ basic	✓ advanced
✓ intermediate	✓ superior

All objects emit heat. The hotter an object, the faster its molecules move and the shorter the wavelength of radiation it emits. This relationship between the heat of a body and wavelength is expressed in Wien's law, which states that the maximum wavelength (in micrometers) of electromagnetic radiation can be determined by dividing 2,897 by the absolute temperature (T). Absolute temperatue T is expressed in kelvins (K). The Kelvin temperature scale begins at absolute 0 (where all particle motion stops), with the freezing point of water (0°C) equivalent to 273 K on this scale (30°C = 303 K).

$$\text{Maximum wavelength} = (2{,}897/T)$$

For example, for the sun, $T = 6{,}000$ K. Substituting the temperature of the sun into the above equation illustrates that incoming solar radiation should have a wavelength of approximately 0.5 micrometer, which lies within the range of visible light.

Use Wien's law to determine the dominant wavelengths emitted from (a) Earth and (b) Venus, and identify the corresponding type of radiation.

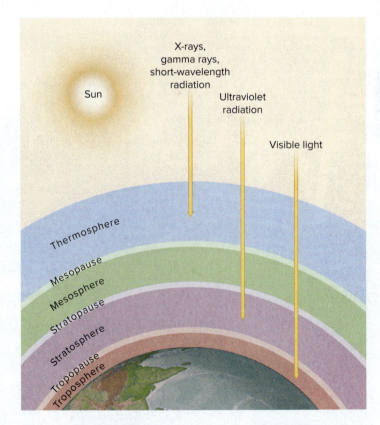

Figure 14.10 Absorption of sun's rays. Most short-wavelength solar radiation, including X-rays, gamma rays, and ultraviolet radiation, is intercepted before it reaches the troposphere. Visible light and radiation associated with heat are the most common wavelengths to reach Earth's surface.

THE EARTH'S ALBEDO

Cirrus clouds
20%–40%

Clouds
10%–90%

Earth's albedo
(average) 31%

Cumulus clouds
75%

Fresh snow
80%–95%

Deciduous forests
13%

Pine forests
9%

Dark roof
8%–18%

Brick, stone
20%–40%

Light roof
35%–50%

Crops, Grasslands
10%–25%

Dry concrete
17%–27%

Asphalt (blacktop)
5%–10%

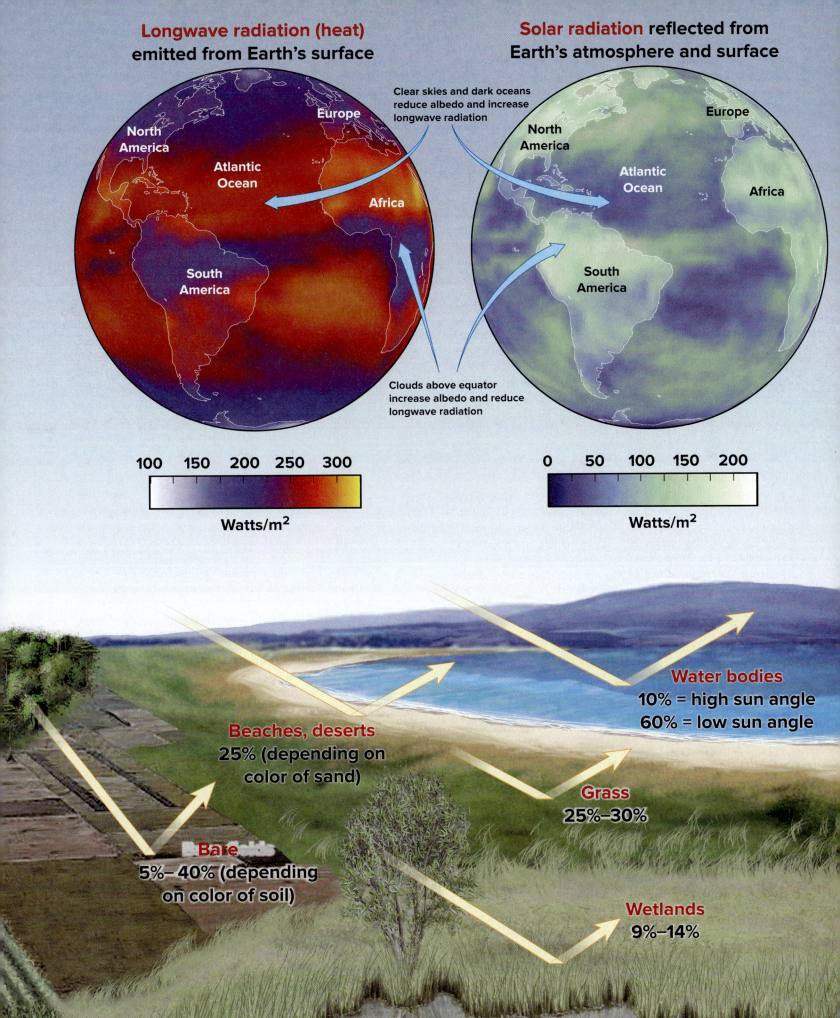

Longwave radiation (heat)
emitted from Earth's surface

North
America

Europe

Atlantic
Ocean

Africa

South
America

Clear skies and dark oceans
reduce albedo and increase
longwave radiation

Clouds above equator
increase albedo and reduce
longwave radiation

100   150   200   250   300

Watts/m²

Solar radiation reflected from
Earth's atmosphere and surface

North
America

Europe

Atlantic
Ocean

Africa

South
America

0   50   100   150   200

Watts/m²

Water bodies
10% = high sun angle
60% = low sun angle

Beaches, deserts
25% (depending on
color of sand)

Grass
25%–30%

Bare fields
5%– 40% (depending
on color of soil)

Wetlands
9%–14%

hotter than it was earlier, because it absorbed both the heat and much of the incident light from the sun. That heat is then reradiated as longer-wavelength infrared radiation.

The Greenhouse Effect. The air temperature in the troposphere decreases steadily as distance increases from the warm Earth's surface, partly because of Earth's heating and partly because the air is thinner (see Section 14.6). Anyone who has ever been at high elevations knows that temperature decreases with altitude (that's why we go to the mountains to ski). **Air temperature declines at a rate of 6.5 C° per kilometer (12 F° per mile, known as the *normal lapse rate*) through the troposphere,** beginning at an average of 15°C (59°F) at sea level.

Earth's surface warms as it absorbs radiation that has not been scattered, reflected, or absorbed elsewhere in the atmosphere. Surfaces with low albedo would warm up and emit energy as heat in the form of long-wavelength infrared radiation (that asphalt parking lot is still warm long after the sun sets). Surfaces with low albedo emit more heat energy than surfaces with high albedo (see Chapter Snapshot). This radiation is absorbed by water vapor, carbon dioxide, and other trace gases (methane, nitrous oxide) in the troposphere, creating a situation that has come to be known as the *greenhouse effect.* **The greenhouse effect increases temperatures at Earth's surface by 33 C° (59 F°),** ensuring that we have a livable planet. Earth would have an average surface temperature of −18°C (0°F) rather than the present average of +15°C (59°F) without the heat-absorbing property of water vapor and other gases in our atmosphere. In contrast, Venus, with a carbon dioxide–rich atmosphere, has an out-of-balance (runaway) greenhouse effect, resulting in surface temperatures of 464°C (867°F; see Chapter 2).

Changing the surface albedo can change how much heat is absorbed and therefore can affect temperatures. A region could experience a warming effect as a result of a decrease in albedo because the surface retains (and then emits) more energy. For example, at high latitudes, warming could occur when white sea ice melts, leaving behind a dark sea surface.

✓ Checkpoint 14.11

| ● basic | ◐ advanced |
| ● intermediate | ◐ superior |

Explain if the interaction of EMR with Earth would cause global temperatures to increase or decrease under the following conditions.

1. The atmosphere is thicker.
2. The atmosphere contains more carbon dioxide.
3. The atmosphere contains more aerosols.
4. The atmosphere contains more black soot.
5. Tree leaves are white.
6. There is no ice on Earth.

✓ Checkpoint 14.12

| ● basic | ◐ advanced |
| ● intermediate | ◐ superior |

Construct a concept map to show the connections between solar radiation and the atmosphere, using at least the following terms: incoming radiation, absorption, reflection, scattering, albedo, and the greenhouse effect.

14.5 The Role of Water in the Atmosphere

Learning Objectives

- Apply concepts of latent heat to explain energy transfer during water phase changes.
- Predict how relative and absolute humidity are affected by temperature changes.
- Apply the concept of latent heat to explain processes behind everyday phenomena.

Can you think of any naturally occurring substance that can exist at Earth's surface as a solid, liquid, and gas? Because of its unique atomic structure, **water is the only natural compound that can exist in all three states on Earth's surface: a liquid (water), a gas (water vapor), and a solid (ice).** We experience these forms of water as rainfall, humidity, or hail, respectively. In Chapter 11, we discussed the various places where water is found in the earth system and how those places are linked together by the hydrologic cycle (Figure 11.3).

The atmosphere contains just a small part of the water present on Earth. Imagine a 1-gallon (3.8-liter) jug of milk sitting on a kitchen table. Now, pour out slightly less than half a cup (118 milliliters) of milk and set it down on the table beside the gallon jug. Next, take just 1 drop of milk and place it on a teaspoon.

- The original 1-gallon jug represents all the water on Earth.
- The milk remaining in the gallon jug represents the 97 percent of Earth's water in the oceans (see Figure 11.4a).
- The half-cup of milk represents all the freshwater in streams, lakes, glaciers, and groundwater on the continents.
- That single drop of milk is equivalent to the tiny fraction of the planet's water that is in the atmosphere.

If we were to add up the annual volume of water falling as precipitation, we would find that it is more than 30 times greater than the moisture stored in the atmosphere at any given moment. To sustain such high precipitation volumes, water must be constantly cycled through the atmosphere. Evaporation and condensation of water are continuous in the atmosphere, and the conversion of water from one state to another transfers energy throughout the troposphere.

Three States of Water

The liquid, solid, and gaseous forms of water differ because of water's atomic structure. One molecule of water is composed of two hydrogen atoms and one oxygen atom. The arrangement

Figure 14.11 Dipolarity of water. Water molecules attract because of the net partial negative charge on the oxygen atom and the net partial positive charge on one of the hydrogen atoms in the two different molecules.

of these atoms results in opposite charges on each end of the molecule; hence, water is termed *dipolar*. The dipolar nature of water molecules results in the formation of a dipole-dipole attraction between their negative and positive ends so that adjacent molecules join together (Figure 14.11). The three states of water are determined by the distance between the individual water molecules and their degree of motion: In solid form, the molecules are joined together in a closely spaced, well-ordered hexagonal structure (Figure 14.12a). Water molecules vibrate in the solid state but do not move around. Ice is the most ordered configuration of water molecules. In liquid form, small groups of molecules remain attached to one another but move rapidly enough to generate semiordered structures (Figure 14.12b). In gaseous form, most individual water molecules move too rapidly to allow the polar attraction to join them together (Figure 14.12c). Water vapor is the least-ordered configuration of water molecules.

Changing States of Water

How does water change from one state to another? Think of a glass of ice water on a hot day. Eventually, the ice will melt and the water will warm up; you know from your own experience that this

happens because the water absorbs heat energy. Conversely, if you put a glass of warm water in the freezer, it will get colder and eventually become solid ice; this happens because the water releases heat energy. As water absorbs or releases heat, it changes state. However, not all heat absorbed is used to raise temperature; some may be required to change the heated material from one state to another.

For example, if a pan containing a mixture of water and ice (initial temperature 0°C [32°F]) were heated on a stove, a thermometer would not record a change in temperature until all the ice had melted. The water-ice mixture would first absorb the heat to convert the ice to water, but there would be no corresponding change in temperature until all the ice had changed state. **The amount of heat absorbed or released as water changes state is termed *latent heat*.**

The absorption of latent heat occurs during melting, evaporation, or sublimation (the change from a solid directly to a gas); the release of latent heat results from freezing, condensation, or deposition (Figure 14.13). The amount of heat gained or lost per gram of water is expressed in calories of latent heat. Thus, the conversion of ice to liquid water absorbs 80 calories of heat for each gram of water converted from a solid to a liquid state. In the reverse reaction, 80 calories of latent heat are released into the atmosphere per gram of

a. Frozen water　　**b.** Liquid water　　**c.** Water vapor

Figure 14.12　Molecular structures of the three states of water. Individual water molecules are composed of one atom of oxygen (red) and two atoms of hydrogen (tan). Molecules are more ordered (that is, more closely spaced and move less) in the **a.** solid state and become less ordered progressing to the **b.** liquid and **c.** gaseous states.
14.12a, 14.12b, 14.12c: Adapted from *Physical Geography Net*

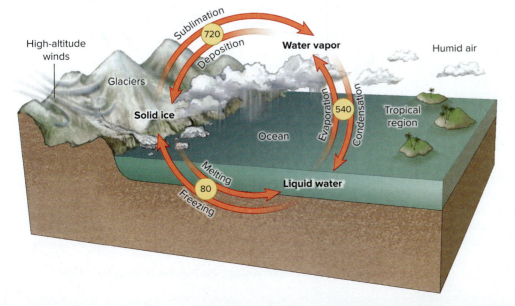

Figure 14.13　Release or absorption of latent heat. Six potential changes in the state of water are possible. The calories per gram of water (heat) that are lost or gained during each change are least for freezing/melting (80) and are greatest for changes between solid and vapor forms (720). The calories released or absorbed during condensation or evaporation are estimated for water at 0°C (32°F). When water changes directly from a solid to a vapor during sublimation, 720 calories per gram of latent heat are required. The latent heat requirements are: 80 calories per gram (solid-liquid) + 100 calories per gram (for liquid-phase temperature rise to 100°C) + 540 cal/g (liquid-vapor).

water that is converted from liquid to solid (Figure 14.13). Much more latent heat is released or absorbed during changes between liquid and gaseous states (evaporation, condensation) than during changes between solid and liquid states. This is a function of the number of bonds that must be broken or modified between water molecules.

Of the six potential changes in the state of water just mentioned, **evaporation and condensation are especially significant.** These two processes occur over large areas of Earth's surface and contribute to weather phenomena and redistribution of heat in the atmosphere.

During evaporation, liquid water is converted to water vapor. You may have noticed this change of state occurring on your skin after you step out of a hot shower into a cool room or when you perspire during exercise. Your body supplies the heat needed for evaporation, and as heat is released, you feel cooler. In the environment, water evaporates from bodies of water and becomes water vapor in the air.

Condensation occurs when a gas is converted to a liquid. During this process, heat is released into the surrounding environment. For example, condensation often causes a film of water to form on the outside of a glass containing an iced drink

✔ Checkpoint 14.13

| ✓ basic | ✓ advanced |
| ✓ intermediate | ✓ superior |

Your body feels cooler when you step out of a warm shower because

   a.   water evaporates on your skin.
   b.   water condenses on your skin.
   c.   water evaporates from the surrounding air.
   d.   water condenses in the surrounding air.

Figure 14.14   Condensation. The lemonade in the glass is initially at room temperature. Ice cubes are added to the glass, causing the temperature of the lemonade and the air surrounding the glass itself to decrease. Moisture forms by condensation on the outside of the glass when the temperature of the air around the glass reaches the dew point.
Adapted from Eric Danielson, *Meteorology 2e* © 2003

(Figure 14.14). The air around the glass becomes slightly warmer as a result of the condensation (but not enough for you to notice). In the environment, water vapor in the air releases heat as it condenses and becomes liquid water (rain).

Humidity

Evaporation and condensation determine **the amount of moisture in the air, termed *humidity*.** Have you ever experienced a hot, humid day? How about a cold, humid day? We rarely experience humid conditions in cold winter weather because cold air is less able to hold moisture than warm air (Figure 14.15). **Air is considered saturated with water if it can hold no more water vapor at a given temperature.** For example, based on Figure 14.15, air at 30°C (86°F) can hold a maximum of 34 grams per cubic meter of water vapor (on the saturation curve), but air with a temperature of 20°C (68°F) can hold only 18 grams per cubic meter.

There are two ways to measure humidity: absolute humidity and relative humidity. **Absolute humidity measures the mass of water (grams) in a volume of air (cubic meters),** essentially the density of water vapor (mass per volume). Absolute humidity is expressed as grams per cubic meter (g/m^3). However, when your local weather forecasters report humidity, they are referring to relative humidity. **Relative humidity measures the amount of water vapor in air in comparison to the maximum mass of water vapor the air could possibly hold (that is, if it were saturated).** Relative humidity is expressed as a percentage. If we were to describe how much water was in a glass, we might say the glass was half full. Relative humidity does the same thing for the atmosphere. It tells us how much water vapor the atmosphere contains and how much more it can hold. Relative humidity is determined by using the following equation, where condensed water in the air and deposited water are not included:

$$\text{Relative humidity} = \frac{\text{actual water vapor density}}{\text{saturated water vapor density}} \times 100\%$$

For instance, according to Figure 14.15, air at 30°C (86°F) can hold a maximum of 34 grams per cubic meter of water vapor (on

Figure 14.15   Humidity and temperature. The amount of water vapor that air can hold per volume of air increases with increasing temperature.
Adapted from Konrad Krauskopf, *The Physical Universe* 11e © 2006

the saturation curve). If the air actually has 17 grams per cubic meter, its relative humidity would be 50 percent. In contrast, if the temperature were 19°C (66°F), air containing 17 grams per cubic meter would have a relative humidity of 100 percent (Figure 14.15).

Changes in Humidity. At any given time, evaporation and condensation are both occurring in air, but one process typically dominates the other. Depending on which process comes out on top, humidity may increase or decrease. **Both absolute and relative humidity can increase when additional water vapor is added to the atmosphere by evaporation if the temperature of the air remains constant.** Absolute and relative humidity may decrease when condensation occurs, converting water vapor to precipitation, which may then leave the atmosphere. Different processes may dominate at different heights in the atmosphere. Therefore, it is possible for condensation to cause precipitation at high altitudes and for the raindrops to evaporate as they fall through a warmer, drier layer of air closer to Earth's surface.

In many instances, air is far enough from sources of water that the actual amount of water vapor in the air does not change as the air moves in the atmosphere. In Section 14.6, we will find that the temperature of air changes when that air rises or falls. What happens to the relative humidity of air if that air gets cooler but no additional water is added to the air? Consider Figure 14.15 again. Imagine a parcel of air with a temperature of 34°C (93°F) and an absolute humidity of 20 grams per cubic meter. If the absolute humidity remains constant and temperature decreases, the relative humidity will increase from 50 percent (move straight left on the chart). The parcel of air would become saturated when the temperature fell to 22°C (72°F). (Note that we are now on the 100 percent saturation curve.) **Relative humidity increases if air temperature decreases** because the saturation level decreases but the amount of water vapor in the air remains unchanged. The reverse is true if the air warms (saturation level increases, relative humidity decreases).

One key difference among the three states of water is the velocity of the water molecules. When cold air moves over a warm water body, evaporation from the water body adds moisture to the air. This is the water vapor (called *steam fog*) that can commonly be seen rising from a warm pond or lake in the early morning during summer, when air temperatures are cooler than water temperatures.

The bonds between the water molecules break as the velocity of the molecules increases and some of the liquid is converted to a gas phase. Reducing air temperature decreases the saturated vapor density because colder air "holds" less moisture than warm air. The *dew point* is the temperature the air must attain for that air to become saturated. When a warm air mass moves over cold water, the cold water cools the air. When the air temperature drops, some moisture condenses and may form fog. **Condensation occurs when the relative humidity of the air increases so much that the air becomes saturated with moisture** (relative humidity = 100 percent). As these fog examples illustrate, relative humidity can increase with the addition of more water vapor or as a result of a decrease in temperature.

✓ **Checkpoint 14.15**

Two identical pans sit on a stove. Pan A contains a mixture of water and ice. Pan B contains an equal volume of water that is chilled to the same temperature. How will the temperatures of the contents change as the pans are heated?

Each pan contains a thermometer that records the temperature of the water and of the water-ice mixture. The stove is turned to high, and both pans are heated until boiling occurs. Heat continues to be applied for several more minutes.

Plot separate curves on the following graph to illustrate how temperature changes with time for both pans throughout the experiment. Label the curves.

✓ **Checkpoint 14.14**

Measurements reveal that 1 cubic meter (35 cubic feet) of air at 12°C (54°F) actually holds 6 grams (0.2 ounce) of water. What happens if the temperature of the air increases? Explain your answer.

a. Absolute and relative humidity both increase.
b. Absolute humidity increases and relative humidity remains constant.
c. Absolute and relative humidity both decrease.
d. Absolute humidity remains constant and relative humidity decreases.

✓ **Checkpoint 14.16**

a. Explain why people can see their breath on a cold winter's day in terms of water changing state and latent heat.
b. Explain why a hair dryer actually dries your hair rather than leaving you with hot, wet hair. Explain the process taking place.

14.6 Air Pressure, Condensation, and Precipitation

Learning Objectives

- Describe the relationship between air pressure and air density.
- Predict how changes to air pressure will affect air temperature.
- Determine the temperature and humidity of air as it rises or falls in the atmosphere.
- Explain connections among lapse rates, cloud formation, and precipitation.

Air Pressure and Air Density

The pressure exerted by the weight of an overlying column of air is atmospheric pressure. For example, the average air pressure on Earth's surface (1,013 millibars, or 14.7 pounds per square inch) is determined by the weight of a 500-kilometer-tall (310-mile) air column. Air pressure declines with increasing altitude (Figure 14.16) because, as altitude increases, there is less and less overlying air.

Air pressure at any point is also influenced by air density. **Air density is determined by measuring the mass of atoms and molecules of nitrogen, oxygen, and other gases per volume of air.** Think of air as a box of gas molecules. The more molecules that are in the box, the greater the density of the air (Figure 14.17a). Air pressure would decrease uniformly with altitude if gases were evenly distributed throughout the atmosphere. However, gravity pulls most gas molecules close to Earth's surface, increasing the air density closer to

the surface in the troposphere and decreasing the density in the outer layers of the atmosphere. Consequently, 50 percent of all air lies below 5.5 kilometers (3.4 miles) of altitude; therefore, air pressure at this altitude is one-half the air pressure at sea level (Figure 14.16). Almost all the air in the atmosphere lies below the height where Felix Baumgartner stepped out of his capsule for his big jump.

✓ Checkpoint 14.17

basic | advanced
intermediate | superior

Which of the following images best approximates the distribution of the two principal gases in Earth's atmosphere? Assume the bottom of each diagram represents relative distributions of gases at Earth's surface and elevation increases toward the top of the page.

● Oxygen   ● Nitrogen

a.   b.   c.   d.

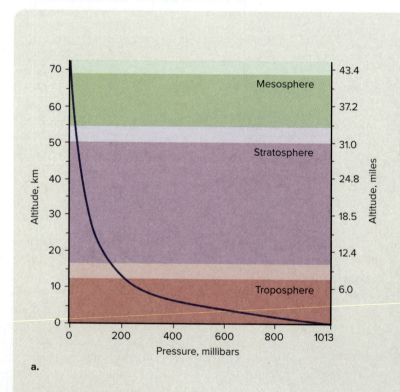

a.

Figure 14.16 Air pressure patterns in the atmosphere. **a.** Air pressure declines with increasing altitude because the bulk of air molecules lie close to Earth's surface. What is the air pressure at an altitude of 15 kilometers (9 miles)? **b.** Air pressure can be measured on a barometer.

14.16a: Adapted from American Chemical Society, *Chemistry in Context: Applying Chemistry to Society* 5e © 2006; 14.16b: © Steve Cole/Getty Images

Oxygen   Nitrogen

More molecules → More density

a.

Volume decreases ← More density

Volume increases → Less density

b.

Figure 14.17 Air density. The number of molecules of gas in a given volume of air determines the air density. **a.** In two equivalent volumes of air, the air mass with the greater number of molecules would have greater density. Which box would represent air closer to Earth's surface? **b.** If the number of molecules remains constant, density increases with a decrease in air volume.

Air density changes as air is warmed or cooled. **Air contracts when cooled,** resulting in gas molecules occupying a smaller volume, consequently increasing density and **generating higher air pressures.** In contrast, **air expands when it is warmed,** resulting in the same gas molecules occupying a larger volume and **generating a decrease in air pressure.** Think of our box of gas molecules: Cooling causes the box to get smaller, but the number of molecules remains the same, so the density increases (Figure 14.17b). In contrast, warming causes the box to expand, so the molecules are farther apart and density is less.

On the basis of temperature alone, where do you predict high- (or low-) pressure air would form on Earth's surface? It should come as no surprise that the highest atmospheric pressures are recorded in cold regions and the lowest pressures are recorded in tropical warm environments (associated with hurricanes and typhoons; Figure 14.18). If we move upward through the atmosphere, **air pressure decreases rapidly at lower altitudes where air density is greatest.** Air pressure decreases more slowly at higher altitudes above 10 kilometers (6 miles) because the air density there is lower (see Figure 14.16). Now back to Joe Kittinger's balloon—why do you think it got larger as he ascended? In his case, the balloon was expandable and filled with helium. As the balloon ascended, it expanded because the air pressure outside was decreasing.

Air pressure varies not only with elevation but also with weather systems. Weather maps typically display air pressure in millibars (mbar). These pressures are measured using a device called a *barometer* (Figure 14.16). Meteorologists must correct for changes in altitude to understand the pressure characteristics of air masses in order to forecast future weather patterns. Pressure data on weather maps are therefore recalculated to reflect pressure measurements for hypothetical sea level. If no effort were made to recalculate air pressures, we would always have low-pressure conditions over mountains due to less-dense air higher in the troposphere and higher pressures along coastlines due to more-dense air at low elevations. This would make accurate weather forecasting nearly impossible because it would not reflect the true difference in the pressures of air masses as they moved to common altitudes.

Effects of Air Pressure on Temperature

How does the pressure of your car tire compare to that of the surrounding atmosphere? Typical tire pressures are about twice that of the outside air. Have you ever noticed that air escaping from a tire is colder than the surrounding air? This temperature change is the result of the drop in pressure as the air expands as it escapes from the high pressure in the tire to the relatively lower pressure in the surrounding atmosphere. **Compressed air becomes warmer; expanding air becomes cooler.** Similar increases or decreases in temperature result from pressure changes in nature. These changes in temperature are called *adiabatic changes* **because they occur without any loss or gain of energy to or from the surrounding air** (as if the air mass were thermally insulated). We can observe a vivid example of this phenomenon when meteors burn up in the

✓ **Checkpoint 14.8**

🔵 basic   🟡 advanced
🔵 intermediate   🟢 superior

In a few sentences, summarize the relationship among density, temperature, and pressure in a mass of air.

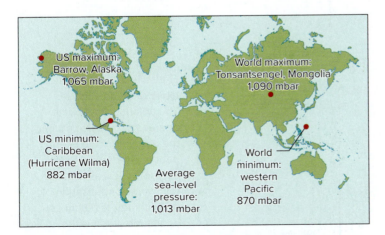

US maximum: Barrow, Alaska 1,065 mbar

World maximum: Tonsantsengel, Mongolia 1,090 mbar

US minimum: Caribbean (Hurricane Wilma) 882 mbar

Average sea-level pressure: 1,013 mbar

World minimum: western Pacific 870 mbar

Figure 14.18 Maximum and minimum global pressure measurements. The highest pressures are found in colder regions, while the lowest values are associated with hurricanes (called typhoons in the Pacific Ocean) in tropical oceans. mbar = millibar.

upper atmosphere (Figure 14.19). A meteor arrives at Earth's atmosphere after traveling through the vacuum of space at more than 50,000 kilometers per hour (31,000 miles per hour). The incoming meteor slows down as it compresses the gases of the upper atmosphere, causing the air temperature around the meteor to rise rapidly. The heat from the compressed gases raises the temperature of the meteor. (*Note:* The heating does not occur because of friction— there are not enough atoms in the upper atmosphere for friction to be that significant!) The extreme heat causes the meteor to glow red hot at temperatures of more than 1,500°C (2,730°F). Faster-moving objects compress more air and generate greater heat; slower-moving objects push against less air and produce less heat.

Some of the mechanical energy needed to force air molecules into a small space in front of the speeding meteor is converted to heat as air is compressed. These temperature changes represent the conversion of mechanical energy into heat energy and are not associated with the addition or loss of heat from outside sources (for example, the sun). In the same way, the air in a car tire warms up as tire pressure is increased, or the barrel of a hand pump becomes warm as a bicycle tire is pumped up. When air expands adiabatically, it converts heat to mechanical energy as air molecules move apart, so the air escaping from a punctured tire feels cooler than the surrounding air. Why are these changes in temperature associated with changes in pressure important? As we will learn in this chapter and Chapter 15, these adiabatic temperature changes help fuel the processes that cause air to rise and promote the development of extreme weather phenomena such as thunderstorms.

Adiabatic Lapse Rates

Perhaps you have been to the mountains and noticed that the air is cooler at higher altitudes than it is at lower elevations. To understand why, we must consider the air in distinct parcels—volumes that can behave adiabatically and independently of the mass of surrounding air. The temperature of the atmosphere is largely determined by the lapse rate, which is the rate at which air cools as it rises

or warms as it falls. **The actual temperature of a rising parcel of air at some altitude is a function of the temperature at which it starts, how high it has risen, and the adiabatic cooling rate. That parcel of air also expands because air pressure is decreasing.** In Section 14.4, we noted that air temperature declines at a rate of 6.5 C° per kilometer through the troposphere (known as the normal or average lapse rate). This is the lapse rate we would expect to find if we measured many locations on Earth and averaged the readings. If a parcel of air in some location begins to rise, an additional cooling effect occurs because of the temperature changes resulting from the expansion of air. The total amount of energy present in the parcel of air will not change, but it can be used either to maintain a constant temperature or to work to expand the size of the parcel of air. As the air expands, the heat energy is converted to mechanical energy and distributed through a larger volume, producing a cooling effect. This results in a **dry adiabatic lapse rate of 10 C° per 1,000 meters (10 C° per kilometer; 29 F° per mile) for unsaturated air** (Figure 14.20).

✓ Checkpoint 14.19

An instructor asked her class to summarize some information about atmospheric processes. Students submitted the following four statements as part of their answers. The instructor returned the statements and told the students that they could correct them for full credit.

Identify what is wrong with each statement, and describe how you would fix these answers to earn full credit.

1. The temperature of a rising parcel of air decreases by the normal lapse rate.
2. The percentage of oxygen in the atmosphere decreases with altitude.
3. When it rains, you have to use the wet adiabatic lapse rate to figure out temperatures at higher elevations.
4. The dry adiabatic lapse rate is higher than the wet adiabatic lapse rate, so air temperatures should be higher in dry air (before condensation occurs) than in wet air (after condensation occurs).

Figure 14.19   A Perseid meteor from the 2002 shower over the southwestern United States. © Joyce Photographics/Science Source

Figure 14.20   Adiabatic lapse rate and altitude. The adiabatic lapse rate decreases because of the addition of the latent heat of condensation for a hypothetical air mass following condensation at 3 kilometers (1.9 miles) of altitude. What is the temperature of a parcel of air rising to 2 kilometers (1.2 miles) of altitude?

As the rising parcel of air cools, its relative humidity increases and the air becomes saturated with water vapor at some elevation. At that point, condensation occurs, and the lapse rate becomes a "wet" adiabatic lapse rate. Latent heat is released in the conversion of water from a vapor to a liquid state, and the addition of the latent heat partially counterbalances the cooling effect of the adiabatic temperature changes. This results in a reduction in the cooling rate. **An average value for the wet adiabatic lapse rate is 6 C° per kilometer (18 F° per mile)** (Figure 14.20). The parcel of air will continue to rise until its temperature (and related density) is the same as that of the surrounding air.

Condensation and Cloud Formation

Rising air gradually cools, slowly increasing the relative humidity of the air mass to the point where saturation occurs, triggering condensation and cloud formation. However, it is not enough just to cool the air for condensation to occur. **To form a cloud, water vapor must also have surfaces upon which to condense.** In the atmosphere, the surfaces come from microscopic aerosol particles (dust, smoke, salt, pollutants) that range in size from 0.001 to 10 micrometers (wider than a billionth, narrower than a thousandth of an inch). For reference, the diameter of human hair is about 50 micrometers. On land, it could be any surface, including grass or your car (dew or frost). In the air, billions of water droplets approximately 20 micrometers in diameter can form on aerosol particles in a single cubic meter of air. The droplets are readily kept airborne by air turbulence until they collide and coalesce (clump) to form larger drops. **Clouds are composed of billions of tiny water droplets that can combine to form rain or hail.**

Precipitation

The larger, heavier **cloud droplets fall, colliding with other droplets and combining with them, a process called *collision coalescence*, to form raindrops** (containing approximately 1 million cloud droplets each). Updrafts in the air can force the droplets upward and recycle them through the cloud, increasing droplet size. Updrafts are relatively common in large clouds as condensation warms the air, causing it to rise. During times of drought, efforts are sometimes made to artificially introduce condensation nuclei into the atmosphere by cloud seeding. Unfortunately, there is little hard evidence that such methods work consistently.

While the collision coalescence process effectively explains rainfall from warm clouds, it is less effective in accounting for the formation of rain from clouds below the freezing point of water. **Air temperatures fall below freezing within a few kilometers of Earth's surface in the lower portion of the troposphere** at normal lapse rates. Consequently, given the typical range of cloud elevations, much of the water in clouds is in the form of supercooled droplets with temperatures ranging from 0°C to −20°C

(32°F to −4°F). You might expect that any water at these altitudes would freeze, but it does not. Supercooled water can freeze only with sufficient agitation (strong vibrations) or when there is a surface upon which to freeze (such as a dust particle or ice crystal). Pure water droplets in high clouds can remain in liquid form down to temperatures as low as −39°C (−38°F).

So how do raindrops form in these supercooled clouds? It turns out that most clouds contain ice crystals and supercooled water droplets. When a water droplet comes close to an ice crystal, supercooled water will transfer onto the ice crystal because air on that surface is unsaturated (compared to air at the surface of the water droplet). When that happens, the air immediately next to the water droplet appears subsaturated, and evaporation occurs. That moisture is also added to the ice crystal, thus growing the crystal. The crystal will fall when it becomes too heavy to be supported by air currents, melting along the way to form cold rain. This **mechanism whereby raindrops are formed is called the *Bergeron process*** after the scientist who first described the process.

How does snow form? Recall that absolute humidity measures the mass of water in a volume of air. The absolute humidity of air required for saturation decreases with temperature (see Figure 14.15). Cold air with temperatures between 0°C (32°F) and −20°C (−4°F) would be saturated with 5 grams per cubic meter to 1 gram per cubic meter of water vapor, respectively. Below temperatures of about −5°C (23°F), air needs a little less water vapor to be saturated for ice than for water. Under these conditions, condensation preferentially produces ice crystals rather than water droplets. **Miniature ice crystals act as condensation surfaces and gradually increase in size to form snowflakes.** Eventually, precipitation in the form of ice or snow occurs, unless the frozen water is warmed as it falls through the atmosphere, resulting in rain.

14.7 Clouds and Frontal Systems

Learning Objectives

- Classify and explain the processes forming clouds of different shape, altitude, and propensity to produce precipitation.
- Compare and contrast the four main mechanisms and/or processes that lift air.

More than 30,000 flights cross the skies of the United States on an average day. As jets pass overhead, their exhausts expel water vapor and a mixture of other gases into the chilly air of the upper troposphere. The water vapor immediately condenses to form a contrail (condensation trail), a skinny line of cloud following the path of the flight.

Climatologist David Travis used weather data from the 3 days following September 11, 2001, when all commercial flights in the United States were canceled, to try to unravel the impact of contrails on surface temperatures. Travis compared average temperature data from that 3-day period with data from the previous three decades. He found that night temperatures were generally cooler and daytime temperatures were warmer than during days with normal flight operations. Travis interpreted the data to show that contrails absorbed heat during the night, causing a warming effect, and had a cooling influence during the day as they reflected

✓ Checkpoint 14.20

⊘ basic ⊘ advanced
⊘ intermediate ⊘ superior

Describe what would happen to a parcel of air that begins to rise. Your answer should discuss the normal lapse rate, dry and wet adiabatic lapse rates, and humidity. Include a sketch showing the parcel of air at different altitudes.

incoming sunlight. When he examined areas with the heaviest air traffic, the temperatures differed by as much as 3 C° (5 F°). The fact that these thin, transitory, upper-level clouds have an effect on surface temperatures illustrates that the type and distribution of all clouds likely affect solar radiation and temperatures across Earth's surface.

Much of the incoming solar radiation is either absorbed by clouds or reflected into space from cloud surfaces (see Figure 14.9). Increasing the thickness or water vapor content of low-level clouds causes more sunlight to be reflected from the cloud surfaces (see Chapter Snapshot) and thus reduces the heat energy that contributes to the greenhouse effect. In contrast, water vapor in clouds is a key greenhouse gas that absorbs heat energy close to Earth's surface; therefore, increasing cloud cover might act to increase the amount of heat absorbed in the lower atmosphere. On any given day, clouds cover about 70% of Earth surface and ongoing atmospheric research is attempting to determine whether clouds provide a net negative or net positive feedback to global warming processes or whether their warming and cooling effects essentially cancel one another out. Unfortunately, the varied lateral and vertical distribution of clouds and their rapidly changing shape make them difficult to incorporate into climate computer models.

Cloud Classification

Clouds are classified on the basis of their altitude, appearance, and associated precipitation (Figure 14.21). High-level clouds are composed of ice; low-level clouds are usually composed of water droplets. Most cloud types occur at a specific range of altitudes, but some, such as cumulonimbus clouds associated with thunderstorms, may span several levels. The prefix *cirr-* indicates a high-level cloud with a wispy shape; *alto-* designates midlevel clouds; and *-cumulus* represents a heap shape. The prefix *nimbo-* or suffix *-nimbus* denotes rain clouds.

Figure 14.21 Types of clouds. Clouds are classified on the basis of their altitude and appearance. Most clouds occur at a specific range of altitudes, but some, such as cumulonimbus clouds associated with thunderstorms, may span several levels.
Adapted from Eric Danielson, Meteorology 2e © 2003

Cirrus clouds are thin, wispy clouds that form in the upper troposphere; they would be equivalent to contrails. Although cirrus clouds appear slight and delicate compared to other cloud types, they are effective heat-absorbing devices, soaking up heat radiating upwards from Earth's surface.

Stratus clouds form sheets that cover the whole sky and may form at any level. Especially thick stratus clouds are termed *nimbostratus*. Looking upward, an air traveler might spy altostratus clouds below high-level cirrostratus clouds.

Puffy, cauliflower-shaped cumulus clouds are the most commonly recognized clouds. If a sufficient number of these clouds form, they may create a low, irregular layer of stratocumulus clouds. Cumulus clouds at higher altitudes are termed *altocumulus* or *cirrocumulus* clouds (Figure 14.21). Cumulus and stratus clouds are more likely to reflect incoming solar radiation, resulting in a relative cooling effect.

Cloud Formation Mechanisms

Varying moisture levels in the air, combined with temperature and related pressure differences, cause imbalances in the Earth system that generate different types of clouds. As described in Section 14.6, clouds form when condensation occurs as rising air cools adiabatically. The tendency of air to rise (or fall) in the atmosphere is closely tied to its density. Just like a hot-air balloon, a parcel of air may rise naturally if it is lighter than the surrounding air masses; this process is called *density* (or *convection*) *lifting*. Otherwise, the parcel of air

may be forced upward by a variety of other processes such as frontal lifting, orographic lifting, or convergence lifting.

Density Lifting. What happens to warm air in your house, and why? As you have probably experienced, warm air rises. It does so because it is less dense (and therefore lighter) than the colder air. In the atmosphere, warm air is usually located near Earth's surface (why?). If air has a tendency to resist movement, it is considered *stable*; if it has a tendency to rise or fall, it is *unstable*. The temperature of air decreases with altitude at the normal lapse rate. So what should happen to a parcel of warm air near the surface? **Warm air will rise as long as the density of the parcel of air remains lower that of the surrounding air.** If the air is unsaturated, warm air will cool according to the dry adiabatic lapse rate (Table 14.1). The air continues to rise until it cools to a temperature equal to that of the surrounding air (Figure 14.22). At that point, the density of the rising air is the same as the surrounding air, and density lifting ends.

Table 14.1	Typical Temperature Changes in Unstable and Stable Air Masses	
Altitude, km (mi)	Warm Air Mass (dry adiabatic lapse rate)	Stable Air Mass (normal lapse rate)
0	20	11.25
0.5 (0.3)	15	8
1.0 (0.6)	10	4.75
1.5 (0.9)	5	1.5
2.0 (1.2)	0	−1.75
2.5 (1.5)	−5	−5

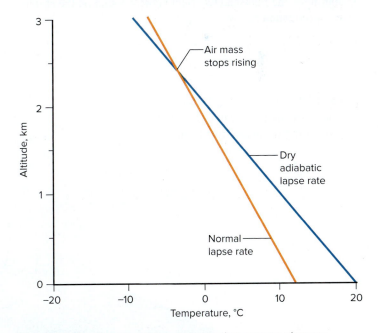

Figure 14.22 Density lifting. Assume that a warm air mass begins to rise at a temperature of 20°C (68°F) and that the surrounding air has a temperature of 11.25°C (52°F). The temperature of the rising air mass and the surrounding air will be equalized at −5°C (23°F) at an altitude of 2.5 kilometers (1.6 miles). At that altitude, all the air has the same temperature; therefore, there is no longer a separate parcel of rising air.

✓ **Checkpoint 14.21**

✓ basic	✓ advanced
✓ intermediate	✓ superior

Classify the clouds in each of the following images.

a.

b.

c.

d.

14.21a, 14.21b, 14.21c & 14.21d: NOAA

Frontal Lifting. Frontal lifting occurs when two large air masses of contrasting density (temperature, moisture content) meet (Figure 14.23). The boundary between the air masses is termed a *front* and may be 10 to 150 kilometers (6 to 93 miles) across and hundreds of kilometers in length.

A warm front forms when a warm air mass advances to meet a cold air mass. Which air mass should rise? The lighter warm air rises above the heavier colder air. Warm air is also forced upward when cold air advances on a warm air mass along a cold front. Cold fronts form when cold air advances over land previously occupied by warm air. Cold fronts are steeper than warm fronts and cause cloud formation and precipitation to occur across a narrower area. Severe weather (heavy rain and snow) usually occurs along cold fronts in the central portion of the continental United States. (See Chapter 15 for more on these weather patterns.)

Orographic Lifting. **Orographic lifting occurs when air is forced to rise over an obstruction in the landscape, typically a mountain range.** The side of a mountain sloping toward the oncoming wind is known as the *windward slope.* The other side of the range slopes in the direction that the wind is blowing and is known as the *leeward slope.* Condensation occurs as the air cools with increasing altitude as it rises up the windward flank of the range, and precipitation may occur at higher elevations. The air warms up and can absorb moisture as it descends the leeward side of the mountains, creating a **rain shadow effect where precipitation is relatively rare.** Where do you predict a rain shadow effect might occur in North America? Because air masses typically move from west to east across the western half of the continent, the rain shadow effect is often found on the eastern slopes of mountain ranges.

✓ Checkpoint 14.22

On July 2, 1982, truck driver Larry Walters decided to attach 45 helium-filled weather balloons to a lawn chair and go for a ride. Lawn Chair Larry rose to an altitude of nearly 5 kilometers (16,400 feet). The high elevations and lack of oxygen made him dizzy, so he decided it was time to deploy his principal altitude control device, a pellet gun. He shot out several balloons and descended back to Earth. Which process was most significant in Lawn Chair Larry's balloon ride? Explain your reasoning.

a. Density lifting
b. Orographic lifting
c. Convergence lifting
d. Frontal lifting

In the following landscape, how would the amount of rainfall change at location X if the mountain eroded down to the dashed line?

a. Rainfall would increase.
b. Rainfall would decrease.
c. Rainfall would stay the same.

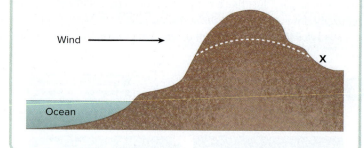

Figure 14.23 Frontal lifting. Different weather conditions and air temperatures are associated with advancing warm and cold fronts.

Convergence Lifting. Convergence lifting occurs when two air masses of similar temperatures collide, forcing some air upward because both air masses cannot occupy the same space (Figure 14.24). This happens regularly over Florida, where air moves westward from the Atlantic Ocean and eastward from the Gulf of Mexico to collide over the Florida peninsula.

✓ **Checkpoint 14.23**

Use the information from the chapter to explain:

a. Which air-lifting mechanism dominates where you live?
b. Which states might provide examples of the four air-lifting mechanisms?

✓ **Checkpoint 14.24**

A parcel of warm air with an average temperature of 29°C (84°F) and 50 percent relative humidity is at sea level. The surrounding air has a temperature of 26.5°C (80°F). The parcel of air begins to rise due to density lifting. The relative humidity of the air mass is 100 percent when it reaches 1-kilometer (0.6-mile) elevation. Analyze how the temperature of the air mass would change as it rises, and determine at what elevation it would become stable.

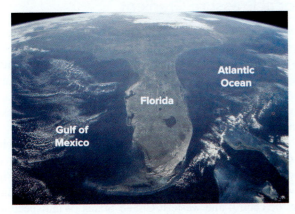

Figure 14.24 Convergence lifting. Condensation above the Florida peninsula may be assisted by density lifting because the air overlying the land mass will be warmer than the air above the adjacent ocean.
NASA

14.8 Winds

Learning Objectives

- Predict relative winds speed and direction, using atmospheric pressure data.

- Compare and contrast processes that form high- and low-pressure air masses.

- Explain how the pressure gradient, Coriolis effect, and friction combine to influence wind speed and direction.

- Illustrate how the atmosphere interacts with other components of the earth system.

A cabin on top of Mount Washington in New Hampshire is secured to the ground by chains to prevent it from blowing away (Figure 14.25). At 1,917 meters (6,290 feet), Mount Washington is the highest peak in the northeastern United States. The highest wind speed ever recorded, 371 kilometers per hour (231 miles per hour), was measured at a weather observatory on the mountain's summit on April 12, 1934. Scientists working at the station had to chip the ice off the anemometer (a device for measuring wind speed) to take the reading. On at least one occasion on more than 100 days each year, the observatory experiences winds with speeds equivalent to those of a typical hurricane. What is it about this location that makes it the home for such ferocious winds? To answer that question, we first have to understand a little more about how air pressure varies on Earth's surface.

The Relationship Between Air Pressure and Wind

We have spent much of the second half of this chapter describing the vertical variations in air pressure—basically, how air pressure decreases with increasing altitude. However, examination of weather

Figure 14.25 Secured building at the Mount Washington Observatory. Note the three sets of chains anchored to the ground to hold the building in place during high winds.
Mt. Washington Observatory

Figure 14.26 Pressure map of North America. Atmospheric pressure for September 25, 2006, is high over the West and low over the Northeast. The lines of equal pressure, called isobars, are spaced 4 millibars apart. What is the range of pressures indicated by the map?
Department of Atmospheric Sciences, University of Illinois

maps reveals that air pressure is not constant across the surface of North America. On a typical day, air pressure over the United States, Canada, and Mexico ranges from lows of around 1,000 millibars (mbar) to highs of more than 1,020 millibars (Figure 14.26).

Why are pressures not the same in different parts of the continent at the same latitude? As we will learn in Chapter 15, **air above Earth's surface inherits the thermal characteristics of the underlying surface.** If the surface is cold compared to the air, the air above the surface is cooled; if the surface is warm, the air is warmed. These temperature differences lead to regional differences in air density and, consequently, differences in atmospheric pressure.

Wind is the horizontal movement of air that arises from differences in air pressure. Winds represent moving air as it **flows from areas of high pressure to areas of low pressure.** Mount Washington's high elevation and location—between the cold air sweeping south from Canada and the air moving north from the Gulf of Mexico or east from the Atlantic Ocean—makes it an ideal place to experience large differences in pressure in the transition zone between pressure systems. Large pressure differences produce fast-moving winds.

Winds are characterized by their speed and direction. **Wind direction refers to the direction from which the wind originates rather than where it is going.** For example, a westerly wind blows from west to east. Three factors influence wind speed and direction: the pressure gradient, the Coriolis effect, and friction with the surface.

Checkpoint 14.25

basic • advanced • intermediate • superior

Find your state on the map in Figure 14.26. On the basis of the isobars, in what direction is the wind blowing in your state?

Regional Pressure Gradient

The pressure gradient is the magnitude of the change in pressure between two points, divided by the distance between those points. **The greater the contrast in pressure, the steeper the gradient will be and the faster the wind will blow.** Isobars join points of equal pressure plotted on a map (Figure 14.26). If no other factors influenced wind direction, winds would blow down the pressure gradient, perpendicular to the isobars.

Since isobars indicate differences in pressure, **the closer the isobars are together, the greater the pressure drop with distance and the faster the wind speed.** For example, in Figure 14.26, wind speed would be greatest over southeastern Canada and least in the central and western United States. Wind speed associated with a pressure gradient between high- and low-pressure systems would increase if the systems moved closer together or if atmospheric pressures became more extreme (Figure 14.27). Initial wind directions will be modified by the influence of the Coriolis effect and friction close to the planet's surface.

Coriolis Effect

Recall from Chapter 13 that atmospheric and oceanic circulation patterns are deflected to the right of their course in the Northern Hemisphere and to the left of their course in the Southern Hemisphere. This pattern, termed the *Coriolis effect,* occurs because objects at different latitudes travel on Earth's surface at different speeds as a result of the rotation of the planet.

> ✓ **Checkpoint 14.26**
>
> | ❷ basic | ❷ advanced |
> | ❷ intermediate | ❷ superior |
>
> Draw arrows indicating the direction of winds in Figure 14.26 on the basis of the pressure gradient alone. Where would the fastest winds be found?

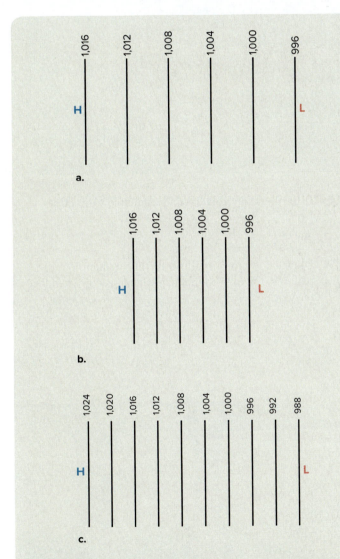

Figure 14.27 Two mechanisms for increasing wind speed between high- and low-pressure systems. Pressure expressed in millibars. **a.** Neighboring high- and low-pressure systems initially have uniform spacing between isobars. **b.** Higher wind speeds are indicated by closer spacing of the isobars as pressure systems converge. **c.** Higher wind speeds are indicated by closer spacing of isobars as pressure systems strengthen.

Winds are deflected to the right of their course in the Northern Hemisphere. These winds continue to be deflected until the pressure gradient effect balances the Coriolis effect and the wind moves parallel to the isobars (Figure 14.28). **Winds blowing parallel to isobars are termed *geostrophic winds.*** Geostrophic winds are disrupted by the influence of friction immediately above Earth's surface but are fully developed a few kilometers above the surface where friction is no longer significant.

Friction

Wind blowing near Earth's surface is slowed by frictional drag from the surface. For instance, friction is most dramatic above rugged land surfaces such as mountain ranges and is least over flat land or open water. An increase in friction reduces the Coriolis effect because the deflection is proportional to wind speed. The diminished Coriolis effect cannot balance the influence of the pressure gradient and typically results in nongeostrophic winds close to Earth's surface (Figure 14.29).

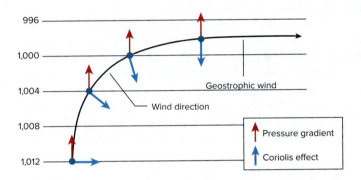

Figure 14.28 Interaction of pressure gradient and Coriolis effect. The pressure gradient causes winds to flow from high pressure toward low pressure. Eventually, the pressure gradient is exactly offset by the Coriolis effect, forming geostrophic winds in the middle to upper troposphere that blow parallel to isobars.

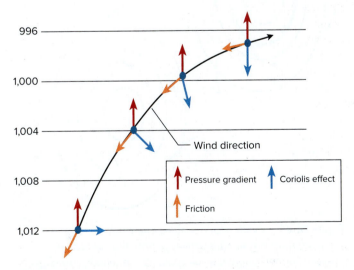

Figure 14.29 Influence of friction. Friction close to Earth's surface reduces pressure gradient and Coriolis effects, resulting in winds that are oriented at an angle to isobars.

Therefore, at the surface, winds are oblique (at an angle, rather than perpendicular) to isobars. The angle between wind direction and the pressure gradient depends on the character of the underlying surface. Generally, wind directions are oriented 40 to 50 degrees from the surface isobars. The more varied the land surface, the greater the friction produced and the larger the angle between wind direction and isobars. Friction is less significant above flat terrain or the ocean surface, but still results in winds that make an angle of 10 to 20 degrees with the regional isobars.

Cyclones and Anticyclones

The combined effect of influences on wind direction is that airflow forms a spiral pattern as it rotates clockwise or counterclockwise around high- or low-pressure systems, respectively. Winds converge toward a low-pressure system from surrounding regions of higher pressure. These converging winds will be deflected to the right of their course, setting up a **counterclockwise airflow, known as a** *cyclone,* **around the low-pressure center** (Figure 14.30a). In contrast, winds blowing outward, away from a high-pressure center, will be deflected to the right of their course to produce a **clockwise flow of air known as an** *anticyclone* (Figure 14.30b).

Low-pressure systems are rapidly dissipated by converging air unless the inrushing air is balanced by a rising air column. **Rising air in low-pressure systems becomes cooler and may reach saturation, resulting in clouds and rain.** At the top of the rising column of air, there must be divergent airflow to balance the convergent flow at the surface (Figure 14.31). Anticyclones must have convergent flow aloft so that air is continually fed into the high-pressure systems to ensure their survival (Figure 14.31).

Figure 14.30 Cyclone and anticyclone in low- and high-pressure systems. **a.** Winds diverge from high-pressure systems, generating a clockwise airflow. **b.** Winds converge in low-pressure systems, creating a counterclockwise airflow at the surface.

✓ Checkpoint 14.27

1. Draw an arrow to show the location and direction of the highest wind velocity on the following air pressure map, on the basis of the isobars alone.

2. Which point on the map is at the center of a low-pressure system?
   a. A   b. B   c. C   d. D   e. E

3. Rising air is present at which point?
   a. A   b. B   c. C   d. E   e. G

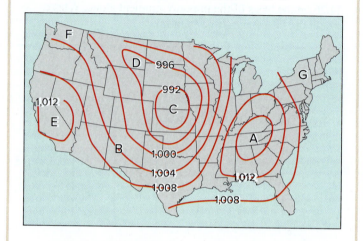

4. If Earth did not rotate, wind would blow directly from _____.
   a. A to C   b. C to A   c. C to B   d. F to E

5. Draw arrows to fully illustrate the circulation patterns of winds on the map when it is corrected for the Coriolis effect.

Figure 14.31 Converging and diverging winds. Rising air produces clouds over low-pressure systems, and descending air results in clear skies above high-pressure systems.

Air descends in high-pressure zones, warming as it approaches Earth's surface. As the air becomes warmer, its relative humidity decreases. Hence, high-pressure anticyclones result in clear skies and dry weather (see Figure 14.26), whereas low-pressure cyclones are often associated with rainfall and cloud formation. We will revisit low- and high-pressure systems in Chapter 15 because they play a significant role in controlling regional weather patterns.

Wind Energy

Wind power accounts for more than 4 percent of US electricity (Figure 14.32a), and its share of national electricity generation is growing faster than any other energy source. Technological advances are making wind power increasingly competitive, with costs ranging from 1 to 10 times those of fossil fuels. Suitable wind velocities (over 20 kilometers per hour; 12 miles per hour) are consistently

a.

Figure 14.32 Wind energy. **a.** Wind farm of multiple turbines, California. **b.** Map of US wind resources with economic potential. Red and blue colors indicate regions with greatest potential for the development of wind energy resources (more consistent winds, higher wind speeds).
14.32a: © Thinkstock/Masterfile; 14.32b: © Department of Energy

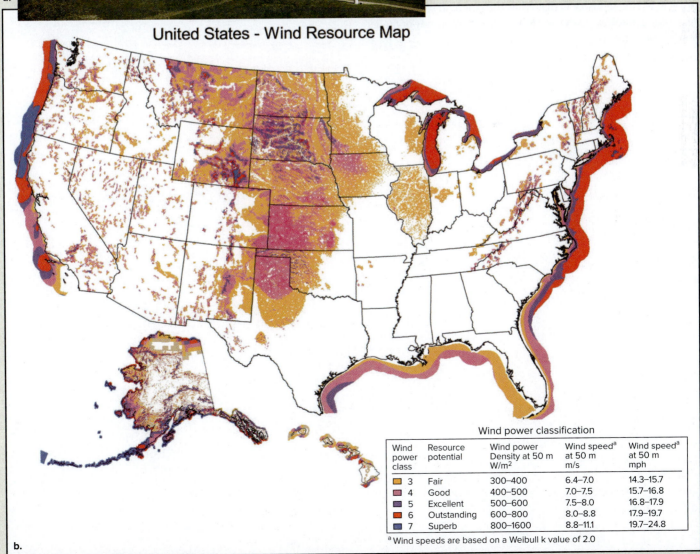

United States - Wind Resource Map

Wind power classification

Wind power class	Resource potential	Wind power Density at 50 m W/m²	Wind speed[a] at 50 m m/s	Wind speed[a] at 50 m mph
3	Fair	300–400	6.4–7.0	14.3–15.7
4	Good	400–500	7.0–7.5	15.7–16.8
5	Excellent	500–600	7.5–8.0	16.8–17.9
6	Outstanding	600–800	8.0–8.8	17.9–19.7
7	Superb	800–1600	8.8–11.1	19.7–24.8

[a] Wind speeds are based on a Weibull k value of 2.0

b.

present over about 13 percent of the United States (Figure 14.32b), and some projections estimate that wind power could generate as much as 20 percent of US energy within two decades. One constraint on the use of wind energy is the amount of land required. Power plants using fossil fuels or nuclear power require a relatively small land area. Depending on conditions, a wind energy installation would require about 20 times more land to generate the same power as a single nuclear power plant.

Many early US wind turbines were located in mountain gaps and other particularly windy parts of California. The land areas with the greatest potential for wind power are determined by the patterns of prevailing winds. These winds consistently exhibit sufficient velocity and reliability only over the the region of the continental US covered by the Great Plains states. In 2015, there were twelve states that generated at least 10 percent of their electricity from wind, and three, all in the Great Plains (South Dakota, Iowa, and Kansas) that have crossed the 20 percent mark. Texas generates more wind energy than any other state. An alternative option would be for coastal states to build offshore wind farms to take advantage of steady coastal winds (Figure 14.32b). However, such plans are often contested on the basis of the potential for lost tourist revenues. It remains unclear if the mix of commercial and government support will be sufficient to continue the rapid growth of wind energy so that it accounts for a more substantial proportion of national electricity generation in years to come. One benefit of the growth in this industry is that it has the potential to reduce the volume of greenhouse gases currently pumped out by power plants driven by fossil fuels.

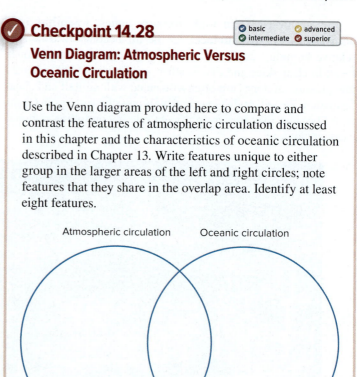

✓ **Checkpoint 14.28**

☑ basic ☑ advanced
☑ intermediate ☒ superior

Venn Diagram: Atmospheric Versus Oceanic Circulation

Use the Venn diagram provided here to compare and contrast the features of atmospheric circulation discussed in this chapter and the characteristics of oceanic circulation described in Chapter 13. Write features unique to either group in the larger areas of the left and right circles; note features that they share in the overlap area. Identify at least eight features.

Atmospheric circulation   Oceanic circulation

the big picture

While at first you may have thought that this chapter would describe a bunch of features in the atmosphere, you should have realized by now that this is really a chapter about energy, especially in the form of heat. The composition of the atmosphere means that enough heat derived from solar radiation is absorbed near the surface to make our planet livable. The amount of heat that land or water can absorb depends on their heat capacity. Water is a terrific place to store heat. It soaks up heat like a sponge, with only a relatively modest increase in temperature. Heat absorbed by water will either slowly raise the water's temperature or fuel changes in the water's state. These changes in state from solid to liquid to gas and back again do not create or destroy energy; they just transfer it from one place to another. The simple processes of evaporation and condensation result in the vertical transfer of heat energy from the lower troposphere to the upper troposphere. The horizontal transfer of warm and cold air is driven by winds produced by pressure differences between bodies of air with different temperatures. So the vertical and horizontal mixing of the atmosphere acts as a heat-transfer system.

Pity the poor pitchers for the Colorado Rockies baseball team. They consistently give up more of their home runs while playing at their home stadium, Coors Field in Denver. It can be hypothesized that this occurs because of the decrease in air density at high altitudes such as Denver (1,610 meters; 5,280 feet). Moving upward through the atmosphere, air pressure decreases rapidly at lower altitudes where air density is greatest. Gas molecules are more widely spaced at Coors Field than at lower-altitude sites such as Wrigley Field in Chicago (about 200 meters [656 feet] of elevation). Consequently, a home-run baseball will push against fewer gas molecules as it flies through the air at Coors Field and will therefore travel about 10 meters (33 feet) farther than at Wrigley Field. So, to answer the question posed with the chapter opening image: It is easier to hit home runs at Coors Field, because it has lower air pressure than at Wrigley Field. Efforts to offset this effect by keeping the baseballs in a humidor before they were used have been shown to have negligible effect. Increasing the humidity of the baseball does nothing to counteract the high altitude. While the ball becomes a little heavier, the moisture actually appears to improve the way it flies through the air.

As a final note, keep in mind that temperature and heat are not the same thing. Higher temperatures do not necessarily correspond to more heat. The few stray gas molecules in the thermosphere are moving much faster than the molecules in the troposphere and result in much higher temperature readings at the top of the atmosphere. But we should all be very grateful that there is a vast reservoir of heat in the slower motions of the billions and billions of nitrogen and oxygen molecules bouncing around the troposphere.

The Atmosphere: Concept Map

Complete the following concept map to evaluate your understanding of the interactions between the earth system and the atmosphere. Label as many interactions as you can, using information from this chapter.

"A great thunderstorm; an extensive flood; a desolating hurricane; a sudden and intense frost; an overwhelming snowstorm; a sultry day—each of these different scenes exhibits singular beauties even in spite of the damage they cause. Often whilst the heart laments the loss to the citizen, the enlightened mind, seeking for the natural causes, and astonished at the effects, awakes itself to surprise and wonder."

Crèvecoeur, M.-G. J., Trent, W. P., & Lewisohn, L. (1904). Letters from an American farmer. New York: Fox, Dutfield & company

Chapter

15

Weather Systems

Chapter Outline

the big picture

Why would someone chase a tornado?

See The Big Picture box at the end of this chapter for the full story on this image.

15.1 The Weather Around Us

Chapter Learning Outcomes

- Students will predict the impact weather has on humans.
- Students will be able to read and interpret a weather map.
- Students will evaluate the risk to humans from weather-related hazards.

At its best, the weather lifts our spirits. At its worst, it kills people. Our society is partially defined by **weather, the state of the atmosphere at any given time and place.** Texas and Oklahoma are two states in a section of the country nick-named "tornado alley" because of severe weather that frequently occurs there. Thunderstorms and rain can affect almost any region of the country. Most of the time, weather simply influences what we do, but occasionally, **extreme weather threatens lives, disrupts transportation, and causes billions of dollars of damage in its destruction** (Figure 15.1). If we analyzed weather records from 1980 to 2010, we would find that an average of between three and four weather disasters per year caused damages valued in excess of a billion dollars. If we analyzed the ten-year period from 2006–2015, we would find the average number more than doubles to be between eight and nine per year. This includes the 2011–2012 period in which the United States was battered by 27 of these costly weather disasters, culminating in the destruction of Hurricane Sandy at an estimated cost of $50 billion. While this recent increase in costs is unusual, it appears there has been an increase in the frequency of these billion-dollar disasters over the last few decades.

Much of this increase is attributed to population growth and development, which are increasing the potential for more

Figure 15.1 US billion-dollar weather/climate disasters. **a.** Disasters from 1980 to 2015 by state (costs of disasters adjusted for 2015 Consumer Price Index). Billion-dollar disasters are more frequent in southern states. **b.** Ten US billion-dollar disasters for 2015.
15.1a, 15.1b: http://www.ncdc.noaa.gov/billions/summary-stats, NOAA

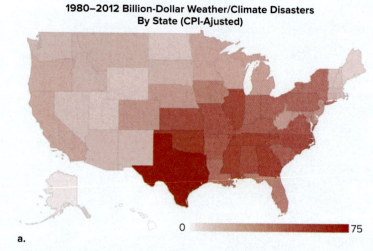

1980–2012 Billion-Dollar Weather/Climate Disasters By State (CPI-Ajusted)

0   75

a.

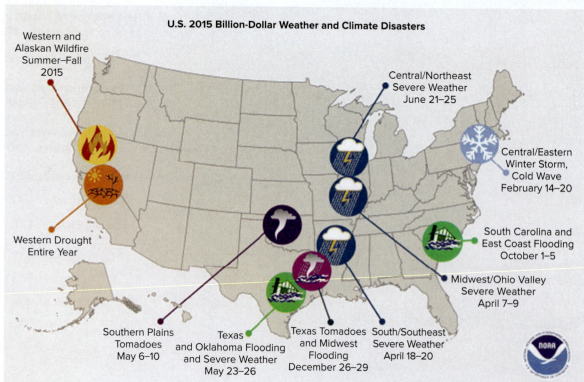

U.S. 2015 Billion-Dollar Weather and Climate Disasters

Western and Alaskan Wildfire Summer–Fall 2015

Western Drought Entire Year

Central/Northeast Severe Weather June 21–25

Central/Eastern Winter Storm, Cold Wave February 14–20

South Carolina and East Coast Flooding October 1–5

Midwest/Ohio Valley Severe Weather April 7–9

Southern Plains Tornadoes May 6–10

Texas and Oklahoma Flooding and Severe Weather May 23–26

Texas Tornadoes and Midwest Flooding December 26–29

South/Southeast Severe Weather April 18–20

b.

costly weather disasters. In 1900, nearly two-thirds of the US population lived in the northeastern and midwestern states. Today, more than 100 million residents make the South the most heavily populated region of the nation. Florida, Georgia, North Carolina, and Texas have seen some of the largest population increases in the last decade. Yet, a cursory glance at the states most likely to experience weather-related disasters shows that these migrants are moving into some of the more disaster-prone regions in America (Figure 15.1a). About one-half of these disasters fall into two types, tropical cyclones (including hurricanes) and severe local storms (Figure 15.1b). Winter storms, wildfires, and droughts together account for another one-third of the total.

Facts About Severe Weather

Which do you think kills more people (on average) every year: lightning, tornadoes, temperature extremes, floods, hurricanes, or snowstorms and ice storms? Data on deaths from natural hazards in the United States for the 25-year span from 1988 to 2012 reveal that the most common cause of death was temperature extremes (summer heat/drought, winter cold), which accounted for more than one-quarter of all deaths (Figure 15.2). On average, more than 500 people a year die from natural hazards, the vast majority of which are related to weather phenomena. Over time, individual high-magnitude events such as Hurricane Sandy are less responsible for deaths than smaller, more frequent events such as storms and lightning. There is some geographical variety to death from natural hazards, as is hinted at by Figure 15.1. Fatalities from natural hazards in southern states are more typically due to severe storms and tornadoes, while extreme heat accounts for more deaths in the northern Great Plains, and winter storms and floods are the greatest hazards in the Mountain West. Midwestern states such as Illinois, Indiana, and Ohio have the lowest mortality rates from all types of natural hazards.

The National Oceanographic and Atmospheric Administration (NOAA) considers **extreme heat "one of the most underrated**

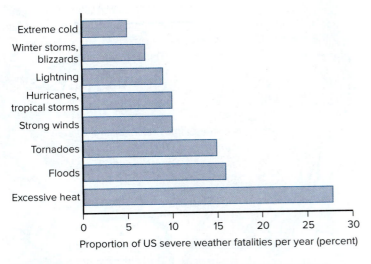

Figure 15.2 US severe weather fatalities, average per year, 1990–2014. The US averaged 515 fatalities per year in the 25-year period from 1990 to 2014. Over one-quarter of those fatalities occurred as a result of heat waves. Phenomena associated with severe types of storms (for example, high winds, lightning, tornadoes) accounted for more than one-half of all deaths.
Adapted from http://www.crh.noaa.gov/mkx/?n=taw-part10-usa_fatality_stats

and least understood of the deadly weather phenomena" because, unlike in the situation with a tornado, the victims are not easily recognized and the hazard is less apparent. When a mass of hot air moves over a region, our bodies respond by losing heat through our skin and through perspiration. However, perspiration works to lower body temperatures only when the humidity is low enough to allow sweat to evaporate (the latent heat of evaporation takes heat away from our bodies) and when we replace lost fluids by drinking water. If the body cannot get rid of the excess heat, the core body temperature will rise above the normal 37°C (98.6°F). Heatstroke occurs when body temperatures rise to around 40.5°C (105°F). More than 55,000 people died in Russia in 2010 when that region experienced its hottest summer on record, with temperatures routinely exceeding 40°C (104°F); a record high temperature of 53.5°C (128.3°F) was recorded in southeastern Russia in June of that year.

Elderly citizens and young children are most at risk from heatstroke because their bodies are less able to make the needed physiological adjustments to extreme heat conditions. It is not uncommon to read of children who were endangered when they were left in a locked car for more than 15 minutes during a hot day. In July 1993, a 3-day period of record high temperatures in Illinois and Wisconsin resulted in more than 500 heat-related deaths; more than two-thirds of those who died were older than 65. Many were found in hot city apartments that lacked air conditioning or had windows that had been nailed shut to protect against burglaries.

Extreme heat, tornadoes, hurricanes, and other dangerous weather phenomena cannot be stopped, but by relying on detailed observations, meteorologists can provide timely warnings to prepare and protect people from these hazards. The most expensive natural disaster in US history occurred in 2005 when Hurricane Katrina wreaked havoc across the Gulf Coast states, causing over $100 billion in damage. Continued coastal development makes a similar future disaster inevitable. Consequently, scientists primarily focus on collecting and analyzing data, and educating those who use that information. Ultimately, it is up to government agencies to make wise decisions that will protect their citizens from harm. It is also important for citizens, like you, to understand and heed the warnings. As we will see with the example of Hurricane Katrina, sometimes the warning signs are not appreciated until it is too late.

The first half of this chapter describes how common weather systems develop across much of the nation. We will discuss how and why air masses range from dry and cold to warm and humid, and what happens when different air masses interact. We will also explore how meteorologists can successfully use these characteristics to forecast weather for 2 to 5 days in the future. The second half of the chapter focuses on dangerous weather phenomena such as thunderstorms, tornadoes, and hurricanes.

Self-Reflection Survey: Section 15.1

Answer the following questions as a means of uncovering what you already know about weather systems.

1. Explain three examples of how the weather influences your daily life.
2. Describe your ideal annual weather conditions.
3. Describe the most extreme weather event you have experienced.

15.3 Air Masses

Learning Objectives

- Explain why the location where an air mass forms affects the temperature, humidity, and pressure of an air mass.
- Predict the physical characteristics of air masses forming at various locations on Earth.
- Predict how the physical properties of air masses would be modified as they traveled from one location to another.

What kind of weather can delay trains, cause traffic lights to malfunction, and result in hundreds of calls to tow-truck companies? All of these conditions occurred on a January morning when the northeastern United States and neighboring parts of Canada were covered by a blanket of bone-chilling cold air. Temperatures in the Canadian cities of Toronto and Québec rivaled those of Siberia as a single large air mass swept south from northern Canada. The cold conditions extended into states as far south as the Carolinas. **An air mass is a large region of the lower troposphere that has relatively uniform temperature and moisture content.** The period from December 2013 through February 2014 was one of the top ten coldest winters on record in the mid-western US. The characteristics of individual air masses depend on where they form (called the *source area*) and any changes that occur as they move into regions having different conditions. Weather is influenced by changes in the air masses over time and interactions at the boundaries between air masses (called *fronts*), where rising air may lead to condensation and precipitation.

Source Areas

Think about the interaction between the atmosphere and the tropical ocean on a sunny day. The ocean exchanges heat with the overlying air, helping to evaporate water. A warm, moist air mass has developed. How would that air mass differ from one over a desert? An air mass develops when the atmosphere above a relatively uniform land or water surface remains stationary for several days and the lower atmosphere takes on the temperature and moisture properties of the underlying surface. **Air masses are identified by the temperature and the moisture characteristics of the underlying surface.** Cold, polar air masses form at high latitudes (more than 50°N, S), and warm, tropical air masses form closer to the equator (Figure 15.5). Maritime air masses develop above oceans and contain much more moisture than dry air masses formed over the continents. However, because temperatures vary with seasons, there are no firm boundaries that define where air masses form. For example, polar air masses creep farther south during winter and retreat northward during summer in the Northern Hemisphere. As we will see later, heavy rains and severe thunderstorms (or snowstorms) can result from the interaction between a continental polar air mass and a maritime tropical air mass that collide over the eastern United States during summer (or winter).

Types of Air Masses

Meteorologists use a form of scientific shorthand to label air masses. Five common types are shown in Figure 15.5 and can be described as follows:

- *cA*. Continental Arctic/Antarctic air forms at high latitudes around the poles above permanently snow-covered ground (or sea ice; Figures 15.5 and 15.6). These air masses are characterized by extremely cold, dry air that may sweep south across Canada and produce days of below-freezing temperatures over much of the central and eastern United States during winter.
- *cP*. Continental polar air forms over the northernmost portions of North America, Europe, and Asia (Figures 15.5 and 15.6). This type of air mass shares its basic characteristics (cold, dry) with cA air but without the exceptionally cold temperatures.
- *cT*. Continental tropical hot, dry air forms over continental interiors such as the dry lands of northern Mexico and the southwestern United States (Figures 15.5 and 15.6). The moisture content and temperature of the air mass are modified as it moves to the east or north.

✓ **Checkpoint 15.5**

basic   advanced
intermediate   superior

Of the five most common types of air masses, which ones most directly affect the area where you live?

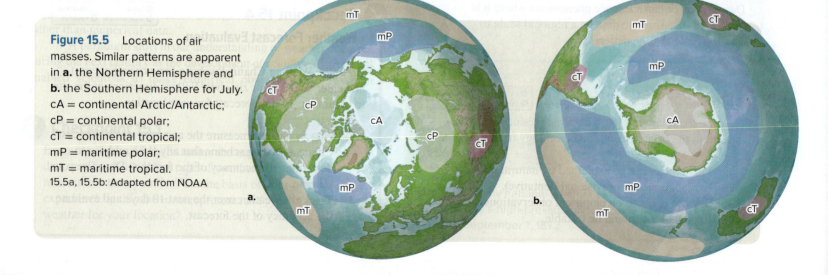

Figure 15.5   Locations of air masses. Similar patterns are apparent in **a.** the Northern Hemisphere and **b.** the Southern Hemisphere for July.
cA = continental Arctic/Antarctic;
cP = continental polar;
cT = continental tropical;
mP = maritime polar;
mT = maritime tropical.
15.5a, 15.5b: Adapted from NOAA

Figure 15.6 North American air masses. Principal source areas and paths for air masses that influence weather patterns across North America. Arrows indicate where the air mass moves to after it forms. For example, cP air masses originate in northern Canada and move to the south and southeast across the United States.
Adapted from NOAA

have created your own personal car-sized air mass. You park and go inside for 20 minutes to stock up on pickles and frozen corn. What will the interior temperature of your car be like when you return? The temperature of the air inside the car will be warmer because the car has been sitting in the sun. In the same way, air masses can gain or lose heat and moisture as they move from one location to another. Can you think of an example of this that we have discussed in a previous chapter?

The principal factors that can modify an air mass are the temperature and topography of the underlying surface. If the air is heated by a warmer land surface (as occurs with cP air moving south), the air will begin to rise. Such air is unstable—that is, it will continue to rise until its temperature matches that of the surrounding atmosphere (see Chapter 14). Cooling (mT air moving north) has the opposite effect. Maritime air, whether polar or tropical, can be forced upward over mountain ranges, leading to cooling and condensation. Condensation removes moisture from the air through precipitation and changes the formerly humid air to a much dryer air mass, producing a rain shadow effect. Washington and Oregon provide a good illustration of these trends. Both states can be divided into wet western highlands and dry eastern deserts (Figure 15.7). Each state experiences plentiful annual rainfall near

- **mP.** Maritime polar air masses form in the northern Atlantic and Pacific oceans and over the Southern Ocean (Figures 15.5 and 15.6). These air masses are characterized by cool, moist air that affects northeastern and northwestern North America. Temperatures at the ocean surface are less extreme (less cold and less hot) than on land, so mP air is warmer than cP air. Such air masses bring rains to the coasts of Washington and Oregon or snows to the mountains inland.
- **mT.** Maritime tropical air masses move inland from the tropical Pacific Ocean, Gulf of Mexico, and tropical Atlantic Ocean (Figures 15.5 and 15.6). They are distinguished by high temperatures and high humidity. The mT air brings hot, humid summers to southeastern states and can form at any time during the year.

Modification of Air Masses

Think about driving to the grocery store on a hot, sunny summer day. On the way there, you run the air conditioning in your car, and the temperature in your vehicle is pleasantly cool. You

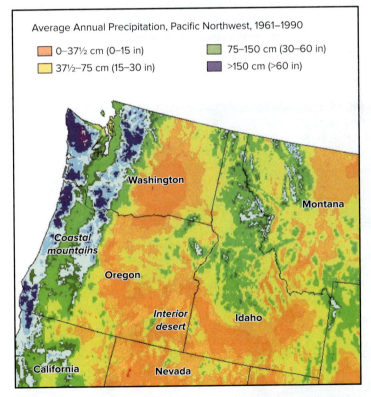

Figure 15.7 Annual precipitation in the Pacific Northwest. Washington and Oregon receive much more rainfall over mountains in the western one-third of the states, where the maritime polar air mass comes onshore. Much of this moisture is lost during orographic lifting, creating a rain shadow effect that contributes to the distribution of desert conditions in the eastern two-thirds of both states.
US Global Change Research Program

✓ Checkpoint 15.6

| ✓ basic | ✓ advanced |
| ✓ intermediate | ✓ superior |

Which air masses would be present during July at each of the locations indicated on the following globe?

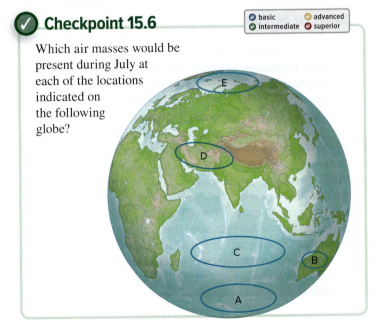

- ***Early stage.*** Friction slows winds at the ground surface, resulting in higher wind velocity moving upward from the surface. These contrasting vertical wind speeds generate winds that tend to rotate counterclockwise about a central, near-horizontal axis, producing a rolling spiral of wind.
- ***Updraft stage.*** Updrafts below a thunderstorm draw the spiraling horizontal winds upward into an arc forming two rotating updrafts on either side called mesocyclones. One mesocyclone is often stronger than the other. In most circumstances, weather radar can observe the mesocyclone, the rotating updraft, several kilometers above the ground within the thunderstorm.
- ***Tornado stage.*** A downdraft from the growing thundercloud pushes a portion of the arc of spiraling winds downward. This acts to concentrate the rotation and stretch it toward the ground, forming the tornado. (Figure 15.18). This is the stage seen by storm chasers.

Sometimes smaller funnels appear to skip across the surface, destroying one home while leaving neighboring properties undisturbed. In reality, scientists hypothesize that tornadoes do not skip—that is, form, dissipate, and then re-form. Instead, they have used film of actual tornadoes to identify small, intense mini-tornadoes embedded in the main funnel. It is these minitornadoes that hit or miss homes.

Improvements in forecasting methods have reduced the number of fatalities associated with tornadoes (Figure 15.19). Nearly one-half of US fatalities occur as a result of tornadoes destroying nonpermanent mobile homes, which are especially susceptible to higher winds as they have little structural strength.

Because cP and mT air masses frequently interact along frontal systems across the central portion of North America, the United States is home to the majority of the world's tornadoes, averaging about 1,200 a year. **Tornadoes occur when thunderstorm activity is at an optimum across much of the nation.** The timing of tornado activity is tied to seasonal movement of the polar front, the boundary between warm, humid air from the Gulf of Mexico and cold, dry air from Canada. Tornadoes are more common in Gulf Coast and southeastern states in early spring. Then tornado activity moves into the Great Plains as the polar front retreats northward in late spring. Tornado activity shifts to the north in summer when these air masses interact in the northern Great Plains and upper Midwest. The highest frequency of tornadoes per area occurs over the Great Plains states (Texas, Oklahoma, Kansas, Nebraska) and parts of the upper Midwest (Iowa, Indiana, Illinois), a region that has come to be known as "tornado alley" (Figure 15.20).

Figure 15.19 US tornado fatalities per recent decade. Fatalities declined from 1950 to 2010 as forecasting technology improved to provide better tornado warnings.

✓ **Checkpoint 15.20**

| ✓ basic | ✓ advanced |
| ✓ intermediate | ✓ superior |

Explain why the number of tornadoes counted each year has increased, while the number of days with at least one tornado sighting has remained essentially unchanged for several decades.

Figure 15.20 Tornado alley. The locations of states with an annual average of more than five strong to violent (F2 to F5) tornadoes, 1999–2008. While Texas experiences more tornadoes, several other states experience more tornadoes per 10,000 square kilometers (for example, Oklahoma, Kansas, Iowa, Louisiana, and Mississippi).
NOAA/NWS Storm Prediction Center, Norman, Okla

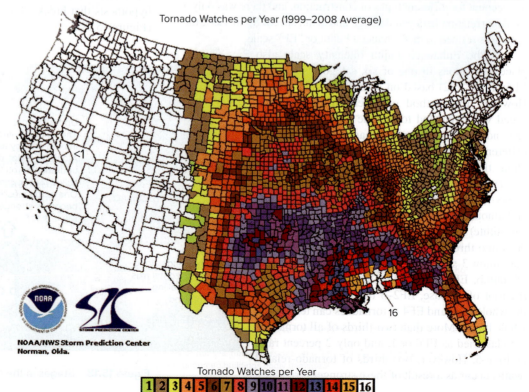

Tornado Watches per Year (1999–2008 Average)

NOAA/NWS Storm Prediction Center
Norman, Okla.

Tornado Watches per Year
1 2 3 4 5 6 7 8 9 10 11 12 13 14 15 16

15.6 Severe Weather: Hurricanes

Learning Objectives

- Describe the life cycle of a hurricane.
- Relate different earth system processes to the formation and decay of hurricanes.
- Compare and contrast hurricanes with other types of low-pressure systems.
- Describe factors used to rank the intensity of a hurricane.
- Evaluate the risk to humans for hurricane scenarios.

Philosopher George Santayana famously remarked, "Those who cannot learn from history are doomed to repeat it." Many scientists are concerned that we will see a repetition of the destruction that

the Gulf Coast experienced from Hurricane Katrina, but perhaps not where we most expect. Records of hurricanes striking the Atlantic and Gulf coasts of North America stretch back over 150 years. This time span is small in geologic terms, but it is long enough to reveal some important patterns. Hurricanes have affected every location along the coasts from Maine to the southern tip of Texas.

Quoted in Perspectives on Milwaukee's Past, Margo J. Anderson, Victor R. Greene, University of Illinois Press, 2009

Analyses of past hurricanes indicate that recent population growth will result in increased loss of life and property when similar storms inevitably occur again in the same locations. Lightning *does* strike the same place twice, and so do hurricanes! This was brought home sharply when Hurricane Katrina barreled into the Gulf Coast on August 29, 2005, submerging most of New Orleans, Louisiana, and blowing away large portions of Biloxi, Mississippi, and other coastal cities (Figure 15.21). More than 1,300 people died, and the

Figure 15.21 Before and after Hurricane Katrina. **a.** NASA's Terra satellite snapped this view of New Orleans, sandwiched between Lake Pontchartrain and the Mississippi River, 2 days prior to the arrival of Hurricane Katrina. **b.** A day after the hurricane, widespread flooding (blue) was apparent. **c.** Biloxi, Mississippi, before Hurricane Katrina. Note the location of the large mansion, pier, and pier house. **d.** After Katrina. The hurricane destroyed these structures.
15.21a & 15.21b: NASA; 15.21c & 15.21d: USGS

a.

b.

c.

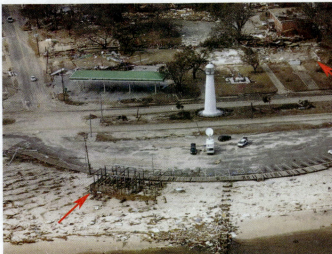

d.

storm caused an estimated $100 billion in damage in communities in Louisiana, Mississippi, and Alabama. **Hurricane Katrina was the worst natural disaster to strike the modern United States.** However, it should not have come as a surprise. Much of the damage from Hurricane Katrina had long been predicted. A dozen major hurricanes had made landfall along the Louisiana coast in the last 100 years; four of those storms resulted in major flooding in New Orleans. It was just a matter of time before the region suffered another hit. Well before Hurricane Katrina struck, government reports, articles in science magazines, and a week-long series of stories in the New Orleans *Times Picayune* newspaper had all predicted in eerie detail what would happen if a major hurricane made landfall near modern New Orleans.

It is one thing to know what might happen; it is another to be able and willing to do something about it. **Measures to protect populations and structures must account for cultural and social factors and are the responsibility of local, regional, and federal governments.** This brings up an important point regarding science and society. Who is best equipped to address the complexity of such hazards? Although scientists and engineers could predict the probability of a disastrous hurricane, they did not have access to funds and resources to do much about it. On the other hand, government officials, who had the power to allocate funds for levee improvements or evacuation plans, may not have known of or understood the seriousness of these predictions. Alternatively, they may have recognized the risks but were unwilling to cut other programs to take measures needed to mitigate these hazards. These contrasting roles for scientists and government officials set up an interesting dialogue. One group seeks to be clear about the potential dangers without being labeled alarmist, while the other tries to protect residents without incurring economic losses. As we discuss the general features of all hurricanes, we will describe the characteristics of Hurricane Katrina and other storms. Ultimately we may conclude, as George Bernard Shaw observed, "We learn from history that we learn nothing from history."

Quoted in When History Teaches Us Nothing: The Recent Reformed Sonship Debate in Context, Tim J. R. Trumper, Wipf and Stock Publishers, 01-May-2008

Building a Hurricane

Before we go any further, take a few minutes to examine the image of Hurricane Katrina in Figure 15.22. What do you observe about this hurricane? Keep looking . . . make at least three separate observations.

OK. You probably noticed the large size of the storm; it was large enough to cover several states at once. You may have observed that the hurricane is composed of a dense mass of clouds surrounding a central clear "eye." Also, you probably noted the swirling, spiral pattern in the clouds. Finally, you may have identified that the hurricane was centered over the ocean—in this case, the Gulf of Mexico. We will explain the development of these features next.

✓ Checkpoint 15.21

The costliest hurricanes are not necessarily the most intense. Explain why two category 1 hurricanes (Agnes, 1972; Diane, 1955) are among are the top 10 most costly US hurricanes when adjusted for inflation.

Figure 15.22 Hurricane Katrina approaching the Louisiana coast, August 28, 2005.
NOAA

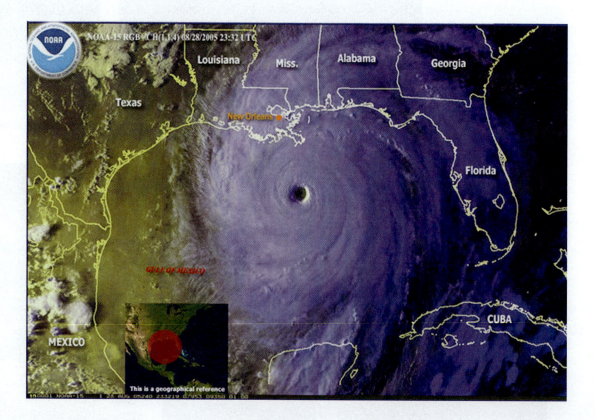

Hurricanes are characterized by high winds, heavy rainfall, and storm surges (elevated water levels) along coastlines. Hurricanes can form under appropriate conditions all over the world but are named differently in different oceans; for example, they are called typhoons in the Pacific Ocean and cyclones in the Indian Ocean. Some hurricanes live their short, dramatic lives exclusively over the oceans, noticed only by meteorologists and a few anxious sailors. It is some of the others, those that make landfall, that grab the attention of millions of coastal residents in the United States during the North Atlantic hurricane season from June 1 to November 30. Let us look at how these storms develop, with special reference to Hurricane Katrina.

Two to Three Weeks Before Landfall. Hurricane development requires warm surface waters and a specific set of conditions in the troposphere. **Surface water temperatures must be at least 27°C (81°F) down to a depth of about 60 meters (200 feet).** Only over warm ocean waters can sufficient evaporation and condensation occur to provide the huge volumes of moisture necessary to form the large masses of clouds present in a hurricane. A deep layer of warm surface water prevents cold water from being drawn to the surface, which would slow the evaporation needed to fuel the developing storm. As we learned in Chapter 13, warm surface waters straddle the equator but move farther north during summer in the Northern Hemisphere, creating optimal conditions for hurricanes with the potential to strike North America (Figure 15.23). Most US hurricanes make landfall from mid-August through September. In the 10 years from 1995 through 2004, the North Atlantic region experienced 20 hurricanes during August months. These storms originated all across the North Atlantic, from just off the east coast of Africa to the west coast of Mexico (Chapter Snapshot). Some of these storms made landfall, while others spiraled harmlessly into the North Atlantic Ocean. The amount of time between formation and landfall varies, depending on the birthplace of the storm. Most storms originating in the eastern Atlantic Ocean require about 3 weeks to reach the coast of North America. Hurricane Katrina developed in the western Atlantic Ocean, near the Bahamas, in a similar location to that of Hurricane Dennis in 1999.

Average annual ocean temperature >27°C \ Hurricane track

Figure 15.23 Hurricanes originate in areas of the world's oceans where water temperatures are greater than 27°C (81°C). Hurricanes in the Northern Hemisphere are most common during summer and early fall. Southern Hemisphere hurricanes are frequent during our winter (their summer). How do you explain the curved track of these storms?

A hurricane begins to grow when warm, humid air is forced aloft. The rising air cools and condenses to form cumulus clouds that will develop into cumulonimbus cells if the air is sufficiently warm and humid. The inflow of air into the developing low-pressure system must be matched with an outflow of air in the upper troposphere to maintain the pressure gradient in the developing hurricane. If not, the pressure contrast decreases and wind speed declines. A developing storm can be dispersed by fast-blowing winds at high altitudes or by a particularly dry layer of air in the upper atmosphere that impedes condensation. The number of storms decreases during El Niño years (see Chapter 13) and when dust storms blow west out of the Sahara desert in North Africa. El Niño alters the path of the jet stream to produce strong wind shears that destroy growing storms. In contrast, the dust storms block sunlight, decreasing ocean temperatures and reducing the heat supply necessary for storm formation.

Earth's rotation, as exhibited by the Coriolis effect (sideways deflection; see Chapters 13 and 14), imparts a clockwise rotation to the growing storm in the Southern Hemisphere and a counterclockwise rotation in the Northern Hemisphere. The magnitude of the Coriolis effect is zero at the equator and increases with increasing latitude. Because there is not sufficient energy in rotation at the equator to produce hurricanes, **the majority of hurricanes originate between 10° and 20°N or S of the equator** (see Figure 15.23).

One to Two Weeks Before Landfall. If the appropriate conditions are met, a tropical depression (wind speed 37–62 kilometers per hour; 23–39 miles per hour) develops from some previously unorganized thunderstorms. If conditions remain favorable, after about 5 days, the tropical depression grows into a tropical storm (wind speed 63–118 kilometers per hour, 39–74 miles per hour) before developing into a hurricane, with wind speeds of at least 119 kilometers per hour (75 miles per hour), a couple of days later. **Air pressure is lowest inside the central eye of the hurricane, where warm air is rising.** Wind speed is related to pressure changes; the lower the pressure inside the hurricane, the higher the wind velocity.

Bands of clouds spiral outward from the eye (Chapter Snapshot), which is surrounded by thick clouds known as the *eye wall.* Beyond the eye of the storm, rising air in the deepening low-pressure system cools and condenses, releasing latent heat. This release of heat helps generate a growing spiral of dense cumulonimbus clouds rotating about the eye. The hurricane continues to grow in size and intensity as long as the underlying water temperature remains above 27°C (81°F). Precipitation is concentrated within a radius of approximately 100 to 200 kilometers (62 to 124 miles) on either side of the eye, releasing up to 20 billion tons of water per day.

Three to Seven Days Before Landfall. Atlantic hurricanes, driven westward by prevailing winds at rates of 10 to 25 kilometers per hour (6 to 16 miles per hour), may turn parallel to the US East Coast or pass south of Florida to strike along the Gulf Coast or the Caribbean islands (Figure 15.23, Chapter Snapshot). Florida and Texas experience more hurricane landfalls than any other states

HURRICANE ANATOMY

Continental US Hurricane Strikes (1950–2005)
- ▲ Category 3
- ▢ Category 4
- ⬤ Category 5

Edna (1954)

Carol (1954)
Donna (1960)
Gloria (1985)

Warm Gulf Stream current fuels hurricanes in the western Atlantic Ocean.

Gloria (1985)
Donna (1960)

Emily (1993)
Connie (1955)
Ione (1955)

Fran (1996)
Hazel (1954)
Hugo (1989)
Gracie (1959)

Katrina (2005)
Betsy (1965)
Carmen (1974)
Andrew (1992)
Hilda (1964)
Audrey (1957)
Rita (2005)
Alicia (1983)
Carla (1961)
Celia (1970)
Bret (1999)
Allen (1980)
Beulah (1967)

Elena (1985)
Frederic (1979)
Juan (2004)
Opal (1995)
Dennis (2005)
Eloise (1975)

Camille (1969)

Easy (1950)

Jeanne (2004)

4

Charley (2004)
Donna (1960)
Wilma (2005)
Donna (1960)

King (1950)
Andrew (1992)
Betsy (1965)

STAGE 3

13

8

10

14

Warm waters in the isolated Caribbean Sea and Gulf of Mexico can cause hurricanes to strengthen approaching the US coast.

1. Additional evaporation of warm water ensures a tropical storm continues to strengthen to hurricane status.
2. Condensation occurs in the saturated air to produce an expanding cloud mass.
3. Latent heat of condensation fuels the rise of air within the storm.
4. The heaviest precipitation and highest winds occur over central part of storm—up to 400 km across. The warmer, less dense air at the storm's center increases the pressure gradient and wind speed.
5. A well-developed spiral pattern is evident in the cloud mass.

1

24

28

28

24

CROSS SECTION THROUGH A HURRICANE

Diverging
upper-level air

Subsiding
air in eye

Air near perimeter of eye is
drawn into eyewall circulation

Eye

Converging
low-level air

Rising air in eyewall,
rotating around eye

Cool waters of the Canary
current mark the northern
limit of hurricane origins.

1. Air moves toward the center of a
developing low pressure system.
2. The low pressure system strengthens,
becomes more organized and forms
a tropical depression. A spiraling wind
pattern develops as the Coriolis effect
imparts a counterclockwise rotation.
3. Continued time over warm waters
causes the depression to strengthen
into a named tropical storm (wind
speeds up to 118 km/h).

STAGE 2

Westerlies drive developing
storm toward the US and
Caribbean.

STAGE 1

Rising, humid air is forced aloft, and
condensation produces unorganized
cumulonimbus clouds. There is no
indication of the near-circular
hurricane-like pattern at this stage.

24

28

Origins of August Hurricanes, 1995–2004

(1) Charley, 2004	(11) Bonnie, 1998
(2) Danielle, 2004	(12) Danielle, 1998
(3) Frances, 2004	(13) Earl, 1998
(4) Erika, 2003	(14) Dolly, 1996
(5) Fabian, 2003	(15) Edouard, 1996
(6) Alberto, 2000	(16) Fran, 1996
(7) Debby, 2000	(17) Felix, 1995
(8) Bret, 1999	(18) Humberto, 1995
(9) Cindy, 1999	(19) Iris, 1995
(10) Dennis, 1999	(20) Luis, 1995

Monthly Mean Sea Surface Temperature for August (°C)

10.0   15.0   20.0   25.0   30.0   35.0

24

"What has happened, can happen; climatic history can repeat itself."

—Reid Bryson, professor, University of Wisconsin

"The paleoclimate record shouts out to us that, far from being self-stabilizing, the Earth's climatic system is an ornery beast which overreacts even to small nudges."

—Broecker, Wallace, "Thermohaline Circulation, The Achilles Heel of Our Climate System: Will Man-Made CO2 Upset the Current Balance?" *Science*, Volume 278 no. 5343, pp. 1582-1588, November 28, 1997

Chapter

16

Earth's Climate System

Chapter Outline

the big picture

Polar bear and tourist. Which is in greater danger?

See The Big Picture box at the end of this chapter for the full story on this image.

sand-sized particles and few examples of larger grains. Iceberg formation is accelerated when the rate of glacial motion increases. Those conditions can occur when glaciers advance more rapidly as thick ice accumulates during cold periods or when ice shelves melt during warming periods. Ocean drilling programs have discovered dropstone layers all across the North Atlantic Ocean, marking cycles of climate changes during the last ice age.

Finally, melting glaciers are a key source of water supplies for farmers and cities around the world. This is especially true in regions with pronounced dry seasons. During the dry season, meltwater streams from the glacier may be the only available source of water. For example, the Gangotri Glacier in northern India (Figure 16.18) supplies about 70 percent of water in the Ganges River during dry periods. Warming climates may result in a short-term increase in stream discharge from glaciers, but as glaciers shrink and recede, water supplies will inevitably decline. This reduction in water supplies will have profound consequences for societies around the world and is one of the most significant consequences of climate change.

Hot Deserts

What does a desert look like? Examine the images in Figure 16.21; which of the images best represents your idea of a desert?

In most hot deserts, temperatures are high, annual rainfall is less than 25 centimeters (10 inches), and evaporation exceeds precipitation. Sand is the dominant feature in one-half of all desert surfaces, while others have a desert pavement composed of larger pebbles and rocks. In reality, most desert surfaces are a combination of sand, desert pavement, and regions dominated by rock outcrops (Figure 16.21). As we learned previously, many deserts are located below subtropical high-pressure systems around 30°N and S. However, deserts also exist in the centers of some continents simply as a result of their greater distance from the ocean. For example, the Gobi Desert (see Chapter Snapshot) is located far inland from the Indian Ocean where the path of moisture-laden winds is blocked by the high peaks of the Himalayan Mountains.

Erosion and Deposition in Deserts. Occasional flash floods and wind action are responsible for eroding, transporting, and depositing material in deserts. Flash floods, though uncommon, rapidly move large volumes of material and play a major role in shaping desert topography. Slower-acting processes associated with winds act over much longer periods and also affect deserts. Most deserts lie in the trade winds belt and experience a prevailing northeast or southeast wind. Winds pick up dry sand

Figure 16.20   Dropstones in fine-grained ocean sediment. The larger grains represent material eroded by glaciers that dropped to the seafloor off the east coast of Greenland when icebergs melted. This core was recovered by scientists of the Ocean Drilling Program (ODP). ODP recovers sediments from the ocean floor around the world as part of a varied research program that analyzes all aspects of the earth system. The numbers represent a centimeter scale.
Integrated Ocean Drilling Program

Figure 16.21   Desert environments. **a.** Sandy desert. **b.** Desert pavement. **c.** Rocky desert. What differences do you notice?
16.21a: © Digital Vision; 16.21b: © C. Allan Morgan/Peter Arnold, Inc/Getty Images; 16.21c: © Digital Vision/PunchStock

a.

Checkpoint 16.11

| ✓ basic | ✓ advanced |
| ✓ intermediate | ✓ superior |

Draw a diagram that illustrates the fate of a snowflake that falls in the accumulation zone of a growing alpine glacier. Begin and end the path of the snowflake in the atmosphere.

Checkpoint 16.12

| ✓ basic | ✓ advanced |
| ✓ intermediate | ✓ superior |

In the introduction to this chapter, we mentioned that glaciers in Alaska are shrinking. Discuss how you would plan a study of glaciers in other places to determine if this was a local, short-term phenomenon or part of a larger, long-term global trend. Include a description of how you would measure changes in the size of a glacier.

grains and sandblast objects in their path. Although the sand is removed, larger pebbles are left behind to form a desert pavement (Figure 16.21b).

The effects of wind vary, depending on the size of the sediment particles. Winds carry the finest particles beyond the boundaries of the desert. These light particles can be picked up and transported in suspension, much as the fine particles of pollution are suspended in the air above some cities. For example, dust particles originating in the Sahara are carried across the Atlantic Ocean by vigorous storms (see Figure 7.24b). Increasingly, China, North Korea, South Korea, and Japan must issue health warnings to warn citizens of approaching dust storms originating from the Gobi Desert. Larger sand grains that are too big to be transported in suspension can be skipped along the desert floor by localized wind eddies, a process known as *saltation* (Figure 16.22). Grains that are too heavy to be picked up by winds creep slowly along the surface because of impacts from grains undergoing saltation.

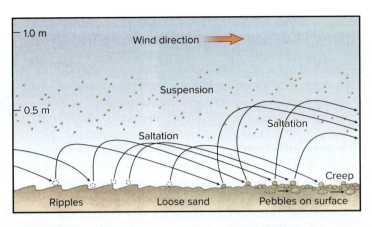

Figure 16.22 Wind action and desert sediment. Winds move sediment by a combination of suspension (fine particles), saltation (sand grains), and creep (larger particles). The wind action results in the formation of ripples in sand and may produce a desert pavement where sand is removed but pebbles are left behind.
Adapted from James Zumberge, *Laboratory Manual for Physical Geology* 12e © 2005

✓ **Checkpoint 16.13**

| ✓ basic | ✓ advanced |
| ✓ intermediate | ✓ superior |

The area around the South Pole receives just a few centimeters of snowfall each year. Is the South Pole a desert? Give reasons to support your answer.

✓ **Checkpoint 16.14**

| ✓ basic | ✓ advanced |
| ✓ intermediate | ✓ superior |

How does the transport of sediment by winds (Figure 16.22) compare to transport in streams?

b.

c.

Chapter

17

Global Change

the big picture

A world inundated by water or gripped by ice. How likely is either scenario depicted in these movies?

See The Big Picture box at the end of this chapter for the full story on this image.

Checkpoint 17.4

Reductions in ozone concentrations involve all aspects of the earth system. UV radiation that interacts with ozone comes from the exosphere (the sun). There is a potential effect on the biosphere, and the hydrosphere, atmosphere, and geosphere all have roles to play. Develop a concept map that illustrates the characteristics of ozone and includes references to different components of the earth system.

ozone measurements is from Arosa, Switzerland, where measurements began in the 1920s. Seasonal variations in the ozone concentration at Arosa range from lows of 280 Dobson units recorded in fall to a maximum around 375 Dobson units during spring. Annual averages at this site have fluctuated between 300 and 360 Dobson units but have declined since the 1970s. That negative trend (about 3 percent per year) results from a reduction in ozone concentrations over Europe.

17.3 CFCs and Ozone Depletion

Learning Objectives

- Connect natural processes and industrial activity to the concentration of stratospheric ozone.
- Explain the significance of time lags in the earth system response to reductions in CFC production.

In Section 17.2, we described the natural fluctuations in ozone concentrations. Now we will explore the effects of human activities on these phenomena. In 1974, scientists F. Sherwood Rowland and Mario Jose Molina published an article suggesting that the use of a manufactured group of gases known as chlorofluorocarbons (CFCs) was resulting in the depletion of ozone. Two decades later (1995), Rowland and Molina (with Paul Crutzen) received the Nobel Prize in Chemistry for their work. **CFCs are volatile organic compounds** that were first patented by Frigidaire in 1928 and were subsequently widely adopted as aerosol propellants and refrigerants. From the 1950s through the 1970s, they were used as propellants in many deodorants and hair sprays. The United States banned aerosol sprays containing CFCs in 1978 and phased out the production of CFCs (replacing them with less ozone-destructive substitutes) by January 1, 1996.

The Nature of CFCs

There are several different types of CFCs. All are inert; that is, they do not react with other gases in the atmosphere. Consequently, they **remain in the atmosphere for periods of 20 to 200 years.** Although they are inert, CFCs can be broken down into their constituent parts by UV radiation (a process known as *photolysis*). Photolysis frees chlorine atoms, which then react with ozone, forming chlorine monoxide (ClO) and oxygen:

(i) $Cl + O_3 \rightarrow ClO + O_2$

The chlorine monoxide (ClO) readily reacts with free oxygen atoms. The result is the formation of another oxygen molecule, leaving the chlorine atom free again:

(ii) $ClO + O \rightarrow O_2 + Cl$

This reaction is significant because an ozone molecule (which is capable of absorbing harmful UV) is destroyed, and a free chlorine atom remains to repeat the process. This sequence is duplicated thousands of times for each chlorine atom. Approximately 80 percent of the chlorine in the stratosphere comes from manufactured compounds (for example, CFCs), and 20 percent comes from natural sources, such as volcanic eruptions and wildfires. In the five decades leading up to the 1990s, the total volume of stratospheric chlorine increased by approximately fourfold.

Reductions in Ozone Concentrations

Remember, the appearance of sunlight starts the reactions that reduce ozone concentrations above Antarctica every spring. However, as we noted in Section 17.1, a reduction in ozone concentrations beyond the expected natural variation was first reported above Antarctica in 1985. Over time, the reductions in ozone at the South Pole gradually became more pronounced. These annual reductions in ozone concentrations over Antarctica have been called the *ozone hole* by the media (see Figure 17.1), even though there is no actual hole—that is, no location in the lower stratosphere where ozone is completely absent. Examination of atmospheric records suggests that ozone levels began to drop in the late 1970s. Recent research has shown that during October (spring in the Southern Hemisphere), ozone concentrations decline to an average of approximately 100 Dobson units above Antarctica (Figure 17.1), with a corresponding increase in the size of the "hole" (Figure 17.5). The levels are down from approximately 320 Dobson units during the late 1950s.

Why Does Ozone Become Depleted over the South Pole?

If the bulk of CFCs are produced and used in the Northern Hemisphere, why are the ozone concentrations so depleted over the unpopulated, ice-covered South Pole? The reasons for the location of the hole are related to the unique weather patterns associated with Antarctica. The ozone hole occupies the polar vortex, a near-circular high-pressure system that isolates the atmosphere over Antarctica during winter in the Southern Hemisphere (June through August). Temperatures in the vortex are below −80°C (−112°F) and lead to the formation of polar stratospheric clouds composed of nitric acid and water. These clouds contain ice particles that provide surfaces upon which chemical reactions can occur. Chlorine compounds in the stratosphere react with cloud particles to release the chlorine that destroy ozone molecules. As spring in Antarctica arrives (September through November), temperatures gradually rise in the stratosphere because of increased solar radiation. Ozone formation (and destruction) is a temperature-dependent phenomenon. As long as the temperatures remain cold enough for stratospheric clouds to exist, reactions that destroy ozone will occur. As temperatures warm, processes that form ozone outpace the destructive cycle, and ozone begins to increase in concentration. Consequently, there is more ozone in the stratosphere in summer than in winter in Antarctica. The absence of polar stratospheric clouds elsewhere results in much less ozone loss over the rest of the planet.

These factors result in a concentration of ozone-depleting chemicals and conditions at the South Pole. The global presence of similar chemicals in the atmosphere raises the possibility for ozone depletion to occur elsewhere. Declines in ozone

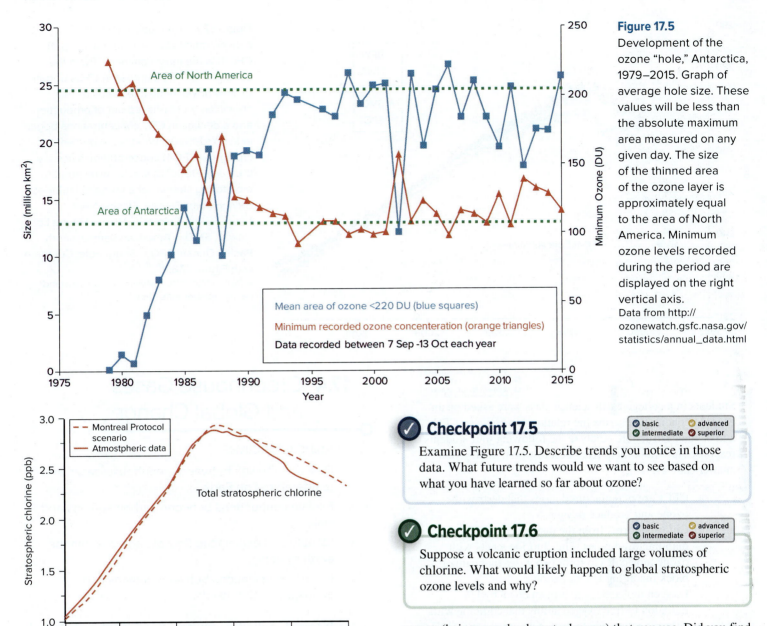

Figure 17.5
Development of the ozone "hole," Antarctica, 1979–2015. Graph of average hole size. These values will be less than the absolute maximum area measured on any given day. The size of the thinned area of the ozone layer is approximately equal to the area of North America. Minimum ozone levels recorded during the period are displayed on the right vertical axis.
Data from http://ozonewatch.gsfc.nasa.gov/statistics/annual_data.html

Figure 17.6 Chlorine concentrations in Earth's atmosphere. The concentrations of chlorine and related elements have steadily declined since the early 1990s due to the gradual phaseout of CFCs. Adapted from http://www.environment.gov.au/soe/2006/publications/commentaries/atmosphere/stratospheric-ozone.html

concentrations are also recognized above North America and Europe. However, the reductions observed at these locations are substantially less than those observed over Antarctica (less than 10 percent compared to 70 percent).

Our Ozone Future

Production of the more common types of CFCs decreased significantly over a decade ago. This decline has been mirrored by a decrease in the concentrations of atmospheric chlorine (Figure 17.6) and various other CFCs. Check some labels to see if CFCs are present in any of the household appliances or pressurized

Checkpoint 17.5

basic · advanced · intermediate · superior

Examine Figure 17.5. Describe trends you notice in those data. What future trends would we want to see based on what you have learned so far about ozone?

Checkpoint 17.6

basic · advanced · intermediate · superior

Suppose a volcanic eruption included large volumes of chlorine. What would likely happen to global stratospheric ozone levels and why?

sprays (hair spray, deodorant, cleaners) that you use. Did you find any? Some older refrigerators and freezers still contain CFC (for example, a product named Freon), as do a few household products. Several new coolants (hydrochlorofluorocarbons, HCFCs; hydrofluorocarbons, HFCs) have largely replaced CFCs. HFCs have no impact on the ozone layer, and HCFCs have only limited ability to deplete ozone because they do not remain in the atmosphere as long as CFCs.

CFC production in the developed world was effectively halted by the Montreal Protocol. Although not all countries adhere to the agreement, most industrialized nations ratified the treaty. Global CFC production has dropped by more than 98 percent from its peak of over 1 million tons (907,000 tonnes) in 1988. However, this huge drop in production has resulted in a much more modest decrease in atmospheric CFC concentrations (Figure 17.7). The long residence time of CFCs in the atmosphere means that it may take a few decades before ozone levels over Antarctica begin to return to normal levels.

Thus, it is possible to forge international agreements to tackle global environmental problems. Next we will see how greenhouse gases compare with CFCs.

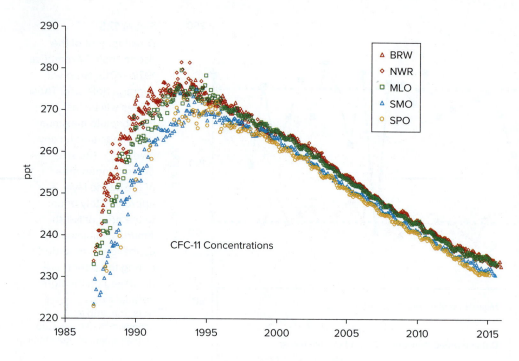

Figure 17.7 Changes in CFC-11 concentrations in parts per trillion (ppt). CFC-11 is the most common CFC in the atmosphere. The increase in CFCs began to slow down in the late 1980s as the chemicals were phased out of production and a decline in CFC concentrations began in the mid-1990s. What is the percentage change in CFC-11 concentration since the early 1990s? Symbols correspond to monitoring stations at a range of latitudes. BRW = Point Barrow, Alaska; NWR = Niwot Ridge, Colorado; MLO = Mauna Loa, Hawaii; SMO = American Samoa, south Pacific Ocean; SPO = South Pole. Compare with Figure 17.6.
Adapted from http://www.esrl.noaa.gov/gmd/hats/graphs/graphs.html

Checkpoint 17.7

Students in a college earth science class were asked on an exam to concisely describe the relationship between CFCs and ozone destruction. Analyze the following four students' responses and rank them from best to worst. Justify your answer choices.

  a. CFCs are manufactured gases that destroy the ozone and produce oxygen.
  b. Chlorine forms from the disintegration of manufactured gases. The chlorine reacts with ozone molecules to form two gases that cannot block incoming UV radiation.
  c. Oxygen molecules are broken down by UV radiation to free oxygen atoms that combine with chlorine monoxide to form new oxygen molecules and free chlorine. Chlorine helps break down CFCs with UV radiation.
  d. UV radiation breaks apart complex manufactured chlorofluorocarbons into its constituent atoms, including chlorine. Chlorine destroys ozone molecules as it cycles through a pair of reactions that produce free oxygen molecules.

Finally, write your own answer to the question.

Checkpoint 17.8

Given the natural variations in climate discussed in Chapter 16, some people argue that governments acted too hastily to ban the use of CFCs just 7 years after the ozone hole had been identified. Summarize three arguments for and three arguments against the ban on CFCs.

17.4 Greenhouse Gases and Global Change

Learning Objectives

- Describe the link between greenhouse gases and temperatures on Earth.
- Explain connections between carbon sources and sinks.
- Identify and explain the flow of carbon within the earth system.
- Discuss associations between humans and atmospheric CO_2 levels.

Infrared radiation emitted from Earth's surface is absorbed by water vapor, carbon dioxide, and other trace gases (methane, nitrous oxide) in the troposphere, creating a situation that has come to be known as the *greenhouse effect*. The heat-absorbing effect of greenhouse gases raises the average surface temperature from a frigid −18°C (−0.4°F) to a pleasant +15°C (59°F). But you can have too much of a good thing. As mentioned before, Venus has a greenhouse effect that results in surface temperatures of 464°C (867°F)—pretty hot! (For more on the relationship between solar radiation and the greenhouse effect, see Chapters 1 and 14.)

Thousands of scientists with the Intergovernmental Panel on Climate Change (IPCC) have reviewed vast amounts of climate data and concluded that human activity has had an unequivocal influence on the global climate. The IPCC was established by the World Meteorological Organization and the United Nations

Checkpoint 17.9

What is the difference between the greenhouse effect and global warming?

Anthropogenic emissions added to atmosphere 3.4 GtC/yr

ATMOSPHERE CARBON RESERVOIR
Approximately half of anthropogenic carbon emissions contribute to an increase in the concentration of atmospheric carbon dioxide.

750 GtC

ANTHROPOGENIC CO₂ EMISSIONS

1.5 GtC/yr

5.5 GtC/yr

Annual flux of carbon between the hydrosphere and atmosphere

Land use changes

Fossil fuels

Biological, chemical processes 90 GtC/yr

Biological, chemical processes 92 GtC/yr

CITIES
Carbon sources from consumption of fossil fuels in industrial processes, electricity generation, and transportation

Anthropogenic emissions added to hydrosphere 2.0 GtC/yr

HYDROSPHERE CARBON RESERVOIR
Most carbon remains in deep ocean water for hundreds of years until brought to the surface by upwelling currents in the oceanic conveyor belt.

39,000 GtC

GtC = Gigatonnes (1,000,000,000 tonnes) of carbon
(1 tonne = 1.1 ton)

= Relative size of carbon reservoirs

we need to understand how these greenhouse gases are produced and absorbed by natural systems. In the discussions that follow, we will focus on carbon dioxide because humans have some control over its production and it has a great potential impact on future warming.

The Global Carbon Cycle

Carbon, one of Earth's most basic components, is present in a variety of forms in all parts of the earth system (Chapter Snapshot). Although this discussion considers each component of the earth system separately, it will become clear that carbon is in constant flux among all the components. Each component acts as a temporary holding place (called a *sink*), storing carbon for short or long time intervals. Each component also acts as a source of carbon that can be released into the atmosphere (or to other earth system components) to increase carbon dioxide concentrations. Carbon dioxide is a product of both natural processes and human activities. While the carbon dioxide generated by human actions is just a small fraction of the volume of gas produced by natural processes, it may just be enough to push our climate past a critical tipping point toward a warmer world.

Atmosphere. The atmosphere is a sink and a source of carbon compounds and carbon-based gases such as carbon dioxide (CO_2; Chapter Snapshot). Carbon enters the atmosphere via the biosphere through respiration (for example, animals exhaling CO_2), the burning of forests, and the decay of dead organisms (plants and animals). Carbon dioxide also enters the atmosphere from the burning of fossil fuels, natural volcanic activity, and the release of dissolved gas from the ocean. The atmosphere is a source of carbon dioxide for plant growth, rock formation (limestone), and other biologic activity. Carbon dioxide from the air can also be absorbed by the ocean. Overall, **more carbon enters than exits the atmosphere** (by more than 3 billion tons [2.7 billion tonnes] per year). Atmospheric carbon dioxide levels also fluctuate annually (notice the sawtooth pattern in Figure 17.9). CO_2 levels are lower during the growing season, when plants absorb gas, and higher in winter, when many plants are

Figure 17.9   Variations in atmospheric carbon dioxide concentrations at Mauna Loa Observatory, Hawaii. The red line represents seasonal carbon dioxide fluctuations and the black line shows annual trends. These data represent the longest record of direct measurements of carbon dioxide in the atmosphere. They were started by C. David Keeling in March 1958.
Adapted from http://www.esrl.noaa.gov/gmd/ccgg/trends/

dormant. The degree of fluctuation in CO_2 levels also varies with latitude. Where do you predict you would find the most pronounced difference in seasonal CO_2 levels due to plant activity?

Hydrosphere. CO_2 is continually exchanged between the atmosphere and the ocean. **Oceans act as a carbon sink by absorbing more gas than they release** to all other components of the earth system (Chapter Snapshot). Oceans retain an additional 2 billion tons (1.8 billion tonnes) of carbon per year. The volume of carbon dioxide that the oceans can absorb increases with decreasing temperature and increasing wind speed. Cooler oceans hold more carbon dioxide. Wave action helps ocean water absorb gas as it produces millions of bubbles that transfer gas from the air to the water. Since carbon dioxide is dissolved in water, it also enters the oceans through precipitation and as runoff from rivers.

Carbon is also carried to the deep ocean in cold, sinking currents such as those in the North Atlantic Ocean that drive

✓ Checkpoint 17.10

✓ basic   ✓ advanced
✓ intermediate   ✓ superior

Carbon is stored in carbon reservoirs, or sinks, just as water is stored in a reservoir behind a dam. The carbon is released from these reservoirs and ultimately ends up in another reservoir. Complete the following table to identify how carbon is stored and released from each component of the earth system.

Sink/source	Carbon is stored when . . .	Carbon is released when . . .
Atmosphere		
Biosphere		
Hydrosphere		
Geosphere		

✓ Checkpoint 17.11

✓ basic   ✓ advanced
✓ intermediate   ✓ superior

Use the concept map provided here to illustrate the natural global carbon cycle by showing how carbon is transferred among the components of the earth system.

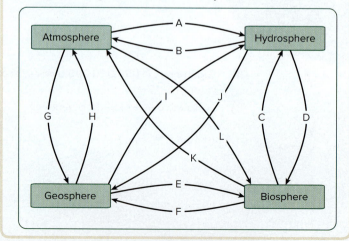

the global ocean conveyor system (Chapter 13 Snapshot). These deep waters remain isolated from the atmosphere for more than 1,000 years, representing an effective, long-term carbon reservoir. Carbon dioxide is brought back to the surface and released by upwelling currents such as those in the northeastern Pacific Ocean. How does it get back into the atmosphere? Think about what happens when you open a warm can of soda. Soda contains dissolved CO_2 to provide bubbles. When you open a warm can, the gas rapidly comes out of solution and sprays liquid all over you. A similar, but less violent, process occurs in the oceans. If water temperatures were warm enough, oceans could release more carbon dioxide to the atmosphere than they absorb.

Biosphere. The biosphere interacts with all other parts of the earth system through the carbon cycle. **Plants extract carbon dioxide from the atmosphere by photosynthesis.** Animals eat the plants and consume the carbon. **Carbon is returned to the atmosphere when the organisms die** or may be withheld from the carbon cycle for millions of years if the organic remains are converted to fossil fuel deposits such as coal, oil, or natural gas. Some carbon is retained in soils. Cultivation can release this carbon, but techniques such as conservation tillage (see Chapter 9) can preserve carbon in the soils. Carbon is also retained by marine organisms such as coccolithophores (see Chapter 7) and foraminifera (see Chapter 16) that extract dissolved carbon dioxide from ocean waters to form their shells. Dead organisms sink to the bottom of the ocean where their remains accumulate to form sedimentary rocks that store the carbon within rock layers.

Geosphere. The **largest sinks for carbon on the planet are rocks and minerals of the solid earth** (Chapter Snapshot). Most of that carbon is present as an element of calcium carbonate (limestone). Limestone releases carbon dioxide to the atmosphere when it undergoes chemical weathering at Earth's surface. These processes are most common in mountainous regions undergoing rapid erosion or where hot, wet climates produce conditions suitable for rapid chemical weathering. Some carbon is stored in fossil fuel deposits formed from decayed organic material and is released when these fuels are consumed to produce energy.

Carbon Produced by Human Activity

The volume of carbon in sinks fluctuates in time and space, but overall, CO_2 in the atmosphere remained constant at 280 parts per million (ppm) for several centuries prior to the Industrial Revolution. However, anthropogenic (human-produced) **emissions of greenhouse gases such as carbon dioxide have increased by one-third since the Industrial Revolution** (Figure 17.10). Most anthropogenic greenhouse gases are produced by the consumption of fossil fuels, such as oil for transportation and coal for electrical power, and through the destruction of forests (deforestation). About one-half of the human-produced carbon dioxide is absorbed in carbon sinks in the biosphere, land, and oceans with the remainder left in the atmosphere (see Chapter Snapshot). Since we know that greenhouse gases absorb energy, it seems apparent that continued emissions will result in additional warming.

How might the recent increase in carbon dioxide concentrations compare to population trends? Over the same time period that carbon dioxide and other greenhouse gases have increased, global population has risen dramatically. There were 1.6 billion people living on Earth in 1900; today, there are more than 7 billion. Not only will there be more people in the future, but also economic growth may increase the volume of greenhouse gases that each of these individuals produces.

We next examine where these gases are produced and consider which ones have the greatest potential to raise global temperatures.

Greenhouse Gas Emissions

The major greenhouse gases are water vapor, carbon dioxide (CO_2), methane, halocarbons (HFCs, including CFCs), and nitrous oxide. Water vapor accounts for most of the natural greenhouse effect, but human production of water vapor is not a major factor in global warming. It is the other greenhouse gases produced by human activities that may have tipped the balance toward global warming. Carbon dioxide accounts for more than three-quarters of anthropogenic global greenhouse gas emissions (Figure 17.11). The bulk of greenhouse gas emissions results from human use of fossil fuels such as oil, natural gas, and coal. Methane and nitrous oxide come from industrial, agricultural, and waste sources, and they account for much of the remaining gases (Figure 17.11). Fossil fuels are burned to produce electricity for residential heat and light, used as gasoline for transportation, or used to power industry. Electricity generates about one-third of US greenhouse gas emissions, and transportation accounts for a little more than one-quarter.

Different fossil fuels produce different amounts of greenhouse gases. For each unit of energy consumed, coal produces 21 percent more CO_2 than oil and 76 percent more than natural gas. Consequently, new power plants may be more likely to burn natural

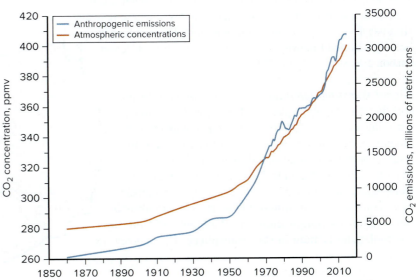

Figure 17.10 Anthropogenic (human-produced) and atmospheric carbon dioxide compared. Human production of carbon dioxide in relatively large concentrations began in the 1800s at the same time as the concentration of carbon dioxide in the atmosphere began to increase. The volume of anthropogenic carbon dioxide can be readily calculated on the basis of the volume of fossil fuels consumed.

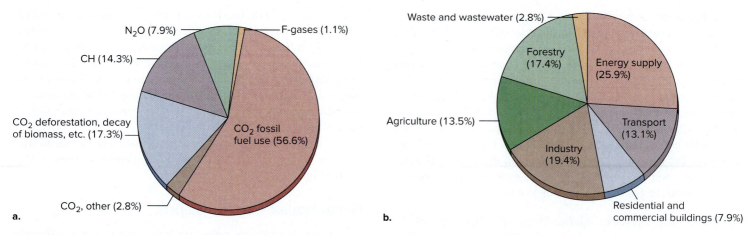

Figure 17.11   Global emissions of anthropogenic greenhouse gases. CH = methane; N₂O = nitrous oxide; F gases = fluorinated gases. **a.** Share of emissions by greenhouse gas; **b.** share of greenhouse gases by sector. 17.11a, 17.11b: Adapted from http://www.ipcc.ch/graphics/graphics/syr/fg2-1.png

gas than coal, even though the United States has one-quarter of the world's known coal reserves. In contrast, nuclear power and alternative power sources, such as solar and wind energy, release no greenhouse gases. The consumption of fossil fuels fluctuates annually. We consume more fossil fuels and produce more greenhouse gases during times of rapid economic growth, low fuel prices, and more extreme summer and winter temperatures, which result in higher demand for domestic cooling and heating. Limits on major nonfossil fuel electricity sources, such as the shutdown of nuclear plants or reduced stream flow decreasing output from hydropower plants, can also result in more fossil fuel use. The United States produces a more than 4 tonnes of carbon per person each year (or about 17 tonnes of carbon dioxide), about four times the global average. US carbon emissions have declined in recent years as natural gas accounts for a larger share of electricity generation. The greatest US carbon emissions occur in cities with high population densities and some combination of concentrated industrial emissions (Houston, Texas), heavy traffic (Los Angeles, California), or dense residential and commercial development (Chicago, Illinois). Even areas of low emissions may be traversed by a heavily traveled highway to generate a thin strip of denser carbon emissions (Figure 17.12).

Global Warming Potential. The ability of gases to absorb heat in the atmosphere is known as their global warming potential (GWP). The GWP compares the amount of heat absorbed by a given mass of a greenhouse gas to the heat absorbed by an equivalent mass of carbon dioxide. Individual molecules of methane, nitrous oxide, and HFCs absorb more heat than a molecule of carbon dioxide. **The amount of heat a gas can absorb depends on its composition and the length of time the gas molecules remain in the atmosphere.** The atmospheric

Table 17.2	Global Warming Potentials for Greenhouse Gases	
Greenhouse Gas	Atmospheric Lifetime, years	Global Warming Potential (GWP for 100 years)
Carbon dioxide	50–200	1
Methane	12	21
Nitrous oxide	120	310
HFCs	1–300	140–11,700

Note: Water (H₂O) is a greenhouse gas that accounts for about 35 to 75 percent of the overall greenhouse effect. Water is not listed here because humans do not produce much water vapor and individual water molecules stay in the atmosphere for only a few years.

lifetimes for key greenhouse gases vary from 1 to 300 years, with carbon dioxide having one of the longer lifetimes (Table 17.2). For example, methane remains in the atmosphere for much less time than carbon dioxide, but it can absorb much more heat, giving it a GWP 21 times greater than that of CO₂. For ease of comparison, all types of greenhouse

Figure 17.12   Annual carbon emissions for the conterminous United States. The map shows annual emissions in 2002 (kilotonnes of carbon, 1 kilotonne = 1,000 tonnes or metric tons). Larger red patches are urban centers; small red dots represent individual point sources (for example, power stations, factories); and linear patterns are major highways. NASA

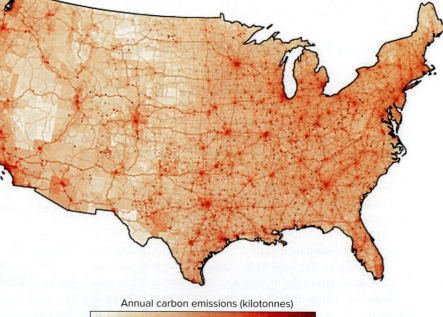

Annual carbon emissions (kilotonnes)
.01   1   100   10,000

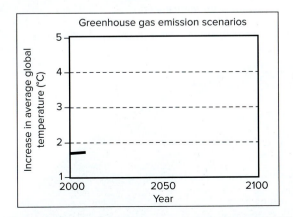

1) China 34.5%	5) Japan 4.4%	9) Saudi Arabia 2.1%	13) South Africa 1.7%	17) Turkey 1.3%
2) USA 19.8%	6) Germany 2.8%	10) South Korea 2.1%	14) Mexico 1.6%	18) Thailand 1.2%
3) India 9.2%	7) Indonesia 2.3%	11) Canada 2.0%	15) United Kingdom 1.5%	19) France 1.2%
4) Russia 5.7%	8) Iran 2.2%	12) Brazil 1.8%	16) Australia 1.4%	20) Italy 1.2%

Figure 17.13   Distribution of global carbon emissions. Data for the 20 nations with greatest emissions in 2014 are listed. Emissions of carbon dioxide vary hugely between places due to differences in lifestyle and ways of producing energy. For example, compare emissions from the different continents. These data will change annually; check online to see recent statistics and discover the most significant changes. How do you think this map will look 50 years from now?
Adapted from http://www.globalcarbonatlas.org/?q=en/emissions

gas emissions may be expressed in terms of "carbon dioxide or carbon equivalents," essentially converting all other gases to their equivalent amount of carbon dioxide or carbon relative to their GWP. (Carbon dioxide equivalents can be converted to carbon equivalents by multiplying by 0.27.)

Concentrations of Greenhouse Gases. Greenhouse gas emissions vary with location, but greenhouse gas concentrations measured in the atmosphere are remarkably uniform worldwide. How do you account for this? Gases are associated with human activities and are therefore initially concentrated where the most people live, but atmospheric circulation patterns thoroughly mix the gases and distribute them throughout the troposphere. Global carbon emissions amount to approximately 32 billion tonnes (gigatonnes) of carbon equivalent each year. The principal global sources of greenhouse gases are nations with the greatest economic growth, large populations, or both. Which countries do you predict have the largest carbon footprint (the amount of carbon dioxide equivalent produced annually)? For example, just four nations produce more than one-half of the world's carbon dioxide emissions (data from 2014)—China, 34.5 percent, United States, 19.8 percent; India, 9.2 percent; and Russia, 5.7 percent (Figure 17.13).

✓ Checkpoint 17.12

✓ basic ✓ advanced
✓ intermediate ✓ superior

Review the following three scenarios and discuss the implications for global greenhouse gas emissions. Which scenario will produce the most or least warming? Complete the graph provided here to predict how temperatures would change over the next century for each scenario. We are more interested in the trends than in the absolute values you identify for each scenario.

Scenario A: Global population peaks around 9 billion in 2050 and declines thereafter. New technologies are rapidly introduced, and economic disparities between regions are substantially reduced. Fossil fuels continue to supply most of the energy used by humans.

Scenario B: Global population continues to increase throughout the twenty-first century (11 billion by 2050). Disparities in the economic growth of regions persist, and technological change occurs slowly. The energy supply gradually changes from fossil fuels toward alternative energy sources.

Scenario C: Global population peaks at around 9 billion in 2050 and declines thereafter. Economies rapidly become service- and information-oriented, income disparities decrease, and energy technologies that do not rely on fossil fuels are introduced.

Greenhouse gas emission scenarios

Increase in average global temperature (°C)

5

4

3

2

1

2000   2050   2100
Year

17.5 Modeling Global Climate Change

Learning Objectives

- Relate climate forcing and feedback mechanisms in the earth system.
- Illustrate how exceeding tipping points may affect the future climate.
- Explain the significance of computer models to understanding the future climate.

The atmospheric concentrations of carbon dioxide and other greenhouse gases will increase, at least for the next few decades. However, only some of the increase in global temperature is directly attributable to rising greenhouse gas concentrations. A larger proportion is related to feedback systems that alter Earth's energy balance and exaggerate the warming trend. In this section, we will discover positive and negative feedbacks that magnify or diminish average global temperature increases. We will also investigate how scientists use their understanding of the earth system to build sophisticated climate models that are used as tools to predict changes in global climate patterns over the next century.

Forcings and Feedbacks

Have you ever experienced audio feedback? Audio feedback can occur when someone speaks into a microphone which transmits that sound to a speaker. The microphone then picks up the sound from the speaker and amplifies part of that sound back to the speaker. This becomes a loop that can result in a high-pitched squeal. If you have heard that happen, you have experienced a forcing and feedback. Feedback results when the output of a system

Figure 17.14 Radiation balance for Earth. Approximately one-third of incoming solar radiation (342 watts per square meter) is reflected back to space (107 watts per square meter). The rest is absorbed by the atmosphere (67 watts per square meter) and the land and water of Earth's surface (168 watts per square meter). Infrared radiation from the heated surface is trapped by greenhouse gases in the atmosphere, adding more warmth to the atmosphere. An increase in greenhouse gases adds 2 to 3 watts per square meter more warming to the atmosphere. Compare with Figure 14.9.
Adapted from UNEP/GRID-Arendal Maps and Graphics Library, http://maps.grida.no/go/graphic/greenhouse-effect

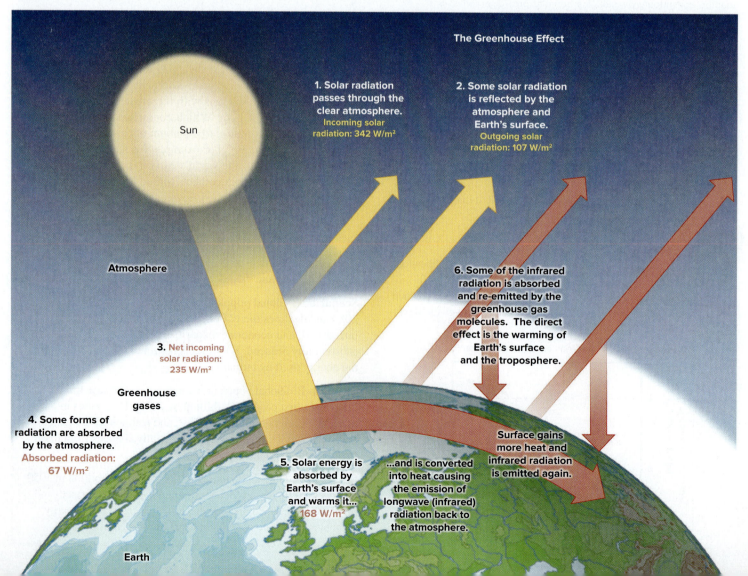

affects the input, in this case the sound. Forcings and feedbacks are very important to our discussion of climate change.

Positive and Negative Climate Forcings. Radiation from the sun is scattered, reflected, or absorbed by Earth (see Chapter 14). The balance between reflected energy and absorbed energy determines the temperature of Earth (Figure 17.14). For instance, the planet cools if more solar radiation is reflected and warms if more radiation is absorbed (see Chapter 14 Snapshot). **Any phenomenon that causes a change in the global solar radiation balance is known as a *climate forcing.*** Forcings are considered positive if they result in global warming and negative if they cause global cooling. We do not mean "good" and "bad" when we use the terms *positive* and *negative* in this context. A positive factor increases an existing trend; a negative factor causes a trend to slow down or reverse. For example, a positive forcing occurs if more solar energy reaches Earth's surface or if more energy is absorbed by the addition of greenhouse gases to the atmosphere. In contrast, a negative forcing occurs if volcanic ash reflects light back into space. The magnitude of a solar forcing mechanism is measured as a change in solar energy in units of watts per square meter (W/m²) passing through the top of the troposphere. Greenhouse gases represent the most significant agent, with an estimated net

forcing of 2.6 watts per square meter (Figure 17.15). Most of this forcing is accounted for by carbon dioxide. Can you think of another forcing that is not related to human activity?

Aerosols: An Example of a Negative Forcing. Aerosols are tiny particles approximately 0.1 micrometer across (about one-thousandth the width of a human hair) produced by the combustion of fossil fuels, burning vegetation, or dust from dry lands. **Most aerosols are particles that reflect sunlight back into space.** Consequently, aerosols cause a negative forcing and diminish the warming effect for some industrialized nations (Figure 17.15).

Checkpoint 17.13

basic | advanced
intermediate | superior

Where would you expect to see the most significant cooling effects due to aerosols? Explain your answer choice.

a. Near the poles
b. Around the equator
c. Between 30° and 60°N latitudes
d. Between 30° and 60°S latitudes

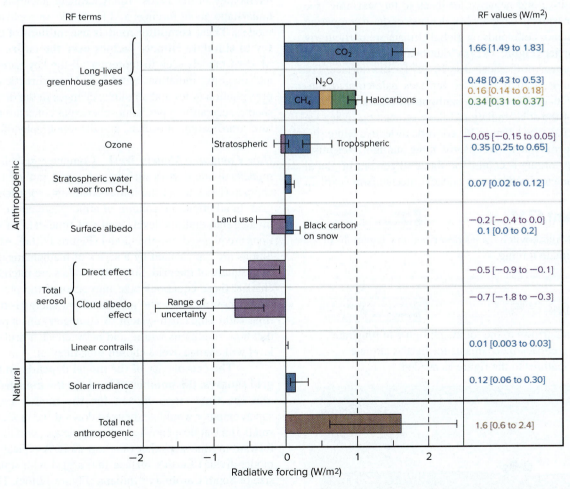

Figure 17.15 Climate forcings. Relative forcings (RFs) are expressed in watts per square meter (W/m²). The warming effect of greenhouse gases was partially offset by forcings due to human activities, such as industrial pollutants in the form of reflective aerosols. Some forcing agents (for example, clouds) have considerable uncertainty as to their degree of forcing. The net effect of forcing is to warm the planet by 1.6 watts per square meter.
Adapted from Data from IPCC

They can even produce a cooling effect in areas of heaviest air pollution (for example, China). Some black particles absorb solar radiation, warming the atmosphere. Ironically, it is only the pollution that produces aerosols that stands between many regions and a more severe warming effect. A surge in pollution associated with rapid industrialization from 1945 to 1975 is thought to have reflected sufficient solar radiation to cause a temporary respite in the steady climb in average global temperatures. Environmental regulations introduced in the early 1970s caused a steep drop in industrial pollution, and the warming trend resumed. Now back to the question at the end of the previous paragraph: What natural geologic event develops aerosols that would provide negative feedback? (See Chapter 6 for a hint.)

Other Feedback Mechanisms. Forcings are only one-half of the story. There are reactions to forcings. The **warming or cooling caused by forcings leads to climatic feedbacks that can exaggerate or reduce temperature changes.** For example, warmer climates add more water vapor to the atmosphere through evaporation. Water vapor is a greenhouse gas that amplifies the warming effect. However, the water vapor may form clouds that reflect more solar radiation. Those same clouds absorb the heat that is radiated from Earth. Consequently, the addition of water vapor may contribute both positive and negative feedback to the warming (see discussion of contrails in Chapter 14). Because of these competing effects, the influences of clouds on global climate are particularly difficult to decipher (Figure 17.15). Additionally, changes in land use result in changes in the heat absorbed at Earth's surface. For instance, warming at high latitudes replaces pale-colored tundra, which reflects much incoming sunlight, with dark forest that absorbs much of the solar radiation. Likewise, melting of sea ice in the Arctic Ocean (see Chapter 16) converts reflective white surfaces to dark ocean water that absorbs more sunlight. Scientists hypothesize that forcing accounts for about 40 percent of global temperature changes and that feedback mechanisms explain

 Checkpoint 17.14

Explain the Milankovitch cycles described in Chapter 16 in terms of a climate forcing.

 Checkpoint 17.15

Add labels and arrows to the following figure to illustrate forcings and feedback mechanisms for Earth's climate system. Add features to the figure as needed.

the rest. Predicting how climate will change is challenging because these multiple forcings and feedbacks are operating simultaneously in a dynamic climate system. To address these complex problems and understand their implications for future climates, scientists have developed powerful computer models.

Climate Models

Climatologists cannot conduct a simple experiment to test the effect of increasing carbon dioxide concentrations on temperature. Even if it were possible to vary one climate factor while keeping all the others constant, it would be of limited value. Instead, climate models must deal with multiple forcings and feedbacks that operate simultaneously. One way to explore how climate-forcing variables interact is by building computer climate simulation models. Climatologists have done just that.

The models scientists use today evolved from the first primitive weather models of the 1950s. Early climate models were very limited, because accurate climate-related global data were not available. Such data include detailed information about winds, temperatures, and moisture for all levels of the atmosphere and for all areas of the globe. This information became more readily accessible and available with the introduction of satellite technology in the 1970s. Today, climate scientists use what they understand about forcings and feedbacks in sophisticated climate models. **These computer models use millions of calculations to try to simulate climate factors over the entire earth system.** Modern models seek to represent all the key factors—incoming and outgoing radiation, wind speed and direction, cloud types, precipitation types and amounts, changes in the dimensions of ice sheets, vegetation types, atmospheric gas concentrations, temperature stratification of oceans, and the topography of continents.

How Computer Models Work. Computer models, like weather models, divide Earth's surface into a grid and use data on surface characteristics to set the initial conditions for each square in the grid. In addition, the column of atmosphere above each square is divided into multiple levels, and real-time data on weather conditions are plugged into the model (Figure 17.16a). All of these conditions are represented by a series of mathematical equations for each square of the grid. Since the models are intended to simulate change, these equations take into account the physical laws that control how air and water move and how energy is transferred over time within the earth system. A computer runs a program to predict how conditions within and between each grid square at each level will change over a chosen time interval.

The complexity of the model depends on the size of the grid squares, the number of levels in the modeled atmosphere, and the time steps chosen for the model. Even the fastest supercomputer would get bogged down if the model grid were too small, if the atmosphere had too many levels, or if the model moved forward in time steps that were too short. A typical global climate model divides Earth's surface into a grid with squares about the size of South Carolina or Indiana (Figure 17.16a). The atmosphere has 15 to 30 levels of differing heights, and most are in the troposphere. Finally, the time steps cannot be longer than the phenomena being modeled. A climate model attempting to predict decades into the future might reexamine the characteristics of each level of each grid square for time intervals as short as 30 minutes (Figure 17.16b).

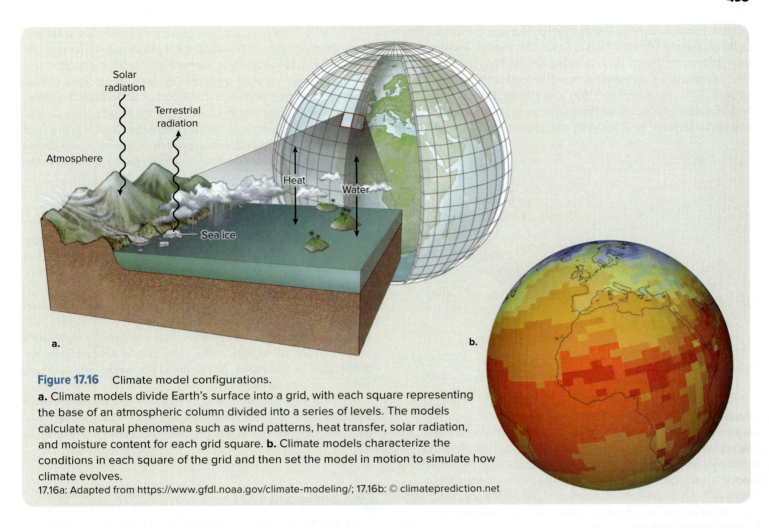

Figure 17.16 Climate model configurations.
a. Climate models divide Earth's surface into a grid, with each square representing the base of an atmospheric column divided into a series of levels. The models calculate natural phenomena such as wind patterns, heat transfer, solar radiation, and moisture content for each grid square. **b.** Climate models characterize the conditions in each square of the grid and then set the model in motion to simulate how climate evolves.
17.16a: Adapted from https://www.gfdl.noaa.gov/climate-modeling/; 17.16b: © climateprediction.net

So, for a single day, the model performs thousands of calculations for each grid square. This process is repeated for each square for thousands of days into the future. No wonder a staggeringly high number of calculations are involved in these models!

Accuracy of Climate Change Prediction Models. Just as there are different computer games to simulate the same activity, there are several computer models designed to simulate climate. The various climate models have many similar features and typically produce consistent predictions. But just how accurate are these models? One way to evaluate their accuracy is to compare their predictions to the actual record of recent climates (1850 to the present). Such comparisons show that most climate models generate the same major trends that are observed in the climate data. For instance, the models correctly predicted that the Arctic would warm faster than the rest of the planet and that the eruption of Mount Pinatubo in 1991 would generate a global cooling event.

Several independent research groups have generated model results that predicted temperature rises if CO_2 increased. Current best estimates are that forcings and feedbacks associated with doubled CO_2 concentrations would produce an average 2 to 4.5 C° (3.6 to 8.1 F°) temperature increase. The range is due to the uncertainties associated with the impact of aerosol forcing as well as social and economic changes that might occur in the future. Today, these models accurately predict the cooling effect that aerosols appear to have had on the climate over the past century. In Section 17.6, we will examine more model predictions regarding the climate in our future.

Checkpoint 17.16

| ⦿ basic | ⦿ advanced |
| ⦿ intermediate | ⦿ superior |

Explain why computer climate models are really "earth system" models.

17.6 A Warmer World

Learning Objectives

- Describe actions anyone could take to reduce her or his greenhouse gas emissions.
- Use results of climate models to compare and contrast the global warming predictions for different continents.
- Show how various components of the earth system will be impacted by global warming.

To return to our opening question: Do you feel lucky? Will our impact on global climate have positive or negative consequences? The people of the Arctic featured in Chapter 16 might tell us that global warming is already here and has brought some serious problems. But what about the rest of the world? How will the remainder of us be affected by additional temperature changes? Well, for one thing, a warmer world holds the promise of a 6-meter-long (20-foot) python on your lawn if you live in the southern half of the United States. Pet owners who dumped their unwanted snakes into the wilds of the Florida Everglades have unwittingly established a breeding population of Burmese pythons. Climate change will turn states as far north as Delaware and Missouri into potential homes for these giant snakes by the end of the century. As if you didn't have enough to worry about.

Figure 17.17 Results of global mean temperature change model for three scenarios, 1900–2100. Each plot corresponds to a different greenhouse gas emissions scenario. Scenario A2 assumes slow economic growth, rapid population growth, and slow technological changes favoring the preservation of fossil fuels as the dominant energy source. Scenario A1B presumes a rapid economic growth, moderate population growth, and a balance between old and new energy sources. Scenario B1 involves moderate economic and population growth with emphasis on new technologies available to all nations. Adapted from Data from IPCC

IPCC climate projections make it clear that the world will warm by an average of 0.2 C° (0.4 F°) per decade for at least the next 15 years. How rapidly it warms and how hot it gets after that will depend on what steps we take in the next two decades. Consequently, it becomes important that we try to understand the potential changes that may occur. Current climate models using various population and economic growth scenarios consistently predict that combinations of **rapid economic growth, significant population increases, and continued reliance of fossil fuels will result in the greatest average temperature increases by 2100** (Table 17.3 and Figure 17.17). The scenarios that produce the greatest temperature increases (A1F1, A2; Table 17.3) assume a future still dependent on fossil fuels with either many more people (15.1 billion in A2) or rapid economic growth that results in peoples of the developing world (for example, China, India, Indonesia) having a standard of living that approaches that of the industrialized nations (A1F1). Moderate average temperature gains (2.4 to 2.8 C°; 4.3 to 5 F°) are expected where a more balanced mix of renewable power and fossil fuels offsets economic or population gains (A1B, A1T, B2; Table 17.3). Finally, the smallest increases in global temperatures occur with the B1 scenario, which describes a sustainable world with "clean" technologies, less demand for goods and resources, and a service-based economy.

It is important to remember that temperature changes predicted by these scenario models are *averages*. These average values mask the fact that global temperatures could increase by as

✓ Checkpoint 17.17

Identify three actions a private citizen could take to reduce the amount of greenhouse gases produced on his or her behalf.

Table 17.3 — Six IPCC Scenarios for Future Global Development Pathways Involving Demographic, Economic, and Technological Factors. Combinations of rapid economic growth, large populations, and continued reliance on fossil fuel technologies result in the greatest average temperature increases by 2100.

Scenario	Scenario Factors			Consequences
	Economic growth	Population growth rates (population in 2100)[a]	Fossil fuel intensive technology[c]	Temperature change, average (range)
A1F1	High	Low[b] (7.1 billion)	High	High, 4.0 C° (2.4–6.4 C°)
A2	Low	High (15.1 billion)	High	High, 3.4 C° (2.0–5.4 C°)
A1B	High	Low[b] (7.1 billion)	Moderate	Moderate, 2.8 C° (1.7–4.4 C°)
A1T	High	Low[b] (7.1 billion)	Low	Moderate, 2.4 C° (1.4–3.8 C°)
B2	Moderate	Moderate (10.4 billion)	Moderate	Moderate, 2.4 C° (1.4–3.8 C°)
B1	Moderate	Low[b] (7.0 billion)	Moderate	Low, 1.8 C° (1.1–2.9 C°)

[a] Global population in 2013 was 7.1 billion.
[b] In the low population growth cases, population is projected to increase to 8.7 billion by 2050 and then decline to 7.1 billion or less by 2100.
[c] High = more than two-thirds of technology relies on fossil fuels; moderate = approximately one-half to two-thirds of technology relies on fossil fuels; low = approximately one-third or less of technology relies on use of fossil fuels.
Source: Data from IPCC

much as 8 C° (14 F°) in some places and see little change in other locations. The severity of climate change will depend on which scenario the world follows. Some consequences are described in the next section.

Effects of Warmer Temperatures

The influences of human emissions on climate are now clear; recent climate changes are having widespread impacts on humans and ecosystems. Carbon dioxide and other greenhouse gasses are at levels higher than has been documented in the last 800,000 years. We now observe unequivocal warming of the planet that is occurring at high rates compared to the past. Changes in extreme weather and climate events have been observed since 1950. All areas of the globe are affected by these changes, though the effects are not uniform. The consequences of rising temperatures vary considerably around the world. What would you do

if the summer temperatures where you live increased by several degrees? Chances are you would use more energy to run your air conditioner. Wealthy nations probably can adapt to small changes in climate that occur over long periods (on human timescales).

Mitigating or adapting to changes will be more difficult in developing nations that do not have the resources of more-affluent countries. Research indicates that changes will be most extreme in the developing world and more moderate in the developed nations that are more responsible for the production of greenhouse gases. Where do you expect that climate change will have the greatest social impact? Projected patterns are similar to those observed over the last few decades. Here are a few predictions from climate models:

- Models predict much greater temperature increases in northernmost latitudes (8 C°; 14 F°) and little change in parts of the Southern Hemisphere and North Atlantic Ocean (Figure 17.18).

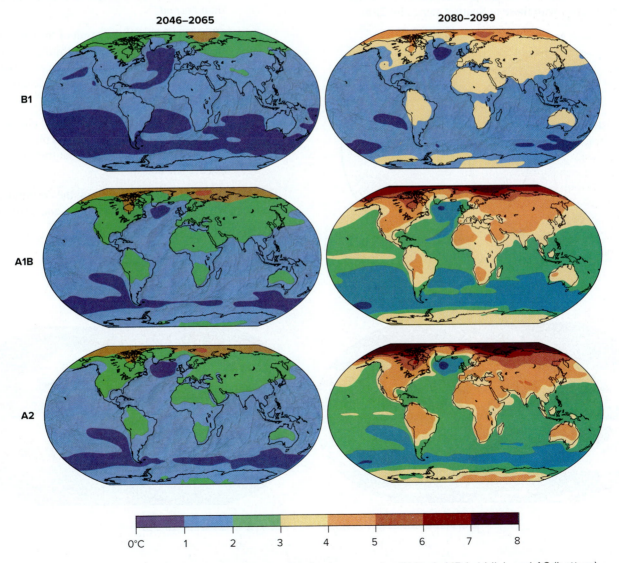

Figure 17.18 Projected annual mean surface warming for the scenarios B1 (top), A1B (middle), and A2 (bottom) for time periods 2046–2065 (left) and 2080–2099 (right). Temperature changes are relative to the average of the period 1980–1999. A warming effect is least for sections of the Southern Ocean and the northern Atlantic Ocean, and greatest for northernmost latitudes. Predicted temperature increases for the United States are between 2 and 6 C° (4 and 11 F°) except for Alaska, where temperatures may increase by 10 C° (18 F°) or more. Adapted from http://www.ipcc.ch/graphics/graphics/ar4-wg1/jpg/fg-10-8.png

- Rising temperatures will increase evaporation and precipitation, but the **extra rainfall will not be distributed evenly around the globe** (Figure 17.19). Precipitation increases will generally be greatest where most rain falls today, around the equator and at high latitudes. A drying trend is evident for latitudes around 20° to 30°N and S that are the sites of modern deserts.
- Regions with no changes or small precipitation increases may actually become drier because the warmer temperatures will result in higher evaporation rates. It is very likely that heat waves will occur more often and last longer.

✓ Checkpoint 17.18

| basic | advanced |
| intermediate | superior |

Venn Diagram: Climate Change in the Americas

Use Figures 17.18, 17.19, and the Venn diagram provided here to compare and contrast the predicted climate changes in North America with those in South America by 2100.

North America   South America

✓ Checkpoint 17.19

| basic | advanced |
| intermediate | superior |

During a homework assignment, some students provided the following descriptions of global warming. Review the definitions and rank them in order of accuracy (1 = most accurate; 5 = least accurate).

_____ Global warming is the gradual warming of Earth due to the greenhouse effect.

_____ Global warming is a gradual, long-term increase in the temperature of Earth due to an increase in the concentration of trace atmospheric gases such as carbon dioxide that absorb heat from Earth's surface.

_____ Global warming occurs when the temperature of Earth slowly increases because a layer of chemicals in the atmosphere absorbs more heat than normal.

_____ Global warming is the result of the ozone layer in the atmosphere becoming thinner because of certain chemicals we use. This causes more solar radiation to reach Earth's surface, leading to higher average temperatures.

_____ Global warming is the warming of climate over many years as a result of natural phenomena, such as volcanic eruptions, that trap heat close to Earth's surface.

Describe the criteria you used to rank the definitions. (For example, why are definitions you rank 1 and 2 better than those you rank 3 and 4?)

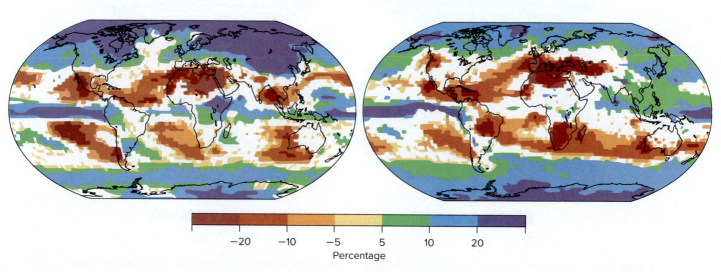

Figure 17.19   Modeled changes in global average precipitation for the period 2090–2099 relative to 1980–1999. Left, changes for the interval December–February; right, changes for July–August. Changes for scenario A1B only. White areas indicate regions where different models do not agree on predictions. Precipitation increases are greatest at equatorial latitudes and for high latitudes. A drying trend is evident for midlatitudes. How will precipitation change over North America?

Adapted from http://www.ipcc.ch/graphics/graphics/syr/fg3-3.png

- Changing climate will **increase the severity of storms,** leading to more structural damage and heat-related deaths. It is very likely that extreme precipitation events will become more intense and frequent in many locations and hurricanes will become more intense (but perhaps not more frequent).
- Snow cover on land will continue to shrink, and sea ice is projected to disappear entirely from the Arctic in late summer.
- Most scenarios predict that **global sea levels will rise by 20 to 50 centimeters** (8 to 20 inches) over the next century. Thermal expansion of ocean waters and alpine glacial melting will continue for centuries, resulting in a slow increase in sea level, even if global temperatures can be maintained at a constant level after 2100. After 2100, models show a switch to Greenland and Antarctica melting as dominant factors in sea level rise.
- Changes in the dynamics of the atmosphere are expected to shift Earth's great pressure belts, causing a poleward migration of subtropical highs in summer and expansion of polar highs in winter.

Differences in heating of land and water will affect monsoons.
- Earth's oceans will become more acidic as CO_2 concentrations increase in the atmosphere. This excess carbon dioxide will change the chemistry of the oceans, facilitating the release of hydrogen ions. This directly affects calcium carbonate minerals in shells of corals, shellfish, and other organisms.

The changing climate has significant consequences for the earth system. The risks from these changes will be unevenly distributed, but are dis-proportionally greater for disadvantaged populations. A hotter, drier climate could **alter the balance of water resources.** Lower stream discharges would lead to less barge traffic and decrease the potential for generating hydroelectric power. Less water available for irrigation and municipal users could also result in regulations limiting the types of crops grown and/or water rationing programs. People relying on snowmelt for their summer water supply may find that warmer climates replace some of the snow with winter rains and cause the snow to melt earlier in the spring. Billions of people are expected to be affected by reductions in water supplies (Table 17.4).

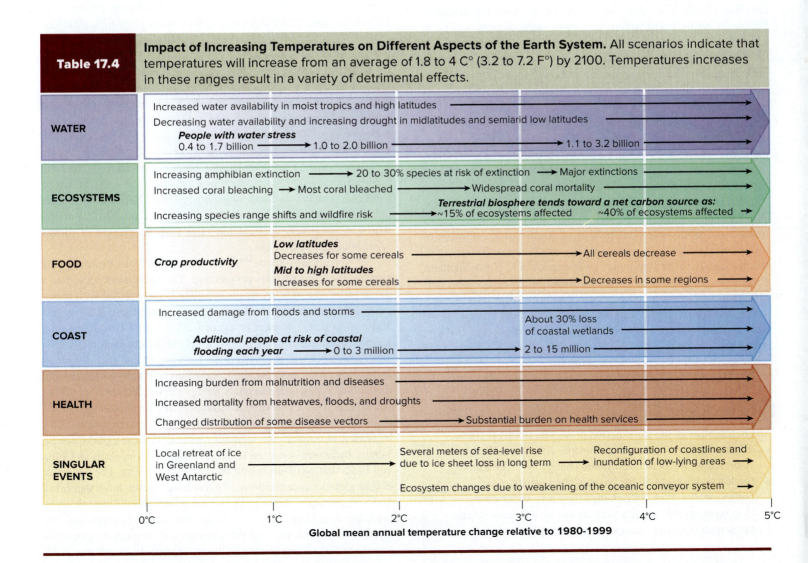

Table 17.4 Impact of Increasing Temperatures on Different Aspects of the Earth System. All scenarios indicate that temperatures will increase from an average of 1.8 to 4 C° (3.2 to 7.2 F°) by 2100. Temperatures increases in these ranges result in a variety of detrimental effects.

Global mean annual temperature change relative to 1980-1999

Agricultural production will increase in some regions and decrease in others (Table 17.4). The impact will be minimal in spatially large countries where losses in productivity in one region are balanced by increases elsewhere. The problem is more threatening for smaller countries. Small, developing nations with large populations face famine and the exodus of refugees because of drought conditions. In contrast, northern nations such as Canada and Russia will benefit from longer growing seasons and increased precipitation (see Figures 17.18 and 17.19).

Changing climate will result in shifting ecosystems. Nonnative plants will migrate northward as climate warms there, displacing existing vegetation. Animals that rely on native plants for sustenance may lose their primary source of food. Trees such as oaks typically migrate at a rate of a few hundred meters per year, about one-tenth of the rate of temperature change over land. Most animals also take generations to adjust to changes in climate. Thus, if ecosystems have insufficient time to adapt to the migration of climate belts, more species could become extinct (Table 17.4). Environmental stresses from climate change are expected to reduce the amount of carbon dioxide extracted by plants, increasing future rates of warming. Nearly one-third of species will be at increased risk of extinction with the smallest projected temperature increases.

Higher sea levels will inevitably place additional millions of people at risk of property damages and loss of life from coastal flooding (Table 17.4), especially in association with hurricanes. People living on small islands and along delta regions will be at greatest risk. While warmer temperatures are expected to reduce the number of deaths due to cold exposure, these will be more than offset by deaths due to malnutrition and infectious diseases related to decreasing agricultural output and expanding range of disease-spreading insects (Table 17.4).

As if these consequences from global warming were not enough, there are some scientists who suggest that various parts of the earth system have tipping points that could change everything. A *tipping point* is a critical threshold above which a small change can cause large changes to the future states of a system in ways sometimes not predicted by models. These tipping points are associated with sub-continental elements of the climate system. Examples include unexpected complete melting of Arctic sea ice, collapse of the Greenland ice sheet, large-scale changes to thermohaline circulation, changes to the amplitude and frequency of El Niño–Southern Oscillation (ENSO), instability of the Indian summer monsoon, permanent tundra loss at high latitudes in Russia, or large-scale changes to rain forests or boreal forest dieback. Any of these changes could have a near-term, regional impact. The greater concern is that any one of these could change other climate-related processes across the entire earth system; hence the term *tipping point*.

 Checkpoint 17.20

| ◉ basic | ◉ advanced |
| ◉ intermediate | ◉ superior |

On the basis of the information presented in this section, discuss how North America, Europe, Africa, and South America will experience climate change by 2100. Rank the four continents on the basis of the severity of the consequences of climate change they may experience (1 = most severe negative consequences; 4 = least severe consequences). Give reasons to support your rankings.

17.7 What Can Be Done?

Learning Objectives

- Calculate your carbon footprint.
- Evaluate the effectiveness of recent carbon emission reduction treaties and agreements.
- Create a plan for reducing carbon emissions.

International Agreements to Improve the Environment

The signing of the Montreal Protocol in 1987 demonstrated that governments can agree on common principles to improve the environment. As a result, the use of compounds that were known to break down ozone was greatly reduced or banned throughout the world. However, the ozone issue was relatively simple and directly threatened people with increased incidences of cancer and cataracts if something was not done. Global warming, by contrast, is complex. People do not appear to be as directly threatened by the predicted consequences, After all, some argue, how important are a couple of degrees average increase? Some countries may even benefit from the warming, and Earth has been much warmer in the past (just ask the dinosaurs!). Consequently, reducing emissions of CO_2 in the atmosphere is proving to be a much more difficult effort.

Many nations signed the Climate Change Convention at the Earth Summit in Rio de Janeiro in 1992. These nations pledged to try to return carbon emissions to 1990 levels by the year 2000. It did not happen. International policymakers met again in Kyoto, Japan, in 1997. The resulting **Kyoto Protocol represented an agreement among the developed nations to reduce greenhouse emissions to 1990 levels by 2008–2012,** and 6 to 8 percent below those levels in the 5 years that follow (6 percent for Japan, 8 percent for nations of the European Union). Some nations have made firm commitments to mitigate potential warming by signing the Kyoto Protocol, which was finally ratified in 2005. Ratification means that the agreement was supported by more than 120 countries, which cumulatively account for more than 55 percent of global greenhouse gas emissions.

Perceptions about the success of the Kyoto Protocol may vary on the basis of your perspective. The agreement served as a major milestone in bringing nations together to grapple with issues related to climate change. Many countries met their 2012 targets and are on track to meet or exceed 2020 goals. However, some people argue this agreement is largely a failure because China and India were excluded from the targets, while other countries were allowed to increase their emissions, Canada withdrew and the United States failed to ratify the agreement. For example, Russia experienced a substantial reduction in greenhouse gas emissions after 1990. Unfortunately, greenhouse gas emissions have increased in other Kyoto signatory nations (for example, Canada, Japan). Increases occurred in developing countries such as China and India that were not bound by the treaty's limits. The net result is that global concentrations of greenhouse gases have increased and will continue to rise in the future unless more dramatic steps are taken to mitigate these emissions. On a more positive note, the Kyoto treaty illustrates that the international community can come together to battle the threat of climate change. In December 2015, efforts to minimize future global warming took a big step forward when the world's nations negotiated the Paris Agreement which seeks to hold the increase in global average temperature to less than 2°C above pre-industrial levels. The agreement will be considered operational if 55 nations that generate at least 55% of the world's greenhouse gas emissions ratify the accord by April, 2017.

Reducing Greenhouse Gas Emissions

Improvements in technology in the decades since the original Earth Summit now present a variety of options for reducing greenhouse gas emissions that were not available in the 1990s. Take a moment to take another look at Figure 17.17. Notice that the difference between the "constant CO_2" and the A1B or A2 plots (the most likely scenarios) forms a triangle. We present an alternative version of this graph in Figure 17.20. This time, we have replaced the temperatures along the left axis with greenhouse gas emissions, we have shortened the timescale from 1955–2055, and we use a single plot of future emissions that projects current trends to double global greenhouse gas emissions by the middle of the century (Figure 17.20). The section of the graph labeled the stabilization triangle represents all the greenhouse gases we want to avoid producing in the next few decades. The more of this

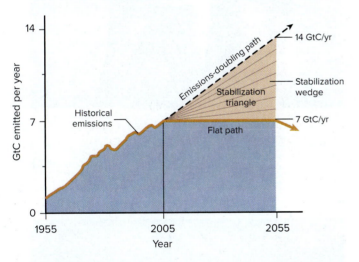

Figure 17.20 Possible carbon emissions through the midcentury. If emissions do not increase, they would follow the "flat path." However, at current trends, it is more likely that we will follow the "emissions-doubling path." The difference in carbon emissions for these two scenarios is represented by the stabilization triangle. Scientists Robert Socolow and Stephen Pacala have suggested that it is useful to divide the triangle into a series of seven stabilization wedges and to pursue a variety of carbon mitigation strategies that could each account for all or part of a wedge, representing 25 gigatonnes of carbon over the time until 2055. Adapted from http://www.princeton.edu/~cmi/resources/cmi_resources_new_fles/cmi_wedge_game_jan_2007.pdf

✓ Checkpoint 17.21

Find an online carbon footprint calculator and determine how much carbon you produce annually. What activities produce the most carbon?

triangle we can get rid of, the less the temperature will increase. We can choose to pursue **mitigation strategies that slow the pace of climate change or adaptation strategies that adjust to its impact.** If effective mitigation strategies are not introduced in the next few decades, then adaptation strategies will play a more significant role by midcentury. We discuss a range of mitigation options and outline adjustment choices in Table 17.5.

No single strategy will "fix" global warming. We will have to employ a variety of approaches to cut back greenhouse gas emissions to reduce future temperature increases. If the stabilization triangle is divided into a series of stabilization wedges, each wedge represents a savings of 25 gigatonnes of carbon (GtC) over the 50 years, from 2005 to 2055 (Figure 17.20). Physicist Robert Socolow and ecologist Stephen Pacala of Princeton University developed the concept of stabilization wedges to encourage consideration of a mix of global warming solutions. They have proposed a number of strategies for saving 25 gigatonnes of carbon over the 50-year time interval that fall into four categories: 1) energy efficiency and conservation; 2) changes in electricity energy sources; 3) changes in transportation energy sources; and 4) increasing natural carbon sinks. Any seven stabilization wedges would achieve the aim of minimizing emission increases, but it is more likely that more than seven will be employed with varying degrees of success in the years ahead. If greenhouse gas emissions increased more rapidly, we would need to use more wedges. Which of the options that follow do you think are most likely to be applied in the future?

Efficiency and Conservation Options.

Improved fuel economy: One stabilization wedge would be saved if the number of miles traveled per gallon (mpg) of fuel were to double from an average of 30 to 60 miles per gallon. This assumes the number of passenger vehicles will increase from 600 million to 2 billion by 2055. Hybrid engines and lighter materials will help increase fuel economy.

Reduced reliance on cars: One wedge would be saved if the annual distance traveled by car was cut in half from

Table 17.5	Adjustment Strategies for Climate Change
Sector	**Adaptation Strategy**
Water	Improve irrigation efficiency; reuse domestic water (Chapter 12); build desalination plants; increase water storage capacity.
Agriculture	Change planting dates; introduce alternative plant varieties; improve land management (for example, conservation tillage methods); change livestock and/or crops.
Infrastructure	Relocate at-risk communities; build/strengthen storm surge barriers and flood levees; expand area of wetland buffers; preserve natural lands; expand climate/weather monitoring systems.
Health	Develop heat-health plans; increase climate-related disease programs; expand health programs in most affected regions.
Transport	Provide incentives for public transportation, telecommuting; relocate at-risk road/rail systems.
Energy	Strengthen electricity transmission lines; provide incentives for renewable resources and energy-efficient technology; expand energy choices.

16,000 kilometers (10,000 miles) to 8,000 kilometers (5,000 miles). Greater use of mass transit and telecommuting can help achieve this goal.

More energy-efficient buildings: A wedge would be saved by employing more efficient technologies for heating, cooling, and lighting to cut emissions from homes and offices by 25 percent. Savings can come from insulation, more-efficient appliances, and passive solar heating of new buildings.

Improved power plant efficiency: Coal power plants operate at about 35 percent efficiency (proportion of coal burned converted to electricity). A wedge would be saved by increasing the efficiency of coal power plants to 60 percent by midcentury. The potential of this strategy is limited as a little over one-third of global electricity is generated from nuclear power and hydropower sources.

Changes in Electricity Energy Sources.

Replacing coal with natural gas: Natural gas power plants produce about one-half as much carbon per unit of electricity as coal power plants. Coal power plants produce about one-quarter of the world's carbon emissions. One wedge would be saved if 1,400 large natural gas plants replaced a similar number of coal power plants.

Carbon capture and storage: Carbon capture and storage (CCS, also termed *carbon sequestration*) technology can trap 90 percent of carbon emissions. The captured carbon dioxide

would have to be stored, probably in underground geologic reservoirs, for hundreds of years (Figure 17.21). Installing CCS technology at 800 large coal power plants (or 1,600 gas plants) would save a wedge, but we would need to vastly expand carbon storage by more than 1,000 times its current capacity.

Nuclear energy: Nuclear power is carbon-free and provides about 17 percent of global electricity (Figure 17.22a). Tripling the number of nuclear power plants while phasing out coal plants with similar capacity would save a wedge of emissions. Nuclear power held great promise as a cheap source of electricity in the 1960s but fell out of favor following a series of plant accidents in the 1970s and 1980s. Additional expansion of nuclear power plants could account for a second wedge. However, problems remain with storage of highly radioactive by-products, with safeguarding materials that could be used in nuclear weapons, and with public concerns over recent nuclear plant failures.

✓ Checkpoint 17.22

| ✓ basic | ✓ advanced |
| ✓ intermediate | ✓ superior |

Review the mitigation strategies described in Section 17.7. Which seven would you choose to add to represent the stabilization wedges in Figure 17.20? Why are these seven the best choices to prevent the most severe future warming?

a.

b.

Figure 17.22 The expansion of carbon-free energy sources such as **a.** nuclear power and **b.** wind energy could account for two or more stabilization wedges of future carbon emissions.
17.22a: © Steve Allen/Getty Images; 17.22b: © McGraw-Hill Education/John Flournoy, photographer

Figure 17.21 Carbon sequestration. An active area of research for sequestering carbon dioxide involves injecting it into oil and gas reservoirs, deep salt formations, or abandoned mines.

Wind energy: As we discussed in Chapter 14, carbon-free wind energy is providing an increasing proportion of global electricity (Figure 17.22b). Electricity from wind is growing at 30 percent per year. To account for a wedge of emissions, the future wind power capacity would have to be 30 times greater than current capacity. That could come in the form of 2 million new 1-megawatt wind turbines that would occupy an area about the size of New Mexico.

Solar energy: As we discussed in Chapter 2, carbon-free, renewable solar energy provides a potential source of future electricity. Photovoltaic (PV) cells covering an area about the size of New Jersey would be sufficient to replace a wedge of carbon emissions generated from coal power plants. PV cells are more expensive than other energy sources, and their deployment would require a dramatic increase in efficiency in order to improve their capacity by several hundred times from current levels.

Changes in Transportation Energy Sources.

Hydrogen fuel and carbon capture: Hydrogen has the potential to be a future energy source for passenger vehicles. Hydrogen can be produced from fossil fuels, but the carbon produced during this process must be captured and stored (see the carbon capture and storage option). A 10-fold increase in hydrogen production would represent one stabilization wedge. One drawback of this option is that there is no distribution system for hydrogen fuels.

Synfuels and carbon storage: Synfuels are liquid fuels produced from coal (which the United States has in abundance) as a replacement for oil (which we import). Synfuel production generates twice the carbon emissions as an equivalent volume of fuel refined from petroleum. The largest single source of carbon dioxide emissions in the world is a South African synfuels plant. A wedge would be saved if carbon can be captured from 180 similar coal-to-synfuel plants.

Biofuels: As discussed in Chapter 16, biofuels in the form of ethanol or biodiesel can be made from crops. If produced sustainably, the carbon dioxide produced by burning these fuels is offset by plant growth. Increasing ethanol production by 30 times its current level would be sufficient to replace one wedge of carbon emissions generated from consuming petroleum fuels. It is unlikely that a sufficient area of global cropland can be dedicated to producing the raw materials needed to produce large volumes of these biofuels.

Increasing Natural Carbon Sinks.

Forest management: At least one wedge is available if current deforestation is halted and forests are managed to serve as carbon sinks. Despite deforestation in some tropical rain forests, there is a net trend for land plants to absorb more carbon than they produce. Stopping deforestation will generate one wedge. Increasing the size of current forests will increase carbon storage; but without an end to deforestation, we would need to cover an unreasonably large area of land in trees to generate a wedge of emissions savings.

Conservation tillage: About one-half of the carbon stored in soils is lost when natural lands are converted to cropland. Up to one wedge would be available from widespread deployment of conservation tillage practices (Chapter 9) that preserve crop waste in the soil and increase the soil carbon.

Socioeconomic Options. We can't forget that we live in a society that is dependent on carbon-based energy. Fossil fuels power the global economy, and that is not likely to change in the foreseeable future. Recognizing that, there are socioeconomic options that could be used to alter consumption patterns. For instance, some argue that fossil fuel usage patterns could be changed if governments were to tax carbon consumption. The tax could be imposed for the production, distribution, or use of fossil fuels with allowances for the amount of CO_2 that fuel emits. The imposing of such a tax is based on an economic principle called *negative externalities*. That principle assumes that the cost of the product to society from future warming, acid rain, or pollution is not currently reflected in its price. The revenues could be used to mitigate those effects or seed alternative energy research. Another idea proposed to address the CO_2 emission problem is called cap-and-trade. This approach sets limits on carbon emissions. Those caps are lowered over time. High CO_2 producers, such as coal-fired power plants, can "trade" carbon credits with low- or non-CO_2 producers, such as gas or nuclear power plants. Over time, new technologies are expected to allow the high producers to transition to different fuels. A variation of this approach is carbon offsetting whereby companies seek to become carbon-neutral by adopting energy efficiency projects, using renewable energy sources, or finding other ways to compensate for their emissions. As is the case with any measures, there will be those who support such measures and those who oppose them.

What Else Can Be Done?

Several technological fixes have been suggested that target different climate-forcing agents. Some are relatively untested and will require considerable time until full deployment. One extreme, relatively unproven option is to alter Earth's energy balance by reflecting incoming solar radiation by using fleets of silver balloons or giant mirrors in the atmosphere. This plan assumes that we could reflect more incoming sunlight, thus reducing the solar energy reaching the surface and the impact of solar forcing. Just how many balloons you would need, where they would be located, and how they would get up there still need to be determined.

Some scientists have suggested adding tiny iron particles to the oceans to spur growth of plankton, organisms that extract carbon dioxide from the ocean. Sounds like a good idea, right? Not so fast. The idea of altering one part of the earth system in an attempt to fix human-induced modifications in another part has received relatively little support, and new research suggests that the story is more complicated than it first appears. Also keep in mind that warmer ocean temperatures might improve biologic activity, but warmer water holds less greenhouse gas. Finally, enormous volumes of iron would be needed to have even a minimal effect.

 Checkpoint 17.23

Explain how the comment "Earth can take care of itself" in the context of global warming relates to the concept of "deep time" discussed in Chapter 8.

Making It Personal. If you want to get started on reducing your individual carbon footprint right now, here are some things you can do, in approximate order of increasing impact:

- Replace your lights with compact fluorescent bulbs.
- Wash with cold water and hang clothes to dry.
- Shut off computers and other appliances when not in use.
- Boost car mileage per gallon by tuning engine, replacing air filter, and inflating tires.
- Turn off the air conditioner and open a window.
- Turn down the thermostat in winter and turn it up in summer.
- Buy locally grown produce to reduce food transportation costs.
- Ride the bus, cycle, or walk instead of driving your car.

✔ Checkpoint 17.24

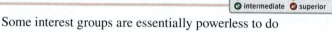

What are some actions that could be taken to diminish the impact of global warming? Different groups of people have different perspectives on this issue. For some, solutions that would avert global warming may not be in their economic interest.

1. List actions each of the following interest groups could take to reduce global warming.
    i. A major oil company
    ii. A utility that burns coal to generate electricity
    iii. A large car manufacturing company
    iv. A company that manufactures wind turbines and solar energy panels
    v. An international insurance company
    vi. The Maldives, a small island nation with an average elevation of 1 meter (3.3 feet)
    vii. A heavily populated developing nation (for example, China, India)
    viii. A heavily populated developed nation with a high standard of living (for example, the United States)
    ix. A family of four in the United States with two vehicles.

Some interest groups are essentially powerless to do anything themselves and must rely on the actions of others. Consider questions 2 and 3 below and evaluate the answer for each group.

2. What would be the impact of global warming on each group?
3. How does the activity of each group impact future global warming?

Choose any three of the groups, and answer these two questions for each group you choose.

4. If applicable, what steps could the group take to diminish its impact on global warming?
5. What incentives would encourage the group to change its habits to reduce its contribution to future warming?

the big picture

Over the last few decades, scientists have learned more about past climates and the factors that influence climate change. It is important to recognize that humans are not the first life-form to affect the planet as a whole. Early organisms, using photosynthesis as a mechanism for sustaining life, gave off oxygen for 2 billion years. Ultimately, that changed the planet. We now have a good idea of the potential range of global temperature increases that human activities could contribute over the next century. Our actions will determine the extent of that increase. In some ways, we are victims of our own success. The faster we grow the global economy, the more greenhouse gases we produce. The more gases we produce, the more extreme future climate changes will be, unless technological advances help us reduce emissions.

In the movie *Waterworld,* the ice caps melted and the planet was inundated with water, flooding almost all the land. We know that won't happen. There just isn't enough ice on the planet to produce the water needed to submerge most of the continents. Increasing rates of melting on Greenland and Antarctica may slowly add a few centimeters a year to global oceans, and in the next few centuries, southern Florida and low-lying parts of coastal nations may slowly submerge. But there is no chance we will be converted to a "waterworld."

In contrast, *The Day After Tomorrow* envisaged much of the Northern Hemisphere locked in the grip of ice sheets. The movie postulated that changes to thermohaline circulation could lead to a shutdown in the global conveyor belt in the North Atlantic that would trigger catastrophic climatic changes in a matter of weeks. Even though current research suggests there may be a slowdown in the conveyor, that process will take many years and the resulting cooling of Europe is likely to be offset by future warming. Like many stories, these movies were each founded on a kernel of truth but quickly lost sight of reality as they spun out their stirring tales. For instance, if you have seen *The Day After Tomorrow,* you know it shows air cooling as it falls from higher altitudes. We learned in Chapter 14 that the opposite actually happens.

Almost all potential concentrations of greenhouse gases are "safe" if you have enough resources to insulate yourself from any negative climate consequences. The greenhouse gases themselves are not a visible threat and pose little direct danger. However, the threat of rising sea levels, decade-long droughts, more severe storms, melting ice caps, bigger floods, and altered ecosystems will make the reality of global change evident in many regions. These changes will be most significant in high latitudes, coastal regions, and areas that already have the potential for severe weather conditions. Computer models of global change are sufficiently sophisticated to give us time to prepare for changes. The clock is ticking. As stewards of planet Earth, it is up to us to propose potential solutions that can diminish the negative effects of global warming. The cure must not be worse than the disease. Nature can provide a solution for this problem, but it may not be to our liking and it probably will not occur on a human timescale.

The signals are clear. Change is inevitable. The question now is just how much we want to control our destiny by playing a role in the solution. Much of what lies ahead for the good Earth is up to us. **Know. Care. Act.**

Global Change: Concept Map

Complete the following concept map to evaluate your understanding of the interactions among the earth system and global change. Label as many interactions as you can. using the information from this chapter.

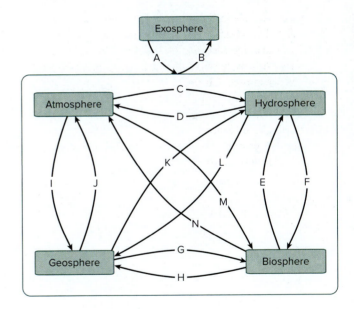

Appendix A
Conversion Factors

To convert	Multiply by	To find
L E N G T H		
Inches	2.54	Centimeters
Centimeters	0.39	Inches
Feet	0.30	Meters
Meters	1.09	Yards
Miles	1.61	Kilometers
Kilometers	0.62	Miles
A R E A		
Square inches	6.45	Square centimeters
Square centimeters	0.15	Square inches
Square feet	0.09	Square meters
Square meters	10.76	Square feet
Square miles	2.59	Square kilometers
Square kilometers	0.39	Square miles
V O L U M E		
Cubic inches	16.38	Cubic centimeters
Cubic centimeters	0.06	Cubic inches
Cubic feet	0.028	Cubic meters
Cubic meters	35.3	Cubic feet
Cubic miles	4.17	Cubic kilometers
Cubic kilometers	0.27	Cubic miles
Liters	1.06	Quarts
Liters	0.26	Gallons
Gallons	3.78	Liters
M A S S E S A N D W E I G H T S		
Ounces	28.35	Grams
Grams	0.035	Ounces
Pounds	0.45	Kilograms
Kilograms	2.205	Pounds
Metric tons (tonnes)	1.102	Tons
Tons	0.907	Metric tons (tonnes)
T E M P E R A T U R E		

To convert degrees Fahrenheit (°F) to degrees Celsius (°C), subtract 32 degrees and divide by 1.8.

To convert degrees Celsius (°C) to degrees Fahrenheit (°F), multiply by 1.8 and add 32 degrees.

Appendix B

The Periodic Table of Elements

The 1–18 group designation has been recommended by the International Union of Pure and Applied Chemistry (IUPAC) but is not yet in wide use. In this text we use the standard U.S. notation for group numbers (1A–8A and 1B–8B). No names have been assigned for elements 112, 114, and 116. Elements 113, 115, 117, and 118 have not yet been synthesized.

Atomic number
Atomic mass

| 9 | F | Fluorine | 19.00 |

Metals
Metalloids
Nonmetals

Group 1 / 1A
- 1 H Hydrogen 1.008
- 3 Li Lithium 6.941
- 11 Na Sodium 22.99
- 19 K Potassium 39.10
- 37 Rb Rubidium 85.47
- 55 Cs Cesium 132.9
- 87 Fr Francium (223)

Group 2 / 2A
- 4 Be Beryllium 9.012
- 12 Mg Magnesium 24.31
- 20 Ca Calcium 40.08
- 38 Sr Strontium 87.62
- 56 Ba Barium 137.3
- 88 Ra Radium (226)

Group 3 / 3B
- 21 Sc Scandium 44.96
- 39 Y Yttrium 88.91
- 57 La Lanthanum 138.9
- 89 Ac Actinium (227)

Group 4 / 4B
- 22 Ti Titanium 47.88
- 40 Zr Zirconium 91.22
- 72 Hf Hafnium 178.5
- 104 Rf Rutherfordium (257)

Group 5 / 5B
- 23 V Vanadium 50.94
- 41 Nb Niobium 92.91
- 73 Ta Tantalum 180.9
- 105 Db Dubnium (260)

Group 6 / 6B
- 24 Cr Chromium 52.00
- 42 Mo Molybdenum 95.94
- 74 W Tungsten 183.9
- 106 Sg Seaborgium (263)

Group 7 / 7B
- 25 Mn Manganese 54.94
- 43 Tc Technetium (98)
- 75 Re Rhenium 186.2
- 107 Bh Bohrium (262)

Group 8 / 8B
- 26 Fe Iron 55.85
- 44 Ru Ruthenium 101.1
- 76 Os Osmium 190.2
- 108 Hs Hassium (265)

Group 9 / 8B
- 27 Co Cobalt 58.93
- 45 Rh Rhodium 102.9
- 77 Ir Iridium 192.2
- 109 Mt Meitnerium (266)

Group 10 / 8B
- 28 Ni Nickel 58.69
- 46 Pd Palladium 106.4
- 78 Pt Platinum 195.1
- 110 Ds Darmstadtium (269)

Group 11 / 1B
- 29 Cu Copper 63.55
- 47 Ag Silver 107.9
- 79 Au Gold 197.0
- 111 Rg Roentgenium (272)

Group 12 / 2B
- 30 Zn Zinc 65.39
- 48 Cd Cadmium 114
- 80 Hg Mercury 200.6
- 112 Cn Copernicium (285)

Group 13 / 3A
- 5 B Boron 10.81
- 13 Al Aluminum 26.98
- 31 Ga Gallium 69.72
- 49 In Indium 114.8
- 81 Tl Thallium 204.4
- 113 Nh Nihonium (286)

Group 14 / 4A
- 6 C Carbon 12.01
- 14 Si Silicon 28.09
- 32 Ge Germanium 72.59
- 50 Sn Tin 118.7
- 82 Pb Lead 207.2
- 114 Fl Flerovium (289)

Group 15 / 5A
- 7 N Nitrogen 14.01
- 15 P Phosphorus 30.97
- 33 As Arsenic 74.92
- 51 Sb Antimony 121.8
- 83 Bi Bismuth 209.0
- 115 Mc Moscovium (289)

Group 16 / 6A
- 8 O Oxygen 16.00
- 16 S Sulfur 32.07
- 34 Se Selenium 78.96
- 52 Te Tellurium 127.6
- 84 Po Polonium (210)
- 116 Lv Livermorium (292)

Group 17 / 7A
- 9 F Fluorine 19.00
- 17 Cl Chlorine 35.45
- 35 Br Bromine 79.90
- 53 I Iodine 126.9
- 85 At Astatine (210)
- 117 Ts Tennessine (294)

Group 18 / 8A
- 2 He Helium 4.003
- 10 Ne Neon 20.18
- 18 Ar Argon 39.95
- 36 Kr Krypton 83.80
- 54 Xe Xenon 131.3
- 86 Rn Radon (222)
- 118 Og Oganesson (294)

Lanthanides
- 58 Ce Cerium 140.1
- 59 Pr Praseodymium 140.9
- 60 Nd Neodymium 144.2
- 61 Pm Promethium (147)
- 62 Sm Samarium 150.4
- 63 Eu Europium 152.0
- 64 Gd Gadolinium 157.3
- 65 Tb Terbium 158.9
- 66 Dy Dysprosium 162.5
- 67 Ho Holmium 164.9
- 68 Er Erbium 167.3
- 69 Tm Thulium 168.9
- 70 Yb Ytterbium 173.0
- 71 Lu Lutetium 175.0

Actinides
- 90 Th Thorium 232.0
- 91 Pa Protactinium (231)
- 92 U Uranium 238.0
- 93 Np Neptunium (237)
- 94 Pu Plutonium (242)
- 95 Am Americium (243)
- 96 Cm Curium (247)
- 97 Bk Berkelium (247)
- 98 Cf Californium (249)
- 99 Es Einsteinium (254)
- 100 Fm Fermium (253)
- 101 Md Mendelevium (256)
- 102 No Nobelium (254)
- 103 Lr Lawrencium (257)

Appendix C
Answers to Selected Checkpoint Questions

Chapter One

1.5 c; **1.6** 1, 2, and 6; **1.10** b.

Chapter Two

2.1 a; **2.2** b; **2.6** b; **2.9** a; **2.10** a, c; **2.13** b; **2.15** Terrestrial characteristics—smaller, rocky surfaces, closer to the sun, shorter orbits, more closely spaced; Jovian—gas giants, strong gravitational fields, larger, have rings, multiple moons; Both—orbit the sun, have atmospheres, spherical, formed at same time; **2.17** a; **2.18** c; **2.21** d; **2.23** a, b; **2.25** Correct answers from top to bottom: cooler, smaller, different, less, more, less, stronger, thinner, colder.

Chapter Three

3.1 d; **3.3** 1c, 2b, 3a, 4b; **3.4** Rank order, best to worst: B, D, A, C; **3.5** b; **3.9** a; **3.10** b; **3.13** a; **3.15** This concept map would probably rate as a "2" on the grading rubric. Some key items are missing (no reference to two types of craters, no reference to asteroids as a cause for impacts) and the organization could be better.

Chapter Four

4.1 a; **4.2** a, b, c, and e; **4.3** #1 fits with pieces 2a, 3a, 5c, and 6b; #2 fits with pieces 1b and 5d; #3 fits with pieces 1a and 5a; #4 fits with pieces 5b and 6a; #5 fits with pieces 1b, 2b, 3b, 4a, and 6a; #6 fits with pieces 1c, 4b, and 5c; **4.5** b; **4.6** d; **4.8** The key contrasts should be that the contracting Earth model should not exhibit any difference in the age of the ocean floor, the ocean floor topography would not include an oceanic ridge system, heat flow would not be concentrated toward the middle of the oceans, volcanic activity should be along the edges of all oceans or none, and there is no logical explanation for the

Wadati-Benioff zone distribution of earthquakes; **4.9** a; **4.10** b; **4.11** c; **4.17** c; **4.19** 1 is divergent, 2 and 3 are convergent; **4.20** part 1, 3.

Chapter Five

5.1 Here are three: Lima, Istanbul, Los Angeles, find the rest; **5.2** d; **5.4** Since general earthquake and engineering principles (for example, frequency of earthquakes, building type and age) are used to derive specific rates for homeowners, the reasoning is general to specific, thus deductive; **5.5** b; **5.7** c; **5.9** 1. a, 2. d; **5.10** 1. b, 2. 5 seconds; **5.12** Hint: 2, 6 for P waves only, 4 for S and P waves only, and 1, 5, 11 for surface waves only; **5.14** a; **5.17** 60–80 kilometers for Northridge, 30–40 kilometers for Whittier.

Chapter Six

6.1 A: milk, B: maple syrup, C: peanut butter, D: frozen yogurt; **6.2** b; **6.3** d; **6.6** Here are five more, find the rest: intermediate silica rock (andesitic), example: Hawaiian Islands (basaltic), contains silica (all three), lowest melting temperature (rhyolitic), most common at convergent plate boundaries (andesitic); **6.7** a. Rainier/Juan de Fuca, Colima/Cocos, Galeras/Nazca, Merapi/Indian-Australian, Sakurajima/Pacific; b. No; c. Mauna Loa or Nyiragongo or Teide; d. Teide; **6.9** a; **6.11** Scientists first became concerned based on observations. Technology was employed to collect more observations used to develop tentative predictions; **6.14** Mount St. Helens: 1, 4, 15, 18, Nyiragongo: 2, 3, 5, 11, 13, 16, Both: 6, 7, 8, 9, 10, 12, 14, 17; **6.17** b; **6.18** 1. b, 2. Rainier, Colima, Santa Maria, Galeras, Taal, Etna, 3. Mauna Loa, Teide; **6.21** b; **6.22** Continental

lithosphere of Australia is not near a plate boundary unlike most of the highest mountain ranges on other continents; **6.25** b; **6.26** 1. Alaska is located adjacent to an active convergent boundary, 2. Rocky Mountains are about 200 million years younger than the Appalachians, 3. Former mountains in Canada are more than a billion years old and have been eroded down to their roots.

Chapter Seven

7.1 a; **7.2** b; **7.3** c; **7.5** d; **7.7** Three cleavage planes; **7.9** a; **7.10** From left to right: granite, basalt, rhyolite, and gabbro. Large individual grains are visible in granite and gabbro but not in basalt and rhyolite; granite and rhyolite are light colored, basalt and gabbro are primarily dark colored; **7.11** Four that are true for both volcanic and plutonic igneous rocks: may form from basaltic magma, have texture, form as a result of melting, contains minerals; **7.14** There are distinct layers with different size materials. The size of the grains would relate to the velocity of the water that deposited them and suggest a depositional history characterized by "normal" flow and flood conditions; **7.15** Three common attributes of chemical and biochemical sedimentary rocks: form under shallow marine conditions, form by precipitation from a solution, examples of sedimentary rocks; **7.17** a; **7.18** Three that are true for both regional and contact metamorphism: formed at temperatures above 200°C, may originally have been an igneous rock, marble is a possible example; **7.21** b, a.

Chapter Eight

8.1 Best order from oldest to youngest: B, C, A, E, D; **8.2** 1b, 2a, 3b; **8.3** Rock units from oldest to youngest: H, F, D, I, B, G, A, C, E, J; **8.6** Fossil 2 is the best index fossil because it is always found in limestone B. Fossil 3 is the least appropriate because it is found in multiple units; **8.7** The image shows corals, crinoids, and cephalopods that were only described in the text from the Redwall Limestone; **8.9** 1b, 2b; **8.10** b, a; **8.11** Answer for part A only: Were insect species affected by the asteroid impact thought to have caused extinction of the dinosaurs? **8.14** a, a, c; **8.15** c; **8.16** a) T, A, C, P, G, N, S, K, L, F; **8.19** Rare, high-magnitude events include asteroid impacts and supervolcano eruptions, and frequent, low-magnitude events include processes such as erosion and plate movements.

Chapter Nine

9.1 a; **9.2** a; **9.7** The rocks that are weathered in each area must have different compositions. Pyroxene and olivine have simpler silicate structures than quartz and would therefore weather more rapidly. Transport distances would be shorter in Hawaii, and therefore, these minerals may not have time to break down; **9.11** Macroprocesses generally result in rock breakage whereas microprocesses operate at the mineral level to change composition; **9.13** Chemical weathering likely dominates with high temperatures and high precipitation at location 1. Location 4 is likely too cold to have any weathering. Assume location 2 is the Sahara desert and location 3 is the Rocky Mountains; **9.14** Changes in the type of material, age, and climate conditions could result in different weathering rates from one site to another; **9.15** Kilwa Kisiwani, has the greatest weathering potential, what would come next? **9.17** b; **9.18** Hint: Regolith could not form until there was rock exposed to weathering and soils need biologic materials (land plants); **9.21** a, b.

Chapter Ten

10.1 b; **10.2** Excessive rainfall, steep slopes, weak soils, plants with shallow root systems; **10.3** Where would you place the following: hundreds affected, town at foot of slope, involved mud or soil, thousands affected, town at mouth of canyon, heavy rain, steep slopes, previous events; **10.5** b, d, d; **10.6** a, c.

Chapter Eleven

11.1 Groundwater and infiltration, stream and runoff, rainfall and precipitation, gas and water vapor, ice and melt water, plants and transpiration; **11.2** c; **11.5** b;

11.8 This region displays dendritic stream systems on the Piedmont and Coastal Plain and trellis patterns in the Valley and Ridge province; **11.10** a; **11.11** a; **11.13** d; **11.14** Deposition will occur in the reservoir behind the dam, reducing the downstream stream load and resulting in more erosion, deepening and widening the channel; **11.17** Increased volume, rate, and/or duration of precipitation, ice dams on north flowing rivers, ground saturation leading to increased runoff, landscape modifications that increase impervious cover, dam collapse, flood control structures; **11.18** d; **11.21** Direct losses of life or property, work-related losses, interruption of commerce, losses to crops, and subsequent land changes; **11.22** Prevention measures are construction of levee, dredging, and dam construction; **11.23** Hint: Flood protection information = 66 points, Open space preservation = 900 points.

Chapter Twelve

12.1 ~0% porosity if sticks set side by side in rows to fill the box, ~40% porosity if sticks arranged in overlapping rows or a latticework pattern; **12.2** Highest permeability = coffee beans, lowest

permeability = flour; **12.3** c; **12.5** d, a; **12.9** b; **12.11** If Bob had won, groundwater could be considered a shared resource that must be managed to the benefit of all. Since Glen won, groundwater would be a commodity to be used as available. As agriculture demands much more water than do people, the implication is that people in cities could lose out to the needs of agriculture; **12.12** Hint: Two other factors could be the porosity of the rocks and the properties of the surface materials; **12.13** c; **12.15** In general, there is no correlation between the location of the maximum and minimum saturation thickness and the locations of streams; **12.17** b; **12.18** Best answers are d and b, then a and c, and the worst answers are e, f, and g. Answers may be ranked in different orders but should be arranged in these groupings; **12.21** Histosols; **12.22** No, wetlands require hydric conditions and hydrophytic plants, not just occasional wet soils; **12.23** Some common features: part of the hydrologic cycle, involve the slow flow of water, may contain freshwater or saltwater, lower levels during drought conditions, store water.

Chapter Thirteen

13.1 Active margin examples: west coast of South America or Central America, south coast of Alaska, western margin of Pacific Ocean; Passive margin examples: east coasts of North and South America, west coasts of Europe and Africa, coast of Antarctica and Australia; **13.3** a; **13.5** Salinity is lower in the Bay of Bengal because of runoff from the Ganges and Brahmaputra rivers (see Chapters 7, 8, and 11); **13.8** Partial answer: Data set 1 is subpolar based on the lower surface temperature. The thermocline for data set 2 at around 100- to 1,500-meter depth and around 300 to 500 meters for data set 1; **13.9** b; **13.10** d;

13.13 b; **13.14** c; **13.17** c; **13.18** b; **13.20** Hint: two factors—wave height, location of human-made water obstacles; **13.22** b; **13.24** Hint: five features— sediment comes from the continental interior, sand is deposited in bars, wide range of grain sizes and sorting possible, occur at a range of elevations, erosion by wave action; **13.27** Prevention strategies include all the human-engineered armoring techniques, adjustment strategies focus on land management practices (for example, limiting coastal development).

Chapter Fourteen

14.1 d; **14.5** a; **14.6** b; **14.7** b; **14.9** The loss of ozone will allow more UV radiation to penetrate the atmosphere; **14.10** For Earth there is a dominant emission wavelength of ~10 micrometers; **14.11** 1. Decrease, 2. Increase, 3. Decrease; **14.13** a; **14.14** d; **14.17** a; **14.19** Responses for parts 1 and 2: 1. An actual parcel of rising air will decrease its temperature by the dry adiabatic lapse rate if it is unsaturated and the saturated lapse rate if saturated; 2. The percentage of oxygen in the atmosphere does not change with altitude. The available number of molecules of all gases will decrease with altitude; **14.21** a. Cirrus, b. Altocumulus, c. Cumulonimbus, d. Cirrostratus; **14.22** a, a; **14.26** Fastest winds will be north of Lake Superior in Canada (near the "L" symbol); **14.27** 2c, 3c, 4a.

Chapter Fifteen

15.2 For example, hazards associated with earthquakes and floods; **15.6** Three answers: A: mP, B: cT, C: mT; **15.7** Midwestern and eastern United States, where cP and mT air masses collide; **15.9** 1b, 2d, 3b, 4d, 5d; **15.10** 1) 10 P.M., October 29, 2)

10 A.M., October 30, 3) Temperature, 4) b; **15.12** Cold front: air becomes cooler after its passing, heavy rainfall with passage, Warm front: air becomes warmer after its passing, increasing humidity after passing, Both: changing wind directions, warm air rises, moves from west to east; **15.13** a; **15.14** c; **15.17** a; **15.18** b; **15.19** Fujita scale: wind speed is a factor, has five levels, Both: related to damage, relies on observations, one level describes nearly complete destruction; **15.23** Tornadoes: winds can be in excess of 480 kilometers per hour, six intensity levels, Midlatitude cyclones: comma-shaped cloud pattern, two different frontal systems involved, Hurricanes: most common in late summer and fall, All three: low-pressure systems, involve counterclockwise rotations.

Chapter Sixteen

16.1 a. 2, b. 1; **16.5** c; **16.6** Regions with the highest population density often occur in association with climate region C, which is in turn characterized by grasslands and deciduous forests; **16.8** James Hansen at NASA established a Common Sense Climate Index that takes into account factors such as seasonal average temperatures, the frequency of extreme daily temperatures, and record high and low temperatures; **16.10** Find moraines, zone of accumulation, zone of ablation and the snowline. The lines in the lower figure are materials that are scoured off of the sides of the valleys; **16.13** Polar regions are deserts because of low precipitation; **16.15** Wind action: forms dunes, Glacial processes: can transport large boulders, occur most frequently at high latitudes, Both: occur on at least five continents, transport sediment in direction of movement, few associated plants and animals;

16.19 A microfossils, B ice cores, C tree rings, D instrumental records, E archeological records; **16.21** Tilt of Earth's axis, distance from the sun, pollution in atmosphere, cloud cover, albedo of features on Earth's surface, surface and deep ocean currents (thermohaline circulation), locations of continents, composition of the atmosphere, presence/absence of ozone in stratosphere, volcanic eruptions; **16.23** 1. Decrease, 2. Small increase, 3. Decrease.

Chapter Seventeen

17.1 d; **17.2** b; **17.3** Melbourne = green squares, San Francisco = red circles, London = blue diamonds, Darwin = brown triangles; **17.7** From best to worst: d, b, c, a; **17.8** One argument for the ban: there was an immediate health threat, One argument against the ban: natural changes were not sufficiently documented; **17.10** Two examples of each: Carbon is stored in the atmosphere when CO_2 is from burning fossil fuels and in the biosphere when CO_2 is used by plants to form biomass. Carbon is released from the geosphere during weathering of limestone and from the biosphere when forests are burned; **17.13** c; **17.17** Possible responses include (but are not limited to): reduce driving (in any way), purchase alternative-fuel vehicle, use less electricity, recycle, reduce home heating or cooling, add solar panels, plant trees; **17.18** South America: greatest temperature increase in warmest locations, drying in northern part of continent, greater range in precipitation changes. North America: sea level increases greatest along northern coasts. Both: greatest temperature increases in northern parts of continents, greater precipitation over east and west coasts; **17.19** Rank of statements from top to bottom: 4, 1, 2, 5, 3.

Glossary

a

ablation zone Area of a glacier that is losing net snow or ice due to melting.

abrasion The erosion of a stream channel as sand or gravel is transported downstream.

absolute humidity A measure of the actual amount of water vapor in the air at a given time—for example, in grams per cubic meter.

abyssal plain Large, flat regions of the seafloor at depths below 3,000 meters.

accuracy The degree to which a set of measured values matches an accepted or known value.

acid mine drainage Runoff resulting when water reacts with metallic ores.

acid rain Rain with a pH value less than the pH of unpolluted precipitation.

active continental margins A margin located adjacent to a plate boundary.

adiabatic lapse rate Rate of temperature change in the atmosphere due to changing atmospheric pressure.

adiabatic temperature change Cooling or warming of air caused when air is allowed to expand or is compressed, not because heat is added or subtracted.

aerosols Tiny solid and liquid particles suspended in the atmosphere.

aftershock Sequence of earthquakes that follow a larger event.

A horizon Top layer of soil defined by the presence of organic matter mixed in with mineral and rock fragments.

air A mixture of many discrete gases, of which nitrogen and oxygen are most abundant.

air density The mass of gases (for example, nitrogen, oxygen) per volume of air.

air mass A large body of air that is characterized by a near-uniform temperature and humidity.

air pollution The presence of airborne particles and gases in concentrations that endanger the health and well-being of organisms or disrupt the orderly functioning of the environment.

albedo The reflectivity of a surface, usually expressed as a fraction of the incident radiation reflected.

alpine glacier A mountain glacier.

altocumulus clouds A principal cloud type that occurs at middle of the troposphere.

altostratus clouds A high cloud type that appears thin and spread out.

amphibole A member of a mineral group in which all members are double-chain silicates.

andesite Fine-grained igneous rock of intermediate composition. Up to one-half of the rock is plagioclase feldspar, with the rest being ferromagnesian minerals (pyroxene, amphibole).

andesitic magma A magma with intermediate silica content.

angle of repose The steepest angle at which loose material remains stationary without sliding downslope.

anticyclones A high-pressure center characterized by a clockwise flow of air in the Northern Hemisphere.

aquifer A body of rock or regolith with sufficient porosity and permeability to provide water in useful quantities to wells or springs.

aquitard Impermeable beds that hinder or prevent groundwater movement.

Archean A term, meaning ancient, which has been applied to the oldest rocks of the Precambrian.

artesian aquifer An aquifer completely filled with pressurized water and separated from the land surface by a relatively impermeable confining bed, such as shale. Also called confined aquifer.

artificial beach nourishment Addition of sand to a beach by mechanical means.

ash Very fine-grained volcanic material erupted from a volcano that commonly forms a cloud.

asteroid A small, generally rocky, solid body orbiting the sun and ranging in diameter from a few meters to hundreds of kilometers.

asteroid belt Region between Mars and Saturn where majority of solar-orbiting asteroids are found.

asthenosphere A weak layer of Earth's mantle below the lithosphere upon which tectonic plates move.

astronomical unit The average distance from the sun to Earth.

atmosphere The gaseous envelope surrounding a planet or celestial body. One of the four components of the earth system.

atmospheric pressure Pressure exerted by weight of the overlying column of air.

atom Smallest possible particle of an element that retains the properties of that element.

atomic mass number The total number of neutrons and protons in an atom.

atomic number The total number of protons in an atom.

aurora A bright display of ever-changing light caused by charged particles interacting with the upper atmosphere in the polar region.

b

barrier island A long, narrow sandy ridge of land that forms parallel to, but separate from, the mainland coast.

barycenter A common center of mass around which two objects revolve.

basalt A fine-grained, mafic, igneous rock composed predominantly of dark ferromagnesian (olivine, pyroxene) minerals and with lesser amounts of plagioclase feldspar.

basaltic magma Mobile rock material with a composition similar to that of basalt.

base level The lowest level to which a stream can erode its bed. The ultimate base level is usually sea level.

batholith A large body of intrusive igneous rocks.

bathymetry The measurement of ocean depths and the charting of the shape or topography of the ocean floor.

bauxite The principal ore of aluminum.

baymouth bar A sandbar that completely crosses a bay, sealing it off from the open ocean.

bed The smallest division of a series of layered rocks, marked by a well-defined divisional plane from its neighbors above and below.

bed load Sand- and gravel-size particles transported downstream by moving along the bottom of a stream channel.

Bergeron process A theory that relates the formation of precipitation to supercooled clouds, freezing nuclei, and the different saturation levels of ice and liquid water.

B horizon The soil horizon notable for colors and textures indicating accumulation of minerals that are not present in the parent material. Typically found below the A horizon and above the C horizon.

Big Bang The theory that the universe began more than 10 to 20 billion years ago with a massive and rapid inflation.

biochemical sedimentary rock A sedimentary rock made by or from organic matter.

biodiversity The number of species in an ecosystem.

biogeochemical cycles The cycling of chemical constituents that pass through geologic and biologic systems.

biological weathering The processes by which the actions of organisms break down rocks and minerals.

biomass The total mass of an organism or group of organisms in a particular area or ecosystem.

biome The community of plants and animals that characterize a region.

biosphere One of the four components of the earth system consisting of plants and animals.

breakwater An offshore structure built to absorb the force of breaking waves and provide quiet water near shore.

body waves Seismic waves that travel through Earth's interior.

boreal forest Forest that grows just south of the tundra.

Bowen's reaction series The sequence by which minerals crystallize from a cooling basaltic magma.

braided stream A stream consisting of numerous intertwining channels.

c

caldera A large depression typically caused by collapse or ejection of the summit area of a volcano.

Cambrian period The oldest period that contains abundant fossils of animals and the first (earliest) geologic period of the **Paleozoic era** from 542 million to 488 million years ago.

Canary current The prevailing southward flow of water along the northwestern coast of Africa.

carbon sequestration The trapping of carbon in natural or artificial storage reservoirs.

catastrophism The concept that some Earth features are shaped by catastrophic events of a short-term nature.

cementation The filling of pore spaces with precipitated minerals that bind individual grains together; a step in lithification of sediment to form sedimentary rock.

Cenozoic era The most recent of the eras; also known as the Age of Mammals.

chalk A very fine-grained biochemical limestone.

channel roughness A measure of the resistance to flow offered by the material in the stream channel.

chemical sedimentary rock A sedimentary rock consisting of material that was precipitated from water by either inorganic or organic processes.

chemical weathering The processes by which the internal structure of a mineral is altered by the removal or addition of elements.

chlorofluorocarbons Volatile organic compounds that remain in the atmosphere for periods of 20 to 200 years.

C horizon The lowest soil horizon, which is the least weathered horizon and commonly represents the parent material for soil formation.

cinder cone volcano A rather small volcano built primarily of pyroclastics ejected from a single vent.

cirrocumulus A high-altitude cloud often with a rippled appearance.

cirrostratus clouds High-altitude, thin clouds made of ice particles that are barely visible.

cirrus clouds One of three basic cloud forms; also one of the three high cloud types. They are thin, delicate ice-crystal clouds often appearing as veil-like patches or thin, wispy fibers.

clastic sedimentary rock A sedimentary rock made of fragments of preexisting rock.

cleavage The tendency of a mineral to break along planes of weak bonding.

climate A description of aggregate weather conditions that helps characterize a place or region; the average of all statistical weather information over a 30-year period.

climate forcing Any phenomenon that causes a change in the global solar radiation balance.

climatic change The long-term fluctuation in rainfall, temperature, and other aspects of Earth's climate.

climatology The scientific study of the climate.

cloud A form of condensation best described as a dense concentration of suspended water droplets or tiny ice crystals.

coastline The coast's seaward edge. The landward limit of the effect of the highest storm waves on the shore.

cohesion A phenomenon in which the surface tension created by a small amount of water holds sand grains together.

cold front A front along which a cold air mass thrusts beneath a warmer air mass.

collision-coalescence process A theory of raindrop formation in warm clouds (above 0°C) in which large cloud droplets collide and join together with smaller droplets to form a raindrop.

comet Small object in space, a few kilometers in diameter, composed of frozen methane, frozen ammonia, and water-ice, with small solid particles and dust embedded in the ices.

Community Internet Intensity Map (CIIM) A map that shows the distribution of intensities associated with an earthquake. A computer program is used to generate an average intensity for each ZIP code affected on a regional map.

compaction A loss in overall volume and pore space of a rock as the particles are packed closer together by the weight of overlying material.

condensation The change of state from a gas to a liquid.

conduction The transfer of heat through matter by molecular activity. Energy is transferred through collisions from one molecule to another.

cone of depression A cone-shaped lowering of the water table that surrounds a well being pumped faster than it recharges.

confined aquifer Source of groundwater capped by an impermeable material.

conservation tillage Soil conservation strategies designed to reduce water and wind erosion.

contact metamorphism Changes in rock caused by the heat from a nearby magma body.

continental Arctic air mass Very cold, dry air masses that form over the Arctic.

continental crust The rocks that make up most continental land masses.

continental drift theory A concept suggesting that continents move over Earth's surface.

continental polar air mass Cold, dry air masses that form over northern Canada and Alaska.

continental rise The sloping surface at the base of the continental slope.

continental shelf The gently sloping submerged portion of the continental margin, extending from the shoreline to the continental slope.

continental slope The steep gradient that leads to the deep-ocean floor and marks the seaward edge of the continental shelf.

continental tropical air mass Warm, dry air masses that form over continents at low latitudes.

contrail Vapor trail in the atmosphere caused by jet exhaust.

convection The transfer of heat by the movement of a mass or substance.

convection cell Atmospheric circulation due to temperature-density differences.

convergence The movement of two objects toward each other.

convergence lifting Occurs when two air masses collide, forcing some air upward, as both air masses cannot occupy the same space.

convergent plate boundary A boundary in which two plates move together.

coquina A limestone consisting of coarse shells.

core Located beneath the mantle, it is the innermost layer of Earth. The core is divided into an outer core and an inner core.

Coriolis Effect The deflection caused by Earth's rotation on all free-moving objects, including the atmosphere and oceans. Deflection is to the right in the Northern Hemisphere and to the left in the Southern Hemisphere.

covalent bond Union formed when two or more atoms mutually share electrons.

creep A very slow, continuous downslope movement of soil or debris.

creep meter An instrument used to record minute changes in distance between two points on Earth's surface.

crest The high point of a wave.

Cretaceous period The last period of the Mesozoic era, between 145 million and 65 million years ago.

Cretaceous-Tertiary extinction (K-T event) Time about 65 million years ago when 70 percent of species became extinct, including dinosaurs.

crevasse Fracture in a glacier caused by tension.

crosscutting relationships, principle of A principle or law stating that a disrupted pattern is older than the cause of the disruption.

crust The thin outermost layer of Earth (also see **continental crust** and **oceanic crust**).

crystal form The arrangement of various faces on a crystal in a definite geometric relationship to one another.

crystal settling During the crystallization of magma, the earlier-formed minerals are denser than the liquid portion and settle to the bottom of the magma chamber.

cumulonimbus clouds A principal cloud type, exceptionally dense and vertically developed, occurring either as isolated clouds or as a line or wall of clouds.

cumulus clouds One of the three basic cloud forms; also the name given to clouds of vertical development. Cumulus clouds are billowy individual cloud masses that often have flat bases.

cut bank The concave bank of a winding stream where erosion takes place.

cyclone A low-pressure center characterized by a counterclockwise flow of air in the Northern Hemisphere.

d

data Information, often involving measurements, used to describe something.

daughter isotope An elemental isotope produced by decay of a radioactive **parent isotope.**

debris flow A relatively rapid type of mass wasting that involves a flow of rock, soil, and regolith containing a large amount of water.

decompression melting Partial melting of hot rock when it moves upward and the pressure is reduced to the extent that the melting point drops to the melting temperature of some minerals in the rock.

deductive reasoning Occurs when scientists draw specific conclusions based on general principles.

delta An accumulation of sediment formed where a stream enters a lake or ocean.

dendritic drainage Irregular stream branching, with tributaries joining the main stream at angles of less than 90°.

dendrochronologist A scientist who measures time intervals and dates events and environmental changes by reading growth layers of trees as demarcated by annual rings.

dendrochronology The science of measuring time intervals and dating environmental changes by reading growth layers of trees as demarcated by the annual rings.

density lifting Rising air due to density differences.

deposition The settling or coming to rest of transported material.

desert Dry environments characterized by evaporation exceeding precipitation.

Devonian period The fourth period of the Paleozoic era, between 416 million and 360 million years ago.

dew point The temperature at which atmospheric water vapor begins to condense into dew as the air cools.

dike A tabular-shaped intrusive igneous feature that cuts through the surrounding rock.

diorite Large-grained igneous rock solidified from intermediatecomposition magma.

dip-slip fault An inclined fault in which the movement is up or down the slope of the fault.

discharge The volume of water in a stream that passes a given point in a period of time.

dissolution Chemical reactions where bonds break between atoms or molecules that then disperse in water.

dissolved load Dissolved materials carried by a stream.

diurnal tide A tidal condition in which only one high tide and one low tide occur each day.

divergent plate boundary A boundary separating two plates moving away from each other.

diversion channel See **floodway.**

Doppler effect An apparent changing frequency of sound or light due to the relative motion of a source and an observer.

Doppler radar In addition to the tasks performed by conventional radar, this new generation of weather radar can directly detect motion and hence greatly improve tornado and severe storm warnings.

downdraft Zone of downward-moving air.

drainage basin The land area that contributes water to a stream.

drainage divide An imaginary line separating the area drained by one stream from the areas drained by adjacent streams.

dropstone deposit Places on the seafloor where rocks carried by melting icebergs are deposited.

e

earthquake A nearly instantaneous release of stored energy resulting from the breaking and sudden movement of stressed rock.

earthquake intensity A measure of the effects of earthquakes on people and buildings.

earthquake magnitude A single, standard measure used to compare the shaking and energy released from different earthquakes.

earth science The investigation of interactions among the four components of the earth system—the atmosphere, hydrosphere, biosphere, and geosphere.

Earth's magnetic field The magnetic field hypothesized to be generated by convection currents in the outer core.

ecology The study of the interaction of organisms with one another and their environment.

ecosystem A community of organisms that inhabit a specific physical environment, defined primarily by its climate and landforms.

E horizon The soil horizon that is the zone of leaching, characterized by the downward movement of water and removal of fine-grained soil components.

electromagnetic radiation The transfer of energy through space by electromagnetic waves.

electromagnetic spectrum The distribution of electromagnetic radiation by wavelength.

electron A negatively charged subatomic particle that has a negligible mass and is found outside an atom's nucleus.

elements Chemical substances that cannot be split into simpler substances.

empirical observations Observations that can be measured and confirmed by others.

eon The largest unit of geological time.

epicenter The location on Earth's surface that lies directly above the focus of an earthquake.

equinox The time when the vertical rays of the sun are striking the equator. The lengths of daylight and darkness are equal at all latitudes during an equinox.

erosion The physical removal of rock by an agent such as running water, glacial ice, or wind.

evaporation The process of converting a liquid to a gas.

evaporite A sedimentary rock formed of material deposited from solution because of evaporation of the water.

exfoliation dome Large, dome-shaped structure formed by physical weathering, usually composed of granite.

exosphere The sun and other components of space that interact with the earth system.

f

fault A break in a rock mass along which movement has occurred.

fault scarp A cliff created by movement along a fault. It represents the exposed surface of the fault prior to modification by weathering and erosion.

fault system A group of faults that move in a similar fashion.

felsic Describes igneous rocks dominated by quartz, sodium-rich plagioclase, potassium feldspars, and the magmas that these rocks crystallize from; derived from the words *feldspar* and *silica*.

Ferrel cell A secondary circulation cell located between Hadley and polar cells.

fetch The distance over which wind blows to produce waves.

flash flood A flood that occurs with little or no warning.

flood A temporary overflow of a river onto adjacent lands not normally covered by water.

flood control Measures taken to reduce the impact of flooding (for example, building levees and dams).

floodplain The flat, low-lying portion of a stream valley subject to periodic inundation.

floodwall Vertical walls constructed to keep water in its channel.

floodway An artificially constructed diversion channel that will transport floodwaters away from inhabited areas.

flow A type of movement common to mass-wasting processes in which water-saturated material moves downslope as a viscous fluid.

focus The location within Earth where rock displacement produces an earthquake.

foliation Planes of minerals formed in response to stress; a feature of many metamorphic rocks.

foraminifera Tiny shelled animals that live or once lived in marine environments. Oxygen isotope data from shells are used as proxy climate indicators.

fossil The preserved trace, imprint, or remains of a plant or animal.

fossil fuel General term for any hydrocarbon that may be used as a fuel, including coal, oil, and natural gas.

fractures Cracks in rocks and materials.

front The boundary between two adjoining air masses having contrasting characteristics.

frontal lifting Forcing aloft of a low-density air mass over a high-density air mass where the two masses meet.

frost wedging A type of physical weathering in which the expansion of freezing water breaks apart a rock.

Fujita intensity scale Scale used to categorize tornadoes by the level of destruction they cause.

fumarole Vents where volcanic gases escape.

g

gabbro Plutonic rock equivalent to basalt.

gaining stream A stream that receives water from the zone of saturation.

geocentric The concept of an Earthcentered universe.

geochemical cycle During geologic changes, the sequence of stages in the migration of elements between the components of the earth system.

geologic time General reference to long time span that predates human records.

geologic timescale The division of Earth history into blocks of time—eons, eras, and periods. The timescale was created by using relative dating principles.

geology The science of the origin, composition, structure, and history of Earth.

geosphere One of the four components of the earth system, consisting of land and rocks.

geostrophic wind A wind, usually above a height of 500 meters, that blows parallel to the isobars.

geothermal gradient The gradual increase in temperature with depth in the crust.

geyser A fountain of hot water ejected periodically.

glacials Cold intervals during the last ice age when temperatures were 5 to 10 C° colder than today.

glaciation Alteration of any part of Earth's surface by passage of a glacier, chiefly by glacial erosion or deposition.

glacier A large, long-lasting mass of ice, formed on land by the compaction and recrystallization of snow, which moves because of its own weight.

granite A felsic, coarse-grained, intrusive igneous rock containing quartz and composed mostly of potassium- and sodium-rich feldspars.

grassland A biome dominated by grasses.

greenhouse effect The transmission of short-wave solar radiation by the atmosphere, coupled with the selective absorption of longer-wavelength terrestrial radiation, especially by water vapor and carbon dioxide.

groin A short wall built at a right angle to the shore to trap moving sand.

ground subsidence Lowering of land surface that can be caused by removal of groundwater.

groundwater Water in the zone of saturation.

groundwater overdraft The withdrawing of groundwater at rates exceeding recharge.

gyre The large circular surface current pattern found in each ocean.

h

Hadley cell An atmospheric convection cell bounded by rising, humid air at the equator and descending, dry air at latitudes 30°N and S.

half-life The time required for onehalf of the atoms of a radioactive substance to decay.

halocline A layer of water in which there is a high rate of change in salinity in the vertical dimension.

heat The kinetic energy of random molecular motion.

heat capacity Change in energy per amount of material divided by the corresponding change in temperature.

heliocentric Describing the view that the sun is at the center of the solar system.

heliosphere Extent to which solar wind from the sun reaches into interstellar space.

highland climate Complex pattern of climate conditions associated with mountains. Highland climates are characterized by large differences that occur over short distances.

Holocene An epoch of the Quaternary period from the end of the Pleistocene, around 12,000 years ago, to the present.

horizon A layer in the soil profile.

hot spot A concentration of heat in the mantle capable of producing magma, which in turn extrudes onto Earth's surface. The intraplate volcanism that produced the Hawaiian Islands is one example.

hot spring Spring with a water temperature warmer than human body temperature.

humidity Term that refers to water vapor in the air but not to liquid droplets of fog, cloud, or rain.

hurricane A tropical cyclonic storm having winds in excess of 119 kilometers (74 miles) per hour.

hydric soils Poorly drained (waterlogged) soils that exhibit anaerobic (oxygen-deficient) conditions during the growing season.

hydrograph A graphical representation of stage, flow, velocity, or other characteristics of water at a given point as a function of time.

hydrologic cycle The complete cycle through which water passes, from the oceans, through the atmosphere, to the land, and back to the ocean. Also referred to as **water cycle.**

hydrologist A scientist who studies the quantity, distribution, and flow of water on a planet.

hydrology The study of water's properties, circulation, and distribution.

hydrolysis Process that occurs when minerals react with water to form products such as clay.

hydrosphere One of the four components of the earth system, consisting of water and ice.

hydrothermal mineral deposits Mineral deposits formed when heated water circulates through rock pores or other openings.

hypothesis An idea shaped from scientific observations.

i

ice age A major interval of geologic time during which extensive ice sheets (continental glaciers) formed over many parts of the world.

igneous rock A rock formed by the crystallization of magma or lava.

inclusion A fragment or rock that is distinct from the body of igneous rock in which it is enclosed.

inclusion, principle of Fragments included in a host rock are older than the host rock.

index fossil A fossil that is associated with a particular span of geologic time.

inductive reasoning Process whereby scientists draw general conclusions from specific observations.

infiltration The movement of surface water into rock or soil through cracks and pore spaces.

infrared radiation Radiation with a wavelength from 0.7 to 200 micrometers.

insolation Amount of solar radiation per area (watts/square meter).

interglacial Descriptive of warm intervals, lasting approximately 20,000 years, during the last ice age when temperatures were up to 3 C° warmer than today.

ion An atom or molecule that possesses an electrical charge.

ionic bond Molecular bond formed when two oppositely charged atoms (ions) attract.

ionosphere A complex zone of ionized gases that coincides with the lower portion of the thermosphere.

isobar A line drawn on a map connecting points of equal atmospheric pressure, usually corrected to sea level.

isostasy The principle that the upper parts of Earth's outer layers are in a state of flotational equilibrium.

isotopes Varieties of the same element that have different mass numbers; their nuclei contain the same number of protons but different numbers of neutrons.

jet stream Regional currents of fast-moving air in the upper troposphere.

Jovian planets The Jupiter-like planets: Jupiter, Saturn, Uranus, and Neptune. These planets have relatively low densities.

Jurassic period The second period of the **Mesozoic era** of geologic time, between 200 million and 145 million years ago.

k

kinetic energy The energy resulting from the constant motion of atoms in a substance.

Köppen-Geiger classification system Climate classification system based on average temperatures, average precipitation, and vegetation.

Kyoto Protocol An agreement among the developed nations to reduce greenhouse emissions to 1990 levels by 2008–2012.

l

lahar Indonesian term describing the rapid flow of water and loose debris down steep volcanic slopes, a volcanic mudflow.

landslide Perceptible downslope movement of earth materials.

latent heat The energy absorbed or released during a change in state of water.

lateral blast A volcanic eruption where force is directed sideways rather than vertically.

lateral migration A process in which a stream migrates across the valley floor in the direction of the erosion.

laterite soil A type of clay-rich soil that has a tendency to harden into a bricklike solid.

lava Magma that reaches the ground surface.

lava dome A steep-sided lava flow commonly almost as high as it is wide and caused by the extrusion of extremely viscous lava.

lava flow Molten material extruded at, and flowing away from, a volcano or fissure.

lava plateau An elevated tableland or flat-topped highland that is several hundreds to several thousands of square kilometers in area and underlain by a thick succession of lava flows.

leachate A liquid that has percolated through soil and dissolved some soil materials in the process.

leaching The depletion of soluble materials from the upper soil horizons by downward-percolating water.

levee An embankment bordering one or both sides of a channel.

limestone A sedimentary rock composed mostly of calcite.

liquefaction The transformation of loosely packed, water-saturated sediment into a fluid mass. Commonly it occurs when grains settle during earthquake ground shaking, which displaces the water in the intervening pore spaces upward such that the water pressure moves the grains apart and the whole sediment-water mixture loses strength.

lithification The process by which unconsolidated sediments become consolidated rocks.

lithosphere The relatively rigid outer layer of Earth that is distinguished from a plastic zone below called the **asthenosphere.**

Little Ice Age The period extending from a.d. 1450 to 1900 during which global temperatures dropped and glaciers advanced.

longshore current A component of motion in the water directed parallel to the shoreline.

losing streams Streams that lose water to the zone of saturation.

Love waves A type of surface seismic wave that moves sideways but with no vertical movement.

luminosity The brightness of a star. The amount of energy radiated by a star.

luster Appearance taken on by a mineral in reflected light.

m

magma Molten rock that may contain dissolved gases as well as some solid minerals.

magmatic mineral deposits Mineral deposits formed when minerals become concentrated as magma cools and solidifies.

magnetic polarity Direction of Earth's magnetic field.

magnetic polarity reversal Direction of Earth's magnetic field when it is the opposite of the present-day field.

mantle A thick shell of rock that separates Earth's crust above from the core below.

marble A coarse-grained metamorphic rock composed of interlocking calcite crystals.

maritime air mass An air mass that originates over the ocean. These air masses are relatively humid.

maritime polar air mass Polar air initially possessing similar properties to those of continental polar air but that, in passing over water, attains a higher moisture content.

maritime tropical air mass The principal type of tropical air, produced over the tropical and subtropical seas; it is warm and humid and is frequently carried poleward on the western flanks of the subtropical highs.

mass number The number of neutrons and protons in the nucleus of an atom.

mass wasting Movement, caused by gravity, in which earth materials move downslope in bulk.

meander A pronounced sinuous curve along a stream's course.

megathrust earthquake An informal term for the largest of earthquakes occurring on long faults associated with subduction zones.

melting The change of state from solid to liquid.

meltwater Melting snow or ice.

mesocyclone A small cyclone formed when updrafts below a thunderstorm draw the spiraling horizontal winds upward.

mesopause The top of the mesosphere.

mesosphere The layer of the atmosphere immediately above the **stratosphere** and characterized by decreasing temperatures with altitude.

Mesozoic era The era that followed the **Paleozoic era** and preceded the **Cenozoic era.**

metamorphic rock A rock produced by metamorphism.

metamorphism The effects of increasing temperature, pressure, or both within Earth that produce solid-state mineralogical, chemical, and textural changes that alter the features of preexisting rocks.

meteor An asteroid that passes through Earth's atmosphere, sometimes incorrectly called a "shooting" or "falling" star.

meteorite An asteroid that strikes Earth's surface.

meteorology The study of the atmosphere and atmospheric phenomena; the study of weather and climate.

midlatitude cyclone Large center of low pressure with an associated cold front and often a warm front. Frequently it is accompanied by abundant precipitation.

Milankovitch cycles Periodic variations in Earth's rotation axis and in its orbit around the sun that affect the amount of solar energy that reaches Earth.

mineral A crystalline substance that is naturally occurring and is chemically and physically distinctive.

mineral resource All discovered and undiscovered deposits of a useful mineral that can be extracted now or at some time in the future.

Mississippian period The time between 360 million and 318 million years ago during the **Paleozoic era.**

mixed tide A tidal condition in which two high and two low tides of different ranges occur each day.

Modified Mercalli scale Scale expressing intensities of earthquakes (judged on amount of damage done) in Roman numerals ranging from I to XII.

Mohs hardness scale Scale on which 10 minerals are designated as standards of hardness.

mud volcano A conical accumulation formed by precipitation of minerals from hot fluids.

mudflow A flowing mixture of finegrained debris and water, usually moving down a channel.

n

neap tide Low tidal range occurring when the moon and sun are 90 degrees apart.

near-Earth objects Asteroids and comets that pass close to Earth's orbit.

Neogene A period of the **Cenozoic era,** between 23 million and 2.5 million years ago.

Neptunism The obsolete theory that all rocks of Earth's crust were deposited from, or crystallized out of, water.

neutron An uncharged subatomic particle found in the nucleus of atoms. Its weight is slightly more than that of a proton.

nimbostratus clouds Precipitation-bearing clouds that form a thick, continuous layer.

nonrenewable resources Resources that form or accumulate over such long time spans that they must be considered as fixed in total quantity.

normal fault A fault in which the rocks above the fault plane have moved down relative to the rock below.

normal lapse rate The temperature of stable air (air that is neither rising nor falling) that decreases with altitude at a rate of 6 C° per 1,000 meters (1.8 C° per 1,000 feet).

numerical time Amount of time elapsed as determined by analyses of the decay of naturally occurring radioactive elements.

o

oasis An isolated fertile area, usually limited in extent and surrounded by desert and marked by vegetation and a water supply.

occluded front A front formed when a cold front overtakes a warm front. It marks the beginning of the end of a middle-latitude cyclone.

ocean A major primary subdivision of the intercommunicating body of saltwater occupying the depressions of Earth's surface, bounded by continents and imaginary lines.

ocean floor The seafloor surface of an ocean basin.

oceanic crust The rocks that make up the uppermost part of the lithosphere on oceanic plates.

oceanic ridge A continuous elevated zone on the floor of the entire major ocean basins, varying in width from 500 to 5,000 kilometers. The rifts at the crests of ridges represent divergent plate boundaries.

oceanic trench A narrow, deep trough parallel to the edge of a continent or an island arc.

O horizon Dark-colored soil layer that is rich in organic material and forms just below surface vegetation.

Oort cloud A massive spherical cloud of cosmic debris where long-period comets originate.

open aquifer A groundwater source that is recharged as water infiltrates through permeable material directly above.

open system One in which both matter and energy flow into and out of the system. Most natural systems are of this type.

Ordovician period The second period of the **Paleozoic era,** after the **Cambrian** and before the **Silurian,** from approximately 488 million to 444 million years ago.

ore Naturally occurring material that can be profitably mined.

original horizontality, principle of Most water-laid sediment was deposited in horizontal or nearhorizontal layers that are essentially parallel to Earth's surface.

orographic lifting The lifting of an air current caused by its passage up and over surface elevations.

oxidation Chemical reaction between a substance and oxygen that produces new substances.

ozone A molecule containing three oxygen atoms.

ozone layer A region in the stratosphere with a greater concentration of ozone that stops 97 to 99 percent of ultraviolet rays from reaching Earth's surface.

p

Pacific Tsunami Warning System (PTWS) A network of stations designed to identify potentially damaging tsunami caused by earthquakes in or around the Pacific Ocean. The PTWS issues warnings or watches that predict tsunami arrival times for coastal areas.

paleoclimate The climate of a given interval of geologic time.

Paleogene The first period of the **Cenozoic era,** between 66 million and 23 million years ago.

paleomagnetism A study of ancient magnetic fields.

paleontology The systematic study of fossils and the history of life on Earth.

Paleozoic era The era that followed the **Precambrian** and began with the appearance of complex life, as indicated by abundant fossils.

Pangaea A supercontinent that broke apart 200 million years ago to form the present continents.

paradigm A theory that is held with a very high degree of confidence and is comprehensive in scope.

paradigm shift A change in a widely held view in science.

parent isotope Radioactive isotope that decays through time to a **daughter isotope.**

partial melting Melting of the components of a rock with the lowest melting temperatures.

passive continental margin A margin that includes a continental shelf, continental slope, and continental rise whose morphology is not affected greatly by active tectonic processes.

peat A brown, lightweight, unconsolidated or semiconsolidated deposit of plant remains.

pelagic community The plants and animals that live in the open ocean, not on the bottom or along shorelines.

Pennsylvanian period The time between 318 million and 299 million years ago, toward the end of the **Paleozoic era.**

percolation Flow of groundwater through the pore spaces in rock or soil.

peridotite Ultramafic rock composed of pyroxene and olivine present in the upper mantle.

period Each era of the standard geologic timescale is subdivided into periods (for example, the **Cretaceous period**).

permeability The capacity of a rock to transmit a fluid such as water or petroleum.

Permian period The time period between 251 million and 299 million years ago; last period of the **Paleozoic era.**

Permian-Triassic extinction (P-T event) Time about 251 million years ago that marked the largest mass extinction event known on Earth.

pH A measure of the acidity or alkalinity of a solution. Explicitly, pH is the logarithm of the hydrogen-ion concentration. A solution with a pH of 7 is neutral, pH values less than 7 are acidic, and pH values greater than 7 are alkaline.

Phanerozoic The part of geologic time represented by rocks containing abundant fossil evidence. The eon extending from the end of the **Proterozoic eon** to the present.

photolysis A chemical reaction caused by the interaction of a material with sunlight.

photosynthesis The process by which plants and algae produce carbohydrates from carbon dioxide and water in the presence of chlorophyll, using light energy and releasing oxygen.

physical weathering Disaggregation of rocks by mechanical processes.

plate tectonics A theory that Earth's surface is divided into a few large, mobile plates that are slowly moving and changing in size. Intense geologic activity occurs at the plate boundaries.

pluton An igneous body that crystallized deep underground.

plutonic igneous rocks Igneous rock formed at great depth.

Plutonism The obsolete theory that the rocks of Earth's crust formed by solidification of a molten material or were heated.

point bar A place where sediment accumulates on the inside of a bend in a stream, where the water flow is slow.

point source pollution Specific, identifiable sources of pollution, such as leaking storage tanks.

polar cell Circulating air mass ascending at high latitudes (~60°N and S) and descending at the poles.

polar front The stormy frontal zone separating air masses of polar origin from air masses of tropical origin.

polar vortex The large-scale cyclonic circulation in the middle and upper troposphere centered generally in the polar regions.

polarity The orientation of Earth's magnetic field, which is described as either normal or reverse polarity.

pore space The space taken up by openings within a rock.

porosity The percentage of a rock's volume that is taken up by openings.

porphyry An igneous rock composed of large crystals embedded in finegrained minerals.

Precambrian The vast amount of time that preceded the **Paleozoic era.**

precession The change in the direction of tilt of Earth's axis.

precipitation Water that falls from the atmosphere in the form of ice, snow, hail, or sleet or in the liquid state.

precision The degree to which measurements in a series match one another.

pressure gradient The amount of pressure change occurring over a given distance.

pressure release A significant type of physical weathering that causes rocks to crack when overburden is removed.

Proterozoic eon The eon following the **Archean** and preceding the **Phanerozoic.**

proton A subatomic particle that contributes mass and a single positive electrical charge to an atom.

proxy A substitute.

P wave A compressional wave (seismic wave) in which rock vibrates parallel to the direction of wave propagation.

pycnocline A layer of water in which there is a rapid change of density with depth.

pyroclastic flow A turbulent mixture of tephra and gases flowing down the flank of a volcano.

pyroxene A group of minerals that are single-chain silicates.

q

quartz A common mineral composed of silicon and oxygen, SiO_2.

quartzite The metamorphic equivalent of quartz sandstones.

Quaternary A division of geologic time from 2.5 million years ago to the present day that represents the last period in the **Cenozoic era.**

r

radial drainage A drainage pattern in which streams diverge outward like spokes of a wheel.

radioactive decay The process by which unstable (radioactive) isotopes transform to new elements by a change in the number of protons (and neutrons) in the nucleus.

radiometric dating The procedure of calculating the absolute ages of rocks and minerals that contain certain radioactive isotopes.

rain shadow A region on the downwind side of mountains that has little or no rain because of the loss of moisture on the upwind side of the mountains.

Rayleigh wave A seismic wave that travels at or close to the surface and involves vertical motions.

recharge zone An area in which water is absorbed that eventually reaches the zone of saturation in one or more aquifers.

rectangular drainage A drainage pattern in which tributaries of a river change direction and join one another at right angles.

recurrence interval The time interval between the occurrences of a type of event. Usually it is used in relation to the probability that a floodplain floods each year; a recurrence interval of 100 years means that there is a 1 in 100 (0.01) probability of a flood with a specified discharge each year.

red shift A shift in the wavelength of electromagnetic radiation to a longer wavelength. For visible light, this implies a shift toward the red end of the spectrum.

regional metamorphism The name applied to those metamorphic alterations that affect rocks over large areas and are indicative of widespread environmental changes rather than localized deformation or magmatism.

regolith A general term applied to soil and other unconsolidated material that lie above bedrock.

relative humidity The ratio of the amount of water vapor in the air to the amount that could be in the air if the air were saturated at the same temperature.

relative time The sequence in which events occurred

renewable resource A resource that is virtually inexhaustible or that can be replenished over relatively short time spans.

reserve Already identified deposits from which minerals can be extracted profitably.

residence time The length of time that a given volume of water remains in part of the hydrologic cycle.

residual mineral deposits Mineral deposit formed from the leaching of soil by water.

resources The total amount of geologic material in all its deposits, discovered and undiscovered.

reverse fault A fault in which the material above the fault plane moves up in relation to the material below.

rhyolitic magma Molten rock with a silica content greater than 68 percent.

Richter scale A scale of earthquake magnitude based on the highest amplitude wave arriving at a seismograph.

rift valley A tensional valley bounded by normal faults. Rift valleys are found at diverging plate boundaries on continents and along the crest of the midoceanic ridge.

rip current Current that moves water directly offshore and away from the coast.

rock cycle A sequence of processes and products that relates each rock type to the others and that describes rocks as continuously forming from preexisting rocks.

rockfall Rock falling freely or bouncing down a cliff.

rock salt An evaporate composed of halite.

rockslide The rapid slide of a mass of rock downslope along planes of weakness.

runoff The portion of precipitation that flows off the ground surface into streams.

S

Saffir-Simpson intensity scale Scale dividing hurricanes into five categories on the basis of wind speed and storm surge.

salinity The proportion of dissolved salts to pure water, usually expressed in parts per thousand or as a percentage.

saltwater intrusion Displacement of fresh groundwater by saltwater due to its greater density.

sand Sediment composed of particles with a diameter between 0.0625 and 2 millimeters.

sand dune A depositional feature consisting of sand.

sandstone A medium-grained sedimentary rock formed by the cementation of sand grains.

saturated zone A subsurface zone in which all rock openings are filled with water.

saturation The maximum quantity of water vapor that the air can hold at any given temperature and pressure.

savanna A tropical biome dominated by grasses but having small trees that do not form a canopy.

schist A metamorphic rock characterized by coarse-grained minerals oriented approximately parallel.

scientific method Process used to systematically and objectively examine and explain a problem or observed phenomenon.

seafloor spreading theory The concept that the ocean floor is moving away from the midoceanic ridge and across the deep-ocean basin, where it is subducted beneath continents and island arcs.

sea level The mean levels of the sea after tidal and wind effects are removed.

seamounts A mountain that rises from the sea floor but does not break the surface of the ocean. Most seamounts are extinct volcanoes.

seawall A wall constructed along the base of retreating cliffs to prevent wave erosion.

sediment Loose, solid particles that can originate by (1) weathering and erosion of preexisting rocks, (2) chemical precipitation from solution, usually in water, and (3) secretion by organisms.

sedimentary rock Rock that has formed from (1) lithification of any type of sediment, (2) precipitation from solution, or (3) consolidation of the remains of plants or animals.

sediment budget The balance between erosion and deposition (can be positive or negative) along a coast.

seismic gap A segment of a fault that has not experienced earthquakes for a long time; such gaps may be the site of large future quakes.

seismic wave The release of energy produced by an earthquake.

seismogram The record made by a seismograph.

seismograph A recording device that produces a permanent record of Earth motion.

seismometer An instrument designed to detect seismic waves or Earth motion.

shale A fine-grained sedimentary rock (grains finer than 0.0039 millimeter in diameter) formed by the cementation of clay (mud). Shale has thin layers (laminations) and an ability to split into small chips.

shield volcano Broad, gently sloping cone constructed of solidified lava flows.

silica A term used for oxygen plus silicon.

silicate A substance that contains silica as part of its chemical formula.

silica tetrahedron The basic building block of the silicate mineral crystal structure, consisting of four oxygen atoms surrounding and bonded to a silicon atom.

sill A tabular intrusive structure concordant with local rock.

Silurian period A period of geologic time of the **Paleozoic era,** covering a time span of between 444 million and 416 million years ago.

sinkhole A closed depression found on land surfaces underlain by limestone.

slope failure The downward and outward movement of earth materials on a natural slope.

slump In mass wasting, movement along a curved surface in which the upper part moves vertically downward while the lower part moves outward.

snowline Hypothetical line marking the transition from the ablation to the accumulation zone on a glacier.

soil A layer of weathered, unconsolidated material on top of bedrock; often also defined as containing organic matter and being capable of supporting plant growth.

soil conservation Practices intended to reduce the loss of soil by erosion.

soil erosion The detachment and movement of topsoil by the action of wind and flowing water.

soil profile A vertical section through a soil showing its succession of horizons and the underlying parent material.

solar radiation Electromagnetic radiation supplied by the sun.

solar system The sun, planets, their moons, and other bodies that orbit the sun.

solstice The time when the vertical rays of the sun are striking either the Tropic of Cancer or the Tropic of Capricorn. Solstice represents the longest or shortest day (length of daylight) of the year.

source area The locality that eroded to provide sediment to form a sedimentary rock or the origin of an air mass.

specific heat Amount of energy required to raise the temperature of a material by 1 C°.

specific retention The volume of water that a given body of rock or soil retains after being drained, divided by its porosity.

specific yield The amount of water that an aquifer will produce as a percentage of the total porosity.

spit A sand bar extending from the shore into an adjoining bay.

spring A place where water flows naturally out of rock onto the land surface.

spring tide Unusually high and low tides that result when the sun and moon lie in line with Earth and the attraction of the two bodies is cumulative.

stage The measured elevation of the water surface in a stream. The stage is defined relative to an arbitrarily selected elevation that is typically close to the elevation of the deepest part of the stream channel.

stalactite Iciclelike structure that hangs from the ceiling of a cave and is formed by precipitation of limestone.

stalagmite Columnlike form that grows upward from the floor of a cavern and is formed by precipitation of limestone.

Rise in the elevation of the sea surface caused by winds associated with hurricanes or other low-pressure storms that form over oceans.

strainmeters Instruments that measure deformation of Earth's crust.

strata Parallel layers of sedimentary rock.

stratification Rock layering.

stratocumulus A low-altitude cloud with a puffy appearance.

stratopause The transition zone between the **stratosphere** and the **mesosphere.**

stratosphere The layer of the atmosphere immediately above the **troposphere,** characterized by increasing temperatures with altitude, owing to the concentration of ozone.

stratovolcano A volcano formed by combined eruptions of cinder and lavas. Eruptions of cinders build the cone high near the vent. Eruptions of lava may extend the base of the volcano.

stratus cloud One of three basic cloud forms; sheets or layers that cover much or all of the sky.

streak Color of a pulverized substance; a useful property for mineral identification.

stream Flowing water that moves through a channel and simultaneously transports dissolved and particulate products of rock weathering.

stream discharge The volume of water flowing through a stream.

stream gauge Device that measures water level at a location in a stream where the channel cross section has been carefully surveyed. Stream gauge data are used to calculate discharge.

stream gradient Downhill slope of a stream's bed or the water surface.

stream load Suspended and dissolved materials that are removed (eroded), carried (transported), and dropped (deposited) by streams.

stream profile The longitudinal profile of a stream.

stream velocity The speed at which water in a stream travels.

strike-slip fault A fault in which movement is horizontal and parallel to the fault surface.

subduction zone Elongated region where one plate descends below another.

sublimation The conversion of a solid directly to a gas without passing through the liquid state.

submarine canyons Valleys that run across the continental shelf and down the continental slope.

subsidence Sinking or downwarping of a part of Earth's surface.

subtropical high A series of highpressure centers characterized by descending air and located roughly between latitudes 25 and 35 degrees north and south of the equator.

superposition, principle of A principle stating that within a sequence of undisturbed sedimentary rocks, the oldest layers are on the bottom, the youngest on the top.

surface wave A seismic wave that travels on Earth's surface.

suspended load Fine particles carried in the water column of a stream.

S wave A seismic wave propagated by a shearing motion, which causes rock to vibrate perpendicular to the direction of wave propagation.

t

talus Rock fragments that accumulate as a heap or sheet at the base of a steep rock surface.

tectonic plate Any of the internally rigid crustal blocks of the lithosphere that move horizontally across Earth's surface relative to one another.

temperate forest A midlatitude biome with trees.

temperature A measure of the average kinetic energy of individual atoms or molecules in a substance.

temperature inversion The atmospheric condition in which a layer of warmer air lies above cooler air.

tephra Rock fragments produced by volcanic explosion.

terminal moraine Depositional feature found at the downslope end of a glacier.

terrestrial planets Any of the Earth-like planets, including Mercury, Venus, Mars, and Earth.

Tertiary period A period of geologic time, from the end of the **Mesozoic era** (65 million years ago) to about 2 million years ago. Authorities have recently suggested it be replaced by the Paleogene and Neogene periods.

texture The general physical appearance of rocks. The size, shape, and arrangement of component minerals in sedimentary rocks determine the texture. In igneous and metamorphic rocks, the crystallinity and fabric of the minerals are major determinants of texture.

theory A well-supported explanation of a natural phenomenon.

thermocline A layer of water in which there is a rapid change in temperature in the vertical dimension.

thermohaline circulation Movements of ocean water caused by density differences brought about by variations in temperature and salinity.

thermosphere The upper layer of the atmosphere located above the **mesosphere,** starting at about an 85-kilometer altitude. Temperature increases with altitude in this layer.

thrust fault A compressional feature causing a younger unit to move over an older unit.

thunderstorm A storm produced by a cumulonimbus cloud and accompanied by lightning and thunder. It is of relatively short duration and usually accompanied by strong wind gusts, heavy rain, and sometimes hail.

tide Daily rising of sea level caused by gravitational attraction of the moon and sun on the oceans.

till Sediment from glaciers that is unsorted.

tillite Sedimentary rock formed from till.

topography The natural elevations of land surfaces in regions, treated collectively as to form.

Torino scale Asteroid and comet impact hazard scale.

tornado A small-scale cyclone with very strong winds. At ground level, most tornados are less than 500 meters (1,640 feet) in diameter.

trace fossil Trail, track, or burrow resulting from animal movement, preserved in sedimentary rock.

trade winds Two belts of winds that blow almost constantly from easterly directions and are located on the equatorial sides of the subtropical highs.

transform plate boundary Boundary between two plates that are sliding past each other.

transpiration Process whereby water in plant leaves and stems is released as water vapor.

transportation Movement of materials from one location to another.

trellis drainage A drainage pattern characterized by parallel main streams and secondary tributaries intersected at right angles by tributaries.

Triassic period The first period of the **Mesozoic era,** occurring after the **Permian** and before the **Jurassic,** 200 million to 251 million years ago.

tributary Any stream that contributes water to another stream.

tropical depression By international agreement, a tropical cyclone with maximum winds that do not exceed 61 kilometers (38 miles) per hour.

tropopause The boundary between the **troposphere** and **stratosphere.**

troposphere The portion of the atmosphere that is in contact with Earth's surface. The top of the troposphere occurs where the temperature drop with altitude levels out and then begins to rise.

trough The low point of a wave.

tsunami A long-period sea wave produced by a submarine earthquake or volcanic eruptions. It may travel for thousands of miles.

typhoon Low-pressure tropical storm.

U

ultraviolet radiation Electromagnetic radiation with a wavelength from 0.2 to 0.4 micrometer.

unconformity A surface that represents a break in the geologic record, with the rock unit immediately above it being considerably younger than the rock beneath.

unconsolidated A state in which sediment grains are loose, separate, or unattached to one another.

uniformitarianism Principle that geologic processes operating at present are the same processes that operated in the past. The principle is stated more succinctly as "The present is the key to the past."

unloading Process of overlying rocks being eroded, pressure being removed, and buried rocks expanding upward.

unsaturated zone A subsurface zone in which rock openings are generally filled partly with water; above the saturated zone.

updraft Rapid upward movement of air associated with severe storms.

V

varve A pair of annual layers deposited in a sequence of sediments.

viscosity Measure of the resistance of a fluid to flow.

visible light Electromagnetic radiation with a wavelength from 0.4 to 0.7 micrometer.

volcanic explosivity index A system describing the relative size of volcanic eruptions.

volcanic gases Volatile matter composed principally of about 90 percent water vapor and carbon dioxide, sulfur dioxide, hydrogen, carbon monoxide, and nitrogen, released during an eruption of a volcano.

volcano A mountain formed of lava, tephra, or both.

W

Wadati-Benioff zone The inclined zone of earthquakes' foci characteristic of subduction zones at convergent plate boundaries.

warm front A front along which a warm air mass overrides a retreating mass of cooler air.

water cycle See **hydrologic cycle.**

water table The upper surface of the zone of saturation.

wave base The depth at which sediments are not stirred by wave action, usually about 33 feet (10 meters).

wave height Vertical distance between the crest and trough of a wave.

wavelength The horizontal distance between two wave crests (or two troughs).

wave refraction Change in direction of waves due to slowing as they interact with the shoreline.

weather The state of the atmosphere at any given time.

weathering The group of processes that change rock at or near Earth's surface.

wedging See **frost wedging.**

westerlies Winds between 30° and 60° latitude that move west to east and greatly influence weather in North America.

wet adiabatic lapse rate The rate of adiabatic temperature change in saturated air. The rate of temperature change is variable, but it is always less than the dry adiabatic rate.

wetland An environment characterized by hydric soils that support development of hydrophytic (water-forming) plants.

wetted perimeter The portion of the perimeter of a stream channel cross section that is in contact with the water.

wind Movement of air.

wind erosion Detachment, transportation, and deposition of loose topsoil or sand by the action of wind.

Y

Younger Dryas A brief interval of cold temperatures at the close of the last major ice age (13,000–11,600 years ago). Temperatures decreased and increased abruptly at start and end of the interval.

Index

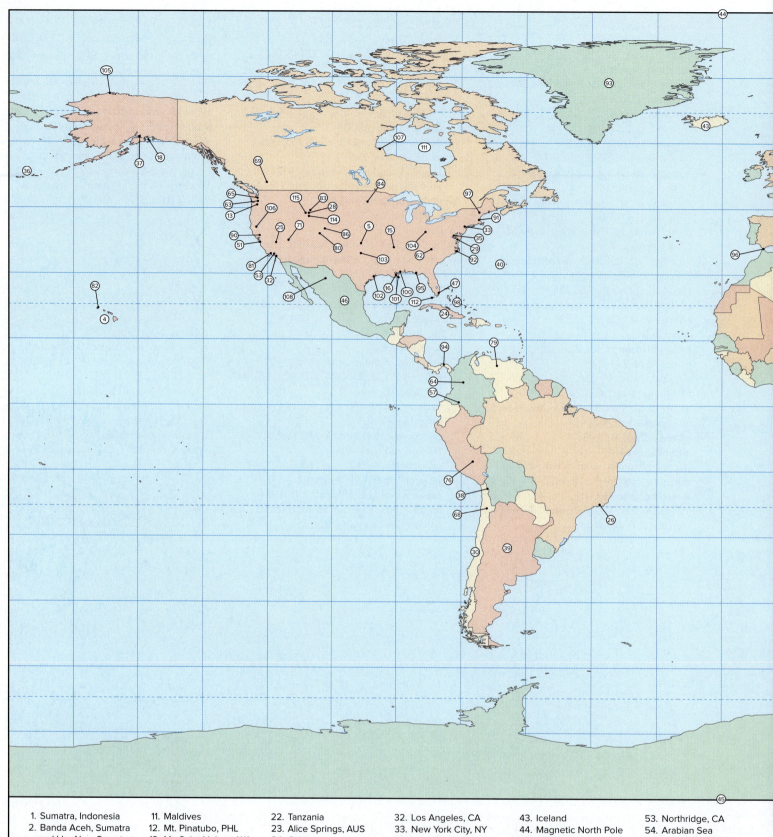

1. Sumatra, Indonesia
2. Banda Aceh, Sumatra and Lho Nga, Sumatra
3. China
4. Hawaii
5. Hutchinson, KS
6. India
7. Japan
8. Kenya
9. Madagascar
10. Malaysia
11. Maldives
12. Mt. Pinatubo, PHL
13. Mt. Saint Helens, WA
14. Myanmar
15. New Madrid, MO
16. New Orleans, LA
17. Phuket Island, Thailand
18. Prince William Sound
19. Seychelles
20. Somalia
21. Sri Lanka
22. Tanzania
23. Alice Springs, AUS
24. Cuba
25. Death Valley, CA
26. Rio de Janiero, Brazil
27. South Africa
28. Yellowstone National Park, WY
29. Chesapeake Bay
30. Chile
31. London, England
32. Los Angeles, CA
33. New York City, NY
34. New Zealand
35. Washington, D.C.
36. Aleutian Islands, AK
37. Anchorage, AK
38. Andes Mountains
39. Argentina
40. Bermuda
41. Gabon
42. Himalayas
43. Iceland
44. Magnetic North Pole
45. Magnetic South Pole
46. Mexico
47. Miami, FL
48. Red Sea
49. Bam, Iran
50. Izmit, Turkey
51. Loma Prieta, CA (epicenter)
52. Kobe, Japan
53. Northridge, CA
54. Arabian Sea
55. Bay of Bengal
56. Democratic Republic of Congo
57. Galeras, Columbia
58. Goma, Congo
59. Lake Nyos, Cameroon
60. Mt. Everest, Nepal
61. Mt. Kilimanjaro, Tanzania
62. Mt. Mitchell, NC

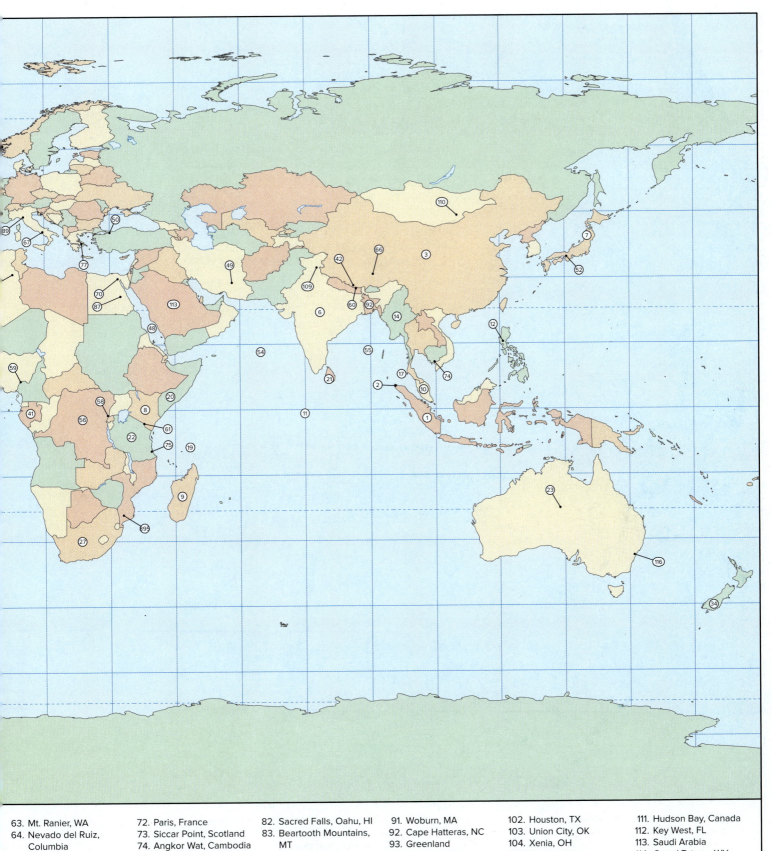

63. Mt. Ranier, WA	72. Paris, France	82. Sacred Falls, Oahu, HI	91. Woburn, MA	102. Houston, TX	111. Hudson Bay, Canada
64. Nevado del Ruiz, Columbia	73. Siccar Point, Scotland	83. Beartooth Mountains, MT	92. Cape Hatteras, NC	103. Union City, OK	112. Key West, FL
65. Seattle, WA	74. Angkor Wat, Cambodia	84. Fargo, ND	93. Greenland	104. Xenia, OH	113. Saudi Arabia
66. Tibetan Plateau	75. Kilwa Kisiwani, Tanzania	85. Limpopo River, Mozambique	94. Isthmus of Panama	105. Barrow, AK	114. Grand Tetons, WY
67. Vesuvius, Italy	76. Machu Picchu, Peru	86. Long's Peak, CO	95. Panama City, FL	106. Blue Canyon, Sierra Nevada Mountains, CA	115. Hebgen Lake, MT (epicenter)
68. Atacma Desert, Chile	77. Parthenon/Athens, Greece	87. Nile River	96. Straits of Gibraltar	107. Churchill, MB	116. Sydney, Australia
69. Banff, Alberta, Canada	78. Sahara Desert	88. Bangladesh	97. Mt. Washington, NH	108. Chihuahuan Desert, Mexico	
70. Egyptian Pyramids	79. Caraballeda, Venezuela	89. Pisa, Italy	98. Bahamas	109. Gangotri Glacier, India	
71. Grand Canyon/ Colorado River, AZ	80. Glenwood Canyon, CO	90. San Joaquin Valley, CA	99. Bergen, Norway	110. Gobi Desert, Mongolia	
	81. La Conchita, CA		100. Biloxi, MS		
			101. Buras, LA		